Vitamin E

Special Issue Editor
Volker Böhm

MDPI • Basel • Beijing • Wuhan • Barcelona • Belgrade

MDPI

Special Issue Editor
Volker Böhm
Friedrich Schiller University Jena
Germany

Editorial Office
MDPI
St. Alban-Anlage 66
Basel, Switzerland

This edition is a reprint of the Special Issue published online in the open access journal *Antioxidants* (ISSN ISSN 2076-3921) in 2017–2018 (available at: http://www.mdpi.com/journal/antioxidants/ special_issues/Vitamin_E).

For citation purposes, cite each article independently as indicated on the article page online and as indicated below:

Lastname, F.M.; Lastname, F.M. Article title. *Journal Name* **Year**, *Article number*, page range.

First Edition 2018

ISBN 978-3-03842-905-0 (Pbk)
ISBN 978-3-03842-906-7 (PDF)

Table of Contents

About the Special Issue Editor   v

Preface to "Vitamin E" vii

Volker Böhm
Vitamin E
doi: 10.3390/antiox7030044   1

Laurent Mène-Saffrané
Vitamin E Biosynthesis and Its Regulation in Plants
doi: 10.3390/antiox7010002   3

Steffi Fritsche, Xingxing Wang and Christian Jung
Recent Advances in our Understanding of Tocopherol Biosynthesis in Plants: An Overview
of Key Genes, Functions, and Breeding of Vitamin E Improved Crops
doi: 10.3390/antiox6040099 20

Ossama Kodad, Rafel Socias i Company and José M. Alonso
Genotypic and Environmental Effects on Tocopherol Content in Almond
doi: 10.3390/antiox7010006 38

Emmanuelle Reboul
Vitamin E Bioavailability: Mechanisms of Intestinal Absorption in the Spotlight
doi: 10.3390/antiox6040095 47

**Martin Schubert, Stefan Kluge, Lisa Schmölz, Maria Wallert, Francesco Galli, Marc Birringer
and Stefan Lorkowski**
Long-Chain Metabolites of Vitamin E: Metabolic Activation as a General Concept for Lipid-
Soluble Vitamins?
doi: 10.3390/antiox7010010 58

Raffaella Comitato, Roberto Ambra and Fabio Virgili
Tocotrienols: A Family of Molecules with Specific Biological Activities
doi: 10.3390/antiox6040093 78

Siti Syairah Mohd Mutalip, Sharaniza Ab-Rahim and Mohd Hamim Rajikin
Vitamin E as an Antioxidant in Female Reproductive Health
doi: 10.3390/antiox7020022 92

Hamza El Hadi, Roberto Vettor and Marco Rossato
Vitamin E as a Treatment for Nonalcoholic Fatty Liver Disease: Reality or Myth?
doi: 10.3390/antiox7010012 107

**Ujwani Nukala, Shraddha Thakkar, Kimberly J. Krager, Philip J. Breen, Cesar M. Compadre
and Nukhet Aykin-Burns**
Antioxidant Tocols as Radiation Countermeasures (Challenges to be Addressed to Use Tocols
as Radiation Countermeasures in Humans)
doi: 10.3390/antiox7020033 120

**Emily S. Mohn, Matthew J. Kuchan, John W. Erdman Jr., Martha Neuringer, Nirupa R. Matthan,
Chung-Yen Oliver Chen and Elizabeth J. Johnson**
The Subcellular Distribution of Alpha-Tocopherol in the Adult Primate Brain and Its
Relationship with Membrane Arachidonic Acid and Its Oxidation Products
doi: 10.3390/antiox6040097 135

About the Special Issue Editor

Volker Böhm, PD Dr., studied "Food Science" at Münster University (Germany). Afterwards, he wrote his Ph.D. thesis on the analysis of residues of coplanar PCB congeners in food and human milk and obtained his Ph.D. in 1992 from Münster University. In 1999, an additional qualification (Antioxidative carotenoids and polyphenols—analysis, contents in food and intestinal absorption) followed as a lecturer at Jena University. Dr. Böhm has been a senior scientific assistant at Jena University since 1999 and a research group leader since 2005. He coordinated the large EU-funded project LYCOCARD (2006–2011). His main research topics are secondary plant products with a focus on carotenoids and polyphenols. His recent research has investigated the contents of carotenoids and polyphenols, their behaviour during food processing, as well as their biological activities and their intestinal absorption.

Preface to "Vitamin E"

Over the last ca. 100 years since the discovery of vitamin E, research has focused on different properties of this molecule, the focus often depending on the specific techniques and scientific knowledge present at each time. Originally discovered as a dietary factor essential for reproduction in rats, vitamin E has since been revealed to have many more important molecular properties, such as the scavenging of reactive oxygen and nitrogen species with consequent prevention of the oxidative damage associated with many diseases, or the modulation of signal transduction and gene expression in antioxidant and non-antioxidant manners.

This Special Issue includes 10 peer-reviewed papers, including one original research paper and nine reviews. They present the most recent advances in vitamin E research, not only looking at its antioxidant properties. Chapters include an overview on the knowledge of tocochromanol biosynthesis in plants and a special look at tocopherols in almonds, which is the nut with the highest level of tocopherols. Other chapters present the fate of vitamin E in the human gastrointestinal lumen during digestion and two long-chain metabolites of vitamin E. Another chapter focuses on the special biological activities of tocotrienols, not shared by tocopherols. Further chapters show the specific effects of vitamin E, e.g. the role of vitamin E as an antioxidant in female reproductive health, the role of vitamin E in nonalcoholic fatty liver disease, and the efficacy of vitamin E as a radiation countermeasure. The last chapter presents the membrane distribution of alpha-tocopherol in brain regions of adult rhesus monkeys.

All authors are acknowledged for their valuable contributions. I would also like to acknowledge all reviewers for their constructive suggestions. Special thanks are to the publishing team of the journal Antioxidants for all their help in the completion of this Special Issue.

Volker Böhm
Special Issue Editor

antioxidants

MDPI

Editorial
Vitamin E

Volker Böhm

Institute of Nutrition, Friedrich Schiller University Jena, Dornburger Str. 25-29, 07743 Jena, Germany;
Volker.Boehm@uni-jena.de

Received: 19 March 2018; Accepted: 20 March 2018; Published: 20 March 2018

Vitamin E is the major lipid-soluble antioxidant in the cell antioxidant system and is exclusively obtained from the diet. Vitamin E protects polyunsaturated fatty acids and other components of cell membranes and low-density lipoproteins from oxidation by free radicals. It is located primarily within the phospholipid bilayer of cell membranes. The most important form is α-tocopherol. Clinical signs of deficiency occur very rarely [1].

During the last ca. 100 years since the discovery of vitamin E, research has focused on different properties of this molecule, the focus often depending on the specific techniques and scientific knowledge present at each time. Originally discovered as a dietary factor essential for reproduction in rats, vitamin E has revealed many more important molecular properties meanwhile, such as the scavenging of reactive oxygen and nitrogen species with consequent prevention of oxidative damage associated with many diseases, or the modulation of signal transduction and gene expression in antioxidant and non-antioxidant manners [2].

This special issue highlights some of the recent advances in vitamin E research, showing the status quo on the one hand and providing new insights in functions and physiological relevance on the other hand. The review of Mène-Saffrané summarizes the current knowledge of tocochromanol biosynthesis in plants and highlights future challenges regarding the understanding of its regulation [3]. Adding to this topic, Fritsche et al. [4] review 30 years of research on tocopherols in model and crop species, with special emphasis on the improvement of vitamin E content using transgenic approaches and classical breeding [4]. Kodad et al. [5] report results for one of the most important nut species worldwide, the almond. Almond kernels show the highest levels of tocopherols among all nuts, being dependent on the genotype and the environment [5].

Reboul [6] describes the fate of vitamin E in the human gastrointestinal lumen during digestion. She focuses on the proteins involved in the intestinal membrane and cellular transport of vitamin E across the enterocyte and discusses factors modulating vitamin E micellarization and absorption. During the metabolism of vitamin E, the long-chain metabolites 13'-hydroxychromanol and 13'-carboxychromanol are formed by oxidative modification of the side-chain [7]. Their occurrence in human serum indicates a physiological relevance. Effects of these metabolites on lipid metabolism, apoptosis, proliferation, and inflammatory actions have been shown. Interestingly, the long-chain metabolites exerted effects different from that of their precursors [7]. Comitato et al. [8] review special biological activities of tocotrienols, not shared by tocopherols. Thus, tocotrienols showed the ability to inhibit cancer cell growth and induce apoptosis thanks to specific mechanisms. In addition, neuroprotective activities of tocotrienols are also presented [8].

Another part of this special issue presents specific effects of vitamin E. Mutalip et al. [9] answers the question: What are the known roles of vitamin E as an antioxidant in female reproductive health? This paper comes back to the initial discovery of vitamin E in 1922 as a substance necessary for reproduction. El Hadi et al. [10] discuss vitamin E as a potent antioxidant being able to reduce oxidative stress in nonalcoholic fatty liver disease. They also present therapeutic efficacy. Nukala et al. [11] discuss another interesting application. They describe the efficacy of tocopherols and tocotrienols as radiation countermeasures and identify the challenges to be addressed to develop them into radiation countermeasures for human use. Mohn et al. [12] describe the membrane distribution of α-tocopherol in brain regions of adult rhesus monkeys. These authors also look for associations between membrane α-tocopherol and content of polyunsaturated fatty acids.

Conflicts of Interest: The authors declare no conflicts of interest.

References

1. World Health Organization (WHO); Food and Agriculture Organization (FAO). *Vitamin and Mineral Requirements in Human Nutrition: Report of a Joint FAO/WHO Expert Consultation, Bangkok, Thailand, 21–30 September 1998*; WHO, FAO: Geneva, Switzerland, 2004; p. 341.
2. Zingg, J.-M. Vitamin E: An overview of major research directions. *Mol. Asp. Med.* **2007**, *28*, 400–422. [CrossRef] [PubMed]
3. Mène-Saffrané, L. Vitamin E biosynthesis and its regulation in plants. *Antioxidants* **2018**, *7*, 2. [CrossRef] [PubMed]
4. Fritsche, S.; Wang, X.; Jung, C. Recent advances in our understanding of tocopherol biosynthesis in plants: An overview of key genes, functions, and breeding of vitamin E improved crops. *Antioxidants* **2017**, *6*, 99. [CrossRef] [PubMed]
5. Kodad, O.; Socias I Company, R.; Alonso, J.M. Genotypic and environmental effects on tocopherol content in almond. *Antioxidants* **2018**, *7*, 6. [CrossRef] [PubMed]
6. Reboul, E. Vitamin E bioavailability: Mechanisms of intestinal absorption in the spotlight. *Antioxidants* **2017**, *6*, 95. [CrossRef] [PubMed]
7. Schubert, M.; Kluge, S.; Schmölz, L.; Wallert, M.; Galli, F.; Birringer, M.; Lorkowski, S. Long-chain metabolites of vitamin E: Metabolic activation as a general concept for lipid-soluble vitamins? *Antioxidants* **2018**, *7*, 10. [CrossRef] [PubMed]
8. Comitato, R.; Ambra, R.; Virgili, F. Tocotrienols: A family of molecules with specific biological activities. *Antioxidants* **2017**, *6*, 93. [CrossRef] [PubMed]
9. Mutalip, S.S.M.; Ab-Rahim, S.; Rajikin, M.H. Vitamin E as an antioxidant in female reproductive health. *Antioxidants* **2018**, *7*, 22. [CrossRef] [PubMed]
10. El Hadi, H.; Vettor, R.; Rossato, M. Vitamin E as a treatment for nonalcoholic fatty liver disease: Reality of myth? *Antioxidants* **2018**, *7*, 12. [CrossRef] [PubMed]
11. Nukala, U.; Thakkar, S.; Krager, K.J.; Breen, P.J.; Compadre, C.M.; Aykin-Burns, N. Antioxidant tocols as radiation countermeasures (Challenges to be addressed to use tocols as radiation countermeasures in humans). *Antioxidants* **2018**, *7*, 33. [CrossRef] [PubMed]
12. Mohn, E.S.; Kuchan, M.J.; Erdman, J.W.; Neuringer, M.; Matthan, N.R.; Chen, C.-Y.O.; Johnson, E.J. The subcellular distribution of alpha-tocopherol in the adult primate brain and its relationship with membrane arachidonic acid and its oxidation products. *Antioxidants* **2017**, *6*, 97. [CrossRef] [PubMed]

antioxidants

MDPI

Review

Vitamin E Biosynthesis and Its Regulation in Plants

Laurent Mène-Saffrané

Department of Biology, University of Fribourg, Chemin du Musée, 10, 1700 Fribourg, Switzerland;
laurent.mene-saffrane@unifr.ch; Tel.: +41-268-008-808

Received: 19 November 2017; Accepted: 21 December 2017; Published: 25 December 2017

Abstract: Vitamin E is one of the 13 vitamins that are essential to animals that do not produce them. To date, six natural organic compounds belonging to the chemical family of tocochromanols—four tocopherols and two tocotrienols—have been demonstrated as exhibiting vitamin E activity in animals. Edible plant-derived products, notably seed oils, are the main sources of vitamin E in the human diet. Although this vitamin is readily available, independent nutritional surveys have shown that human populations do not consume enough vitamin E, and suffer from mild to severe deficiency. Tocochromanols are mostly produced by plants, algae, and some cyanobacteria. Tocochromanol metabolism has been mainly studied in higher plants that produce tocopherols, tocotrienols, plastochromanol-8, and tocomonoenols. In contrast to the tocochromanol biosynthetic pathways that are well characterized, our understanding of the physiological and molecular mechanisms regulating tocochromanol biosynthesis is in its infancy. Although it is known that tocochromanol biosynthesis is strongly conditioned by the availability in homogentisate and polyprenyl pyrophosphate, its polar and lipophilic biosynthetic precursors, respectively, the mechanisms regulating their biosyntheses are barely known. This review summarizes our current knowledge of tocochromanol biosynthesis in plants, and highlights future challenges regarding the understanding of its regulation.

Keywords: vitamin E; tocochromanol; tocopherol; tocotrienol; plastochromanol-8; tocomonoenol; homogentisate; polyprenyl pyrophosphate; nutrigenomics

1. Introduction

Natural compounds exhibiting vitamin E activity in animal cells belong to the chemical family of tocochromanols, a group of organic molecules with a polar chromanol ring and lipophilic polyprenyl side chain that varies according to the type of tocochromanol [1]. The polyprenyl precursor of tocopherols, tocotrienols, plastochromanol-8 (PC-8), and tocomonoenols is phytyl pyrophosphate (PPP), geranylgeranyl pyrophosphate (GGPP), solanesyl pyrophosphate (SPP), and tetrahydrogeranylgeranyl pyrophosphate (THGGPP), respectively (Figure 1). According to the degree of methylation of the chromanol ring, each type of tocochromanol exhibits up to four different forms, named α- (three-methyl groups), β- and γ- (two-methyl groups), and δ-tocochromanol (one-methyl group; Figure 1), respectively. While the four forms of tocopherol, tocotrienol, and tocomonoenol have been identified in wild-type plant extracts, only the γ- form of the solanesyl-derived tocochromanol PC-8 naturally exists.

The first article published on vitamin E relates the finding of an unknown nutritional factor that prevents embryo resorption during rodent gestation [2]. Several years later, this *"factor X"*, as it was originally named, was identified as α-tocopherol [3]. In addition to its essential role in animal reproduction, vitamin E has been shown to have beneficial roles in human health [4], notably by inhibiting lung cancer [5], by delaying brain aging and reducing the risk of developing Alzheimer's disease [6], and by suppressing cholesterogenesis [7,8]. Although vitamin E is present in many edible plant-derived products, converging nutritional surveys conducted in both poor and developed countries, respectively, have shown that a significant proportion of human populations exhibit mild to severe vitamin E deficiencies [9–13]. It has been shown, for instance, that 23% of the Seoul metropolitan population exhibited plasma α-tocopherol concentrations below 12 μmol/L, the threshold defining vitamin E deficiency in humans [11]. The consequences of vitamin E deficiency on human health have not yet been fully documented nor investigated. However, it has been clearly established that low plasma vitamin E is strongly associated with miscarriage during the first trimester of woman pregnancy [12]. Moreover, it has been shown that diet supplementation with vitamin E decreased the miscarriage rate of pregnant woman by approximately 50% [12]. Collectively, these results demonstrate the importance of adequate vitamin E intake for proper reproduction.

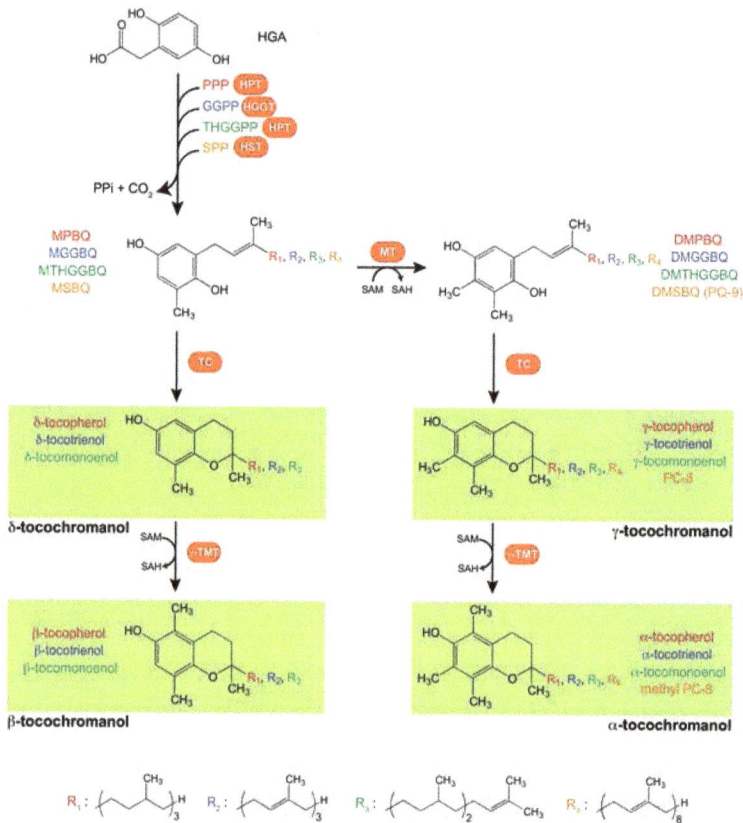

Figure 1. Tocochromanol biosynthetic pathways in plants. Tocochromanol and prenyl benzoquinol chemical structures and biosynthetic enzymes (highlighted in orange). Tocochromanol and prenyl benzoquinol names are color-coded to distinguish each tocochromanol pathway: red for the tocopherol pathway, blue for the tocotrienol pathway, green for the tocomonoenol pathway, and orange for the PC-8 and methyl PC-8 pathway. The α-, β-, γ-, and δ-forms of tocopherols, tocotrienols, and tocomonoenols have been identified in plants. For solanesyl-derived tocochromanols, only PC-8 has been identified in wild-type plants, and only methyl PC-8 has been identified in transgenic Arabidopsis overexpressing the γ-TMT/VTE4 gene. Abbreviations: HGA, homogentisate; HGGT, homogentisate geranylgeranyltransferase; HPT, homogentisate phytyltransferase; HST, homogentisate solanesyltransferase; GGPP, geranylgeranyl pyrophosphate; γ-TMT, γ-tocopherol methyltransferase; MT, methyltransferase; PC-8, plastochromanol-8; PPi, pyrophosphate; PPP, phytyl pyrophosphate; PQ-9, plastoquinol-9; SAM, S-adenosyl-L-methionine; SAH, S-adenosyl-L-homocysteine; SPP, solanesyl pyrophosphate; TC, tocopherol cyclase; THGGPP, tetrahydrogeranylgeranyl pyrophosphate. Prenyl benzoquinol acronyms are detailed in the main text.

It has been assumed for a long time that tocochromanol biosynthesis was the exclusive appanage of plants, algae, and some cyanobacteria that are all photosynthetic organisms. However, a recent study showed that *Plasmodium falciparum*, a non-photosynthetic parasite that causes malaria, synthesizes both α- and γ-tocopherols during its intraerythrocytic stages to avoid oxidative stress [14–16]. Besides this exception, tocochromanol metabolism has been primarily studied in angiosperms and in the vitamin E-producing cyanobacteria *Synechocystis*. The present review summarizes our current knowledge on tocochromanol metabolism in plants, including the core tocochromanol biosynthetic pathway that is now well delineated (Section 2); the transcriptional regulation of γ-*TMT* expression, which is the gene encoding the biosynthetic enzyme of the most potent vitamin E form of α-tocopherol in animals (Section 3); the regulation of homogentisate (HGA) biosynthesis, which is the polar precursor of tocochromanols (Section 4); and the regulation of polyprenyl pyrophosphate biosynthesis, which is the lipophilic precursor of tocochromanols (Section 5). Thus, this work complements the very recent review that highlighted the biosynthetic origins and transports of polar and lipophilic tocochromanol biosynthetic precursors in plants [17].

2. Tocochromanol Biosynthetic Pathways

Tocochromanol biosynthesis is initiated by the condensation of the polar aromatic head HGA with various lipophilic polyprenyl pyrophosphates that determine the type of tocochromanol. PPP is the lipophilic biosynthetic precursor of tocopherols, while GGPP is the one for tocotrienols, as is SPP for PC-8, and THGGPP for tocomonoenols (Pellaud and Mène-Saffrané, 2017 [17] and Figure 1). The condensation reaction is catalyzed by three types of HGA prenyltransferases that possess each their substrate specificities. Tocopherol synthesis is initiated by HGA phytyltransferases (HPTs) that condense HGA and PPP. This enzyme has been identified in both Arabidopsis and the cyanobacterium *Synechocystis* [18,19]. While the Arabidopsis HPT (AtHPT), also named VTE2, preferentially utilizes PPP as a prenyl donor, its *Synechocystis* counterpart uses both PPP and GGPP [18]. Although AtHPT poorly utilizes GGPP as a substrate in vitro, and wild-type Arabidopsis accessions do not naturally accumulate tocotrienols, it has been shown that AtHPT catalyzes the synthesis of tocotrienols in transgenic Arabidopsis plants that are overaccumulating HGA [20]. Recently, AtHPT was also shown to be catalyzing the accumulation of γ-tocomonoenol in Arabidopsis seeds, indicating that THGGPP is a substrate for this enzyme [21]. Collectively, these data show that the substrate specificity of HGA phytyl transferases is wider than that originally described, and is notably modulated by HGA availability. In Poaceae, tocotrienol synthesis is initiated by HGA geranylgeranyltransferase (HGGT), which utilizes both GGPP and PPP as prenyl donors in vitro. In addition, overexpression of the barley *HGGT* gene in the Arabidopsis *vte2* mutant induces the quantitative accumulation of both tocotrienols and tocopherols in seeds and leaves, indicating that HGGT utilizes both prenyl donors in vivo as well [22]. The Arabidopsis *vte2-1* mutant lacks tocopherols, but still accumulates PC-8 in both seeds and leaves, confirming in vitro data showing that VTE2 is not involved in the condensation of HGA with the solanesyl derivative SPP [18,23]. This reaction is catalyzed by the HGA solanesyltransferase (HST) that has been identified in both *Chlamydomonas* and Arabidopsis [24,25]. The condensation reaction between HGA and polyprenyl pyrophosphates produces 2-methyl-6-phytyl-1,4-benzoquinol (MPBQ) for tocopherols, 2-methyl-6-geranylgeranyl-1,4-benzoquinol (MGGBQ) for tocotrienols, 2-methyl-6-solanesyl-1,4-benzoquinol (MSBQ) for PC-8, and 2-methyl-6-tetrahydrogeranylgeranyl-1,4-benzoquinol (MTHGGBQ) for tocomonoenols (Figure 1). These compounds are either direct precursors of δ- and β-tocochromanols, or can alternatively be methylated by a methyltransferase (MT/VTE3) that uses *S*-adenosyl-L-methionine (SAM) as a methyl donor [26,27]. The products of the later reaction are prenylated dimethyl-benzoquinols, namely 2,3-dimethyl-6-phytyl-1,4-benzoquinol (DMPBQ) for tocopherols, 2,3-dimethyl-6-geranylgeranyl-1,4-benzoquinol (DMGGBQ) for tocotrienols, 2,3-dimethyl-6-solanesyl-1,4-benzoquinol better known as plastoquinol-9 (PQ-9) for PC-8, and 2,3-dimethyl-6-tetrahydrogeranylgeranyl-1,4-benzoquinol (DMTHGGBQ) for tocomonoenols

(Figure 1). Both methyl- and dimethyl-benzoquinols are then further cyclized by the tocopherol cyclase (TC/VTE1) into δ- and γ-tocochromanols, respectively. Seeds of the Arabidopsis *vte1-1* mutant lack all tocopherols, PC-8, and γ-tocomonoenol, indicating that the cyclase indiscriminately uses mono- and dimethyl-benzoquinols carrying either a phytyl, a solanesyl, or a tetrahydrogeranylgeranyl side chain as substrates [21,23,28,29]. The fourth and final step of tocochromanol biosynthesis consists of the methylation of γ- and δ-tocochromanols into α- and β-tocochromanols, respectively [30,31]. This reaction is catalyzed by the γ-tocopherol methyltransferase (γ-TMT/VTE4) that utilizes SAM as methyl donor (Figure 1). In Arabidopsis leaves and seeds, VTE4 converts γ- and δ-tocopherols into α- and β-tocopherol, respectively, indicating that the methyl transferase efficiently methylates phytyl-derived tocochromanols [30]. In addition, transgenic Arabidopsis lines overexpressing the barley *HGGT* gene notably produce α-tocotrienol [32]. This indicates that the γ-tocopherol methyltransferase VTE4 catalyzes that methylation of geranylgeranyl-derived tocochromanol such as γ-tocotrienol as well.

3. Regulation of the *γ-tocopherol methyltransferase* Expression

The different tocochromanol forms identified in plants do not exhibit the same vitamin E activity in animal cells [1]. Based on the rat fetal resorption assay, α-tocopherol (1.49 IU/mg; 100%) is by far the most potent vitamin E form, followed by β-tocopherol (0.75 IU/mg; 50%), α-tocotrienol (0.45–0.75 IU/mg; 30–50%), γ-tocopherol (0.15 IU/mg; 10%), β-tocotrienol (0.08 IU/mg; 5%), and δ-tocopherol (0.05 IU/mg; 3%). The vitamin E activities of γ- and δ-tocotrienol have been tested and were below the limit of detection, while those of other tocochromanols such as PC-8 and tocomonoenols have not been tested yet. The difference in vitamin E activity among tocochromanol forms results from their differential affinities to the α-tocopherol transfer protein, a cytoplasmic protein that transports tocochromanols from the endosomal fraction of hepatocytes into the bloodstream of animals [33,34]. This implies that at equal molarity, a plant tissue accumulating α-tocopherol instead of γ-tocopherol for instance, exhibits 10 times more vitamin E activity. Based on this, a relevant question regarding the regulation of tocochromanol biosynthesis in plants and the improvement of crops vitamin E activity is: what are the molecular mechanisms regulating *γ-TMT* expression?

The main tocochromanol identified in wild-type leaves analyzed to date is α-tocopherol, indicating that leaf γ-TMT/VTE4 activity is sufficient to quantitatively methylate the γ-tocopherol pool present in this tissue. In contrast, despite the presence of PC-8 (γ-tocochromanol) in wild-type Arabidopsis and maize leaves, methyl PC-8 (α-tocochromanol) has not been detected in any wild-type tissue analyzed to date [23]. In contrast, methyl PC-8 accumulates in transgenic Arabidopsis leaves overexpressing *γ-TMT/VTE4*, demonstrating that PC-8 is a substrate for γ-TMT/VTE4, and that endogenous γ-TMT/VTE4 activity is not sufficient to convert all γ-tocochromanols into α-tocochromanols [35]. This later conclusion is further supported by the tocochromanol composition of transgenic Arabidopsis leaves overexpressing the barley *HGGT* gene. Indeed, while these transgenic lines accumulate tocopherols mostly under the form of α-tocopherol, their tocotrienol pool is mainly under the form of γ-tocotrienol [32]. In contrast to this, another metabolic engineering study in which tocochromanol biosynthesis has been increased by enhancing the production of HPP in chloroplasts suggests that γ-TMT activity might be limiting for α-tocopherol synthesis in leaves. In comparison to controls, these Arabidopsis transgenic lines overaccumulated both γ-tocopherol and γ-tocotrienol, but exhibited control levels for α-tocopherol in leaves [36]. In addition, they did not accumulate any detectable α-tocotrienol, despite the strong accumulation of γ-tocotrienol in leaf tissues. In alfalfa, it has been shown that the overexpression of a Medicago *γ-TMT* modestly but significantly increased α-tocopherol accumulation in leaves [37]. Collectively, these data show that the leaf γ-TMT/VTE4 activity is clearly not sufficient to fully methylate both γ-tocotrienol and PC-8, and is sometimes limiting α-tocopherol biosynthesis. The former conclusion might indicate that γ-TMT/VTE4 has a lower affinity for γ-tocotrienol and PC-8 than for α-tocopherol.

In contrast to leaves, the tocochromanol composition of wild-type seeds is much more variable, and three groups of plant species can be currently distinguished according to this criterion. The seeds of some plant species accumulate mostly γ-tocochromanols and very little or no α-tocochromanols. This is notably the case of wild-type Arabidopsis seeds, which accumulate mostly γ-tocopherol and PC-8, very low levels of α-tocopherol, and no methyl PC-8 [23]. The overexpression of γ-*TMT* genes in Arabidopsis and *Brassica napus* is sufficient to trigger the quantitative accumulation of both α-tocopherol and methyl PC-8 in transgenic seeds, thus demonstrating that the lack of α-tocochromanols in the seeds of these species results from the very low transcriptional activity of the γ-*TMT/VTE4* gene in seeds [30,35,38]. In contrast to Arabidopsis, the seeds of other plant species mostly accumulate α-tocochromanols, and very few γ-tocochromanols. This is notably the case of sunflower and wheat, which accumulate mostly α-tocopherol ± α-tocotrienol, and very low levels of γ-tocochromanols [39,40]. This indicates that the VTE4 activity in seeds of these species is sufficient to quantitatively convert almost all of the γ-tocochromanols into α-tocochromanols. The third type of plants accumulate both α- and γ-tocochromanols in seeds, suggesting that although their VTE4 activity is sufficient to support the synthesis of α-tocochromanols, it is not strong enough to fully methylate the pools of γ-tocochromanols. This group includes species such as rapeseed and maize [41–43]. Our current knowledge of the transcriptional mechanism(s) regulating γ-*TMT/VTE4* expression, notably in seeds, is currently limited to a single study performed in soybean. Tocopherol analysis in seeds of 1109 wild and cultivated soybean varieties showed that while almost all of them produce primarily γ-tocopherol, three varieties accumulate up to seven times more α-tocopherol [44]. Quantitative trait loci (QTL) analysis in high α-tocopherol soybean varieties showed that polymorphism located in the γ-*TMT3* promoter correlated with the higher expression of the methyl transferase, and thus that α-tocopherol level is transcriptionally regulated in soybean seeds as well [45]. Collectively, these data demonstrate that the synthesis of α-tocochromanols, such as α-tocopherol in seeds and leaves, is determined by the mechanism(s) regulating the expression of the γ-*TMT/VTE4* gene. To date, despite many independent research initiatives that led to the identification of numerous α-tocopherol QTLs in Arabidopsis [46], soybean [45,47,48], maize [42,43,49,50], winter oilseed rape [51,52], and barley [53], the transcriptional machinery regulating the expression of γ-*TMT* genes in plants is still unknown.

4. Plastidic Availability in Homogentisate Regulates Tocochromanol Synthesis

Early feeding experiments of safflower, sunflower, and soybean cell cultures with HGA have shown that an exogenous HGA supply significantly increased tocochromanol biosynthesis in plant cells [54–56]. These indicate that the endogenous mechanism(s) regulating HGA biosynthesis and availability in plant cells directly determines the final amount of tocochromanols. The tocochromanol biosynthetic precursor HGA comes from the degradation of the aromatic amino acid L-tyrosine (L-tyr), which is produced by the plastidic shikimate pathway (Figure 2). L-tyr is first converted into 4-hydroxyphenylpyruvate (HPP) by tyrosine aminotransferases (TATs). Among the six to 10 *TAT* genes identified in the genetic model Arabidopsis, the enzymatic activity has been experimentally confirmed only for TAT1 (also named TAT7) and TAT2 [57–60]. Regarding tocochromanol synthesis, it has been shown that TAT1/TAT7 controls 35–50% of α-tocopherol biosynthesis in Arabidopsis leaves [59]. The other TAT(s) involved in the TAT1-independent tocochromanol biosynthesis, as well as the ones involved in seed tocochromanol biosynthesis, have not been identified yet. Following tyrosine transamination, HPP is converted by 4-hydroxyphenylpyruvate dioxygenase (HPPD) into HGA (Figure 2). A corpus of biochemical, genetic, and cell biology data indicate that the cellular compartment hosting HGA biosynthesis is different among plant species. This aspect of tocochromanol biosynthesis has been recently reviewed in detail [17]. In the genetic model Arabidopsis, several studies have provided evidence that HGA biosynthesis is confined to the cytoplasm. The sequence analysis of the Arabidopsis *TAT1/TAT7* and *HPPD* genes have shown that both lack the typical sequence encoding for a chloroplast transit peptide, suggesting that TAT1/AtTAT7 and HPPD are localized in the cytoplasm [60]. This has been confirmed by Western blot for HPPD, and with TAT1:GFP and HPPD:GFP fusion proteins, which were both localized in the cytoplasm of Arabidopsis cells [60,61].

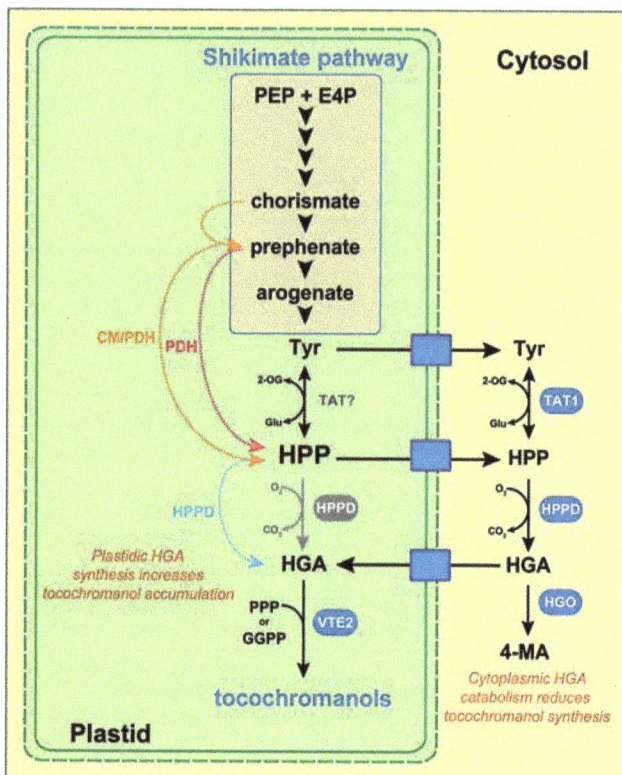

Figure 2. Biosynthesis and transport of homogentisate in plants. Red, orange, and blue dotted lines correspond to transgenic plants overexpressing the coding sequence of yeast *PDH*, bacterial *CM/PDH*, and plant *HPPD*, respectively, all fused to a sequence encoding a chloroplast transit peptide. Biosynthetic enzymes demonstrated to be involved in tocochromanol synthesis are highlighted in blue or gray. In species such as Arabidopsis, HPPD is localized in the cytoplasm. In contrast, in maize, tomato, and cotton, *HPPD* genes exhibit a typical chloroplast transit signal, suggesting that this enzyme is localized in the chloroplasts of these species (HPPD in grey). In soybean, HPPD have been localized in both compartments. Plant species in which HGA biosynthesis is localized in the cytosol must have chloroplast membrane transporters (blue boxes) exporting Tyr and HPP into the cytosol, and importing HGA back into chloroplasts. Abbreviations: CM/PDH, bacterial bi-functional chorismate mutase/prephenate dehydrogenase; E4P, erythrose 4-phosphate; GGPP, geranylgeranyl pyrophosphate; Glu, glutamate; HGA, homogentisate; HGO, homogentisate dioxygenase; HPP, 4-hydroxyphenylpyruvate; HPPD, 4-hydroxyphenylpyruvate dioxygenase; 4-MA, 4-maleylacetoacetate; 2-OG, 2-oxoglutarate; PDH, prephenate dehydrogenase; PEP, phosphoenolpyruvate; PPP, phytyl pyrophosphate; TAT, tyrosine aminotransferase; Tyr, L-tyrosine.

In addition, *HPPD* genes isolated from barley, carrot, and coleus also lack chloroplast transit peptide sequences [62]. Collectively, these demonstrate that HGA biosynthesis, at least in these species, is exclusively localized in the cytoplasm. This data has major consequences regarding the regulation of tocochromanol biosynthesis in these species, since the cytoplasmic HGA biosynthesis implies the existence of chloroplast membrane transporters that export L-tyr/HPP into the cytoplasm, and other one(s) that import HGA back into chloroplasts (Figure 2). Thus, beyond HGA biosynthesis *per se*, these chloroplast membrane transporters might constitute additional levels of regulation of plastidic HGA availability, and thus determine the final amount of tocochromanol produced by a given tissue.

Genetic data obtained from soybean further illustrate how the cytoplasmic HGA biosynthesis and the exchanges of tocochromanol biosynthetic precursors between chloroplasts and the cytoplasm participate in the regulation of tocochromanol biosynthesis in plants. In this species, the HPPD activity is encoded by a single gene that exhibits two transcription start sites. It has been shown that the long transcript encodes a polypeptide imported into the chloroplast, while the short one encodes a polypeptide that remains in the cytoplasm [63]. This indicates that soybean HPPD activity is localized in both the cytoplasm and chloroplasts. In addition, seeds of the soybean *hgo1* mutant, which carries a mutated *HOMOGENTISATE DIOXYGENASE 1* gene encoding an enzyme that catabolizes HGA into 4-maleylacetoacetate (Figure 2), exhibits higher amounts of both HGA and tocochromanols [64]. Interestingly, the HGO1 activity catabolizing HGA is localized in the cytoplasm of soybean cells, indicating that the cytoplasmic HGA catabolism negatively impacts tocochromanol biosynthesis in this species. Similarly, the Arabidopsis HGO gene (At5g54080) encodes a protein expected to be localized in the cytoplasm, according to the The Arabidopsis Information Resource website. Collectively, these data demonstrate that the cytoplasmic HGA synthesis and catabolism in species such as Arabidopsis and soybean are two mechanisms that determine the final amount of tocochromanols in seeds. Moreover, they reveal that yet-unknown chloroplast membrane transporters exporting L-tyr/HPP into the cytoplasm, and other one(s) importing HGA back into chloroplasts, are potentially limiting factors for tocochromanol accumulation. To date, neither the membrane chloroplast transporters exporting L-tyr and HPP in the cytoplasm, nor the one importing HGA back into chloroplasts, have been identified in Arabidopsis. In contrast, a cationic amino acid transporter localized in the chloroplast membrane and capable of exporting phenylalanine, tryptophan, and L-tyr from the stroma into the cytosol has been identified in petunia [65].

The major role of plastidic HGA availability and its transport from the cytoplasm into plastids in the regulation of tocochromanol biosynthesis can be deduced from four metabolic engineering studies that aimed at increasing plant vitamin E activity. In higher plants, HPP derives from the sequential catabolism of L-tyr into chorismate, prephenate, and arogenate (Figure 2). In plants, L-tyr restricts its own biosynthesis, and thus the one of tocochromanols, by feedback inhibition of both arogenate dehydrogenase and prephenate dehydrogenase [20]. In contrast, prokaryotes and yeast produce HPP directly from chorismate via bi-functional chorismate mutase (CM)/prephenate dehydrogenase (PDH) activity, and from prephenate via PDH activity, respectively (Figure 2). It has been previously shown that the overexpression of bacterial *CM/PDH* or yeast *PDH* genes in transgenic plants, including Arabidopsis, strongly increases the accumulation of both HGA and tocochromanols [20,56,66]. Interestingly, both *CM/PDH* and *PDH* coding sequences were fused to a sequence encoding a chloroplast transit peptide and expressed in combination with *HPPD* genes that were also fused to a sequence encoding a chloroplast transit peptide. Collectively, these studies demonstrate that bypassing the plant L-tyr feedback inhibition and functionalizing chloroplasts with the HPPD activity significantly stimulate tocochromanol metabolism in plants.

Interestingly, the overexpression of bacterial *CM/PDH* or yeast *PDH* genes alone is more effective in terms of tocochromanol biosynthesis in Arabidopsis leaves than when these genes are co-expressed in combination with *HPPD* [20,36]. Compared with transgenic Arabidopsis overexpressing *CM/PDH* alone, the lower tocochromanol accumulation in transgenic Arabidopsis leaves carrying both transgenes (*CM/PDH* or *PDH* with *HPPD*) correlates with a much higher accumulation of free HGA [20]. This indicates that whereas HGA availability is limiting for tocochromanol accumulation, a high HGA concentration likely inhibits their biosyntheses. In addition to revealing the potentially toxic effect of HGA on tocochromanol metabolism, these two later studies indicate that the HPPD activity and HPP/HGA trafficking across chloroplast membranes are both not limiting in Arabidopsis leaves. Indeed, the overaccumulation of tocochromanols resulting from the sole overexpression of *CM/PDH* or *PDH* in chloroplasts implies that HPP is efficiently exported into the cytosol where the HPPD activity is localized, and that the cytoplasmic HGA is imported back into chloroplasts to sustain tocochromanol biosynthesis (Figure 2). Since tocochromanol biosynthesis is strongly increased by the overexpression of a bacterial *CM/PDH* gene alone in Arabidopsis leaves, one can conclude that the endogenous HPPD activity (cytoplasmic) and HPP/HGA trafficking are both not so limiting for leaf tocochromanol metabolism in Arabidopsis. In contrast to this, the situation seems very different in tobacco leaves; while the overexpression of the yeast *PDH* gene does not significantly alter tocochromanol accumulation in transgenic tobacco leaves, its overexpression in combination with *HPPD* fused to a chloroplast targeting sequence strongly increases tocochromanol accumulation [66]. The reasons explaining these differences between Arabidopsis and tobacco leaves are currently unknown but suggest that HPPD activity and/or HPP/HGA trafficking in tobacco is limiting compared to Arabidopsis leaves.

In Arabidopsis seeds, re-analysis of previously published data indicates that HPP/HGA trafficking across the chloroplast membranes is much less efficient than in leaves despite the significant tocochromanol metabolism in this tissue. Co-expression of *CM/PDH* in combination with *HPPD* in Arabidopsis stimulates seed tocochromanol accumulation, whereas the expression of *CM/PDH* alone does not significantly alter it (construct pMON36596 versus pMON36520 in the Supplementary data of Karunanandaa et al. [56], 2005, respectively). In this study, a sequence encoding a chloroplast transit peptide was fused to both *CM/PDH* and *HPPD* sequences as well, showing that seed tocochromanol might be enhanced in seeds if seed chloroplasts are functionalized with both activities. This indicates that the endogenous HPPD activity and/or HPP/HGA trafficking is much less efficient in Arabidopsis seeds than in leaves. However, the *HPPD* gene is strongly expressed in Arabidopsis seeds, a tissues in which *HPPD* expression is among its highest during plant development (Figure 3).

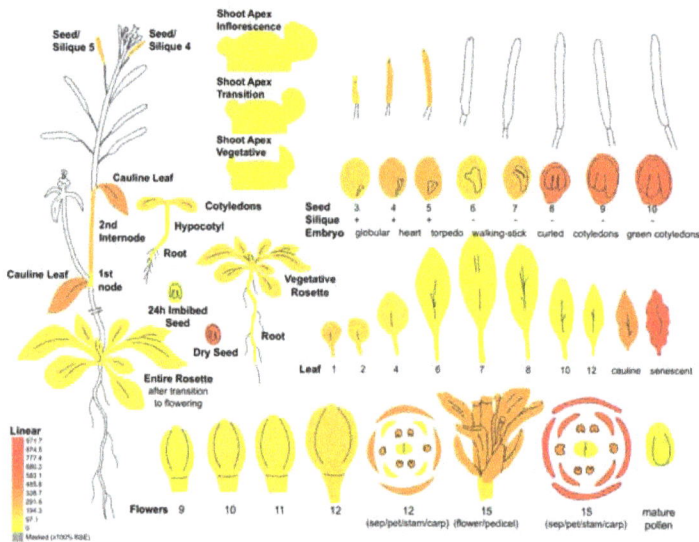

Figure 3. *HPPD* expression pattern during Arabidopsis development. The expression of the Arabidopsis *HPPD* gene (At1g06570) was assessed with ePlant (http://bar.utoronto.ca/eplant). *HPPD* expression level is equal to 697.5 in green cotyledons (SD = 36.3; n = 3), and is the third highest *HPPD* expression after sepals (level = 971.7; SD = 84.2; n = 3) and senescent leaves (level = 797.7; SD = 6.9; n = 3).

This indicates that HPPD activity is likely not limiting tocochromanol metabolism in Arabidopsis seeds, and that in contrast, HPP/HGA trafficking restricts tocochromanol biosynthesis in these tissues. This novel interpretation of old data also suggests that HPP/HGA trafficking across the chloroplast membrane is likely a target of choice to increase tocochromanol biosynthesis in seeds.

The tocochromanol patterns of transgenic plants expressing bacterial *CM/PDH* and yeast *PDH* alone or in combination with *HPPD* vary according to the plant species, organ, and developmental stage. Thus, while young tobacco and Arabidopsis leaves overexpressing *CM/PDH* or *PDH* genes overaccumulate primarily tocotrienols, mature Arabidopsis transgenic leaves expressing these constructs accumulate mostly tocopherols [20,36,66]. In contrast to leaves, Arabidopsis and soybean transgenic seeds expressing *CM/PDH* or *PDH* genes primarily overaccumulate tocotrienols [56]. Since the type of tocochromanol results from the polyprenyl pyrophosphate substrate, these results indicate that the tocopherol precursor PPP is likely limiting in seeds and young leaf tissues, whereas GGPP is not.

5. Regulation of Polyprenyl Pyrophosphate Availability

Since the type of tocochromanol (tocopherol, tocotrienol, PC-8, tocomonoenol) is determined by the polyprenyl pyrophosphate substrate condensed to HGA, the biosynthetic pathways producing the lipophilic tocochromanol precursors (PPP, GGPP, SPP, and THGGPP) are key components in the regulation of tocochromanol biosynthesis. To date, although the tocotrienol biosynthetic pathway is well characterized, and key tocotrienol biosynthetic genes have been identified in plants, there are currently no data describing the mechanisms regulating the biosynthesis of GGPP used for tocotrienol biosynthesis [32]. This may result from the fact that the wild accessions of the genetic model Arabidopsis do not accumulate the tocotrienols that are mostly prevalent in the seeds of Poaceae. As the GGPP that is used for tocotrienol synthesis likely originates from the plastidic methyl erythritol phosphate pathway (MEP), one can assume that the regulation of tocotrienol biosynthesis might be directly linked to this pathway. In addition, the mapping of the tocotrienol QTLs identified in maize and barley grains might improve our understanding of the mechanisms regulating its accumulation [67,68]. It is known that tocotrienols accumulate in plants or tissues that do not usually produce them when HGA is overaccumulated, such as in transgenic Arabidopsis plants expressing *CM/PDH* or *PDH* genes, or in the leaves of transgenic tobacco carrying similar constructs [20,36,56,66]. The current interpretation of these data is that GGPP is likely not limiting in these species or tissues, and that HGA overaccumulation alters the specificity of the tocopherol prenyltransferase. These results might indicate that HGA availability in Poaceae seeds might contribute to the regulation of tocotrienol biosynthesis in these species.

The biosynthesis of PC-8 and its solanesyl benzoquinone precursor plastoquinone-9 (PQ-9) is initiated by the homogentisate solanesyltransferase-dependent condensation of HGA with SPP that comes from the preferential concatenation of GGPP and isopentenyl pyrophosphate [25,69–71]. To date, four genes have been found to specifically regulate PQ-9/PC-8 accumulation in plants. The *Nicotinamide Adenine Dinucleotide Phosphate (NADPH) dehydrogenase C1 (NDC1;* AT5G08740) is a type II NAD(P)H quinone oxidoreductase that reduces PQ-9 into PQH$_2$-9, using ferredoxin as an electron donor [72]. This enzyme is localized in plastoglobuli together with prenyl quinones/quinols and tocochromanols such as PC-8. In Arabidopsis *ndc1* mutants, the leaf PQ pool is more oxidized than in wild-type controls, while PC-8 amounts are reduced by ca 66% [72]. In contrast, leaf tocopherol levels were not affected by the mutation. The proposed model explaining the role of NDC1 in PC-8 accumulation is based on tocopherol cyclase—the enzyme that notably catalyzes the cyclization of PQH$_2$-9 into PC-8—preferentially using reduced prenyl quinones [73]. Two ABC1 atypical kinases, *ABC1K1* (AT4G31390) and *K3* (AT1G79600), have been shown to regulate PQ-9/PC-8 metabolism in Arabidopsis [74–76]. Both kinases are abundant in plastoglobuli, and single and double *abc1k1* and *abc1k3* mutants exhibit altered PQ-9/PC-8 metabolism. Although the function of these atypical kinases in PQ-9/PC-8 metabolism has been clearly demonstrated, their mode(s) of action remains to be determined. It has been shown that both ABC1K1 and K3 phosphorylate the tocopherol cyclase VTE1. Since the biosynthesis of tocopherols, which requires VTE1 activity as well, was not affected in *abc1k3* leaves, it is unlikely that the ABC1K3 mode of action is mediated by VTE1 phosphorylation. In contrast, tocopherol amounts were significantly reduced in *abc1k1* leaves, suggesting that the ABC1K1 mode of action might involve the regulation of the tocopherol cyclase activity. The role of ABC1K1 and K3 in tocochromanol metabolism might also be mediated by the synthesis of fribrillin, a group of lipid-associated proteins that are localized in chloroplasts and essential for the biosynthesis of PQ-9, and thus PC-8 [77,78]. Indeed, Arabidopsis *fibrillin-5* mutants exhibit lower PQ-9/PC-8 amounts, and it was shown that mutation in *abc1k3* strongly reduces the accumulation of fibrillin proteins [74,77].

Regarding the regulation of tocopherol biosynthesis, it is known that the availability of phytol is a major parameter governing the final amount of tocopherols in a plant tissue. It has been shown, for instance, that the supplementation of plant cell suspensions with free phytol is sufficient to significantly increase tocopherol accumulation in these cells [54–56]. Moreover, feeding Arabidopsis seedlings with radiolabeled phytol has been associated with the biosynthesis of radiolabeled α-tocopherol, thus suggesting that free phytol is likely recycled into tocopherols [79]. The first genetic evidence supporting this biosynthetic model was provided by the Arabidopsis *vte5* that lacks 80% and 65% of the tocopherols in seeds and leaves, respectively [80]. Recently, the VTE5-dependent biosynthesis of tocopherols has also been reported in tomato leaves and fruits [81]. VTE5 is a phytol kinase that phosphorylates the free phytol that is released during chlorophyll catabolism into phytyl phosphate [80]. The chlorophyll-dependent tocopherol biosynthesis is further supported by genetic data provided by the Arabidopsis *g4/chlsyn1* mutant, which carries a mutated chlorophyll synthase that esterifies geranylgeranyl pyrophosphate on chlorophyllide *a* [82]. The leaves of this mutant grown in vitro on a media supplemented with sucrose are albino, and do not accumulate tocopherols. In addition, *g4/chlsyn1* seeds lack 75% of the tocopherols, thus indicating that tocopherol biosynthesis is also mostly chlorophyll-dependent in Arabidopsis seeds as well. Collectively, these data demonstrate that the PPP used for tocopherol biosynthesis primarily originates from the recycling of phytol released during chlorophyll catabolism in both leaves and seeds. This biosynthetic model, including details about enzymes such as Mg-dechelatase, chlorophyllases, and pheophytinase, has been very recently reviewed, and will not be developed here [17].

If one can easily understand the origin of PPP in the seeds of chloroembryophytes such as Arabidopsis or rapeseed, which accumulate significant amounts of chlorophylls in developing embryos, how do leucoembryophyte seeds produce tocopherols, since they *a priori* do not synthetize chlorophylls? A comprehensive analysis of vitamin E natural variation in maize grain recently provided an unexpected answer to this puzzling question. Among the 52 QTLs identified for maize grain tocochromanols, QTL5 and QTL24 exhibited scores for phenotypic variance explained among the highest of all identified QTLs [68]. Interestingly, DNA loci covered by QTL5 andQTL24 both carry a gene encoding protochlorophyllide reductase, an enzyme of the chlorophyll biosynthetic pathway [68]. This result suggests that despite their lack of green coloration, leucoembryophyte seeds such as maize grain might produce chlorophylls that, once degraded and recycled, might provide PPP for tocopherol biosynthesis. A biochemical analysis of developing maize grain showed that, whereas they lack any macroscopic green coloration, embryos accumulate very low but detectable traces of the chlorophyll immediate precursor chlorophyllide *a*, of chlorophylls *a* and *b*, and of the first chlorophyll catabolite that lacks the Mg atom pheophytin *a*. In addition, endosperm also accumulated detectable traces of chlorophyll *a*. Although further studies—notably with genetic evidence—are required to definitively link chlorophyll metabolism and tocopherol biosynthesis in leucoembryophyte seeds, these data indicate that leucoembryophyte seeds might produce tocopherols *via* the same mechanism originally described in chloroembryophyte seeds [80,83]. This exiting perspective would definitively link tocopherol metabolism to the one of chlorophylls in plants, and open news questions regarding tocopherol metabolism in non-photosynthetic organisms such as in *Plasmodium falciparum* [14–16].

Although the chlorophyllic origin of tocopherols is clearly established in the seeds and leaves of chloroembryophyte species such as Arabidopsis, little is known about the regulation of PPP biosynthesis linked to tocopherol accumulation. Recently, two novel Arabidopsis ethane methylsulfonate mutants, the *enhanced vitamin e* (*eve*) *1* and *4* mutants, have been isolated by forward genetics [21]. The seeds of *eve1* and *eve4* mutants overaccumulate tocopherols, PC-8, and γ-tocomonoenol, a monounsaturated form of tocochromanols. Interestingly, both *eve1* and *4* seeds did not accumulate any tocotrienols, such as transgenic Arabidopsis plants overaccumulating HGA, suggesting that although HGA metabolism is likely increased in these mutants to sustain tocochromanol metabolism, HGA overaccumulation is not the primary source of its enhancement. In addition, γ-tocomonoenol accumulation has never been reported in Arabidopsis HGA overaccumulating lines, further suggesting that the misregulation of HGA biosynthesis is likely not the primary source of the enhancement of tocochromanol accumulation in *eve1* and *eve4* seeds. The fact that all tocochromanol forms are overaccumulated in *eve1* and *eve4* seeds likely suggests that both mutations affect the chloroplastic isoprenoid metabolism produced by the methyl erythritol phosphate pathway (Figure 4). The mutation responsible for the *eve1* seed phenotype affects the *WRINKLED1* gene that encodes for a transcription factor containing two APETALA2/ethylene response element DNA-binding proteins [84]. This gene is mostly expressed during seed development, although moderate expression has been also detected in roots and flowers [85,86].

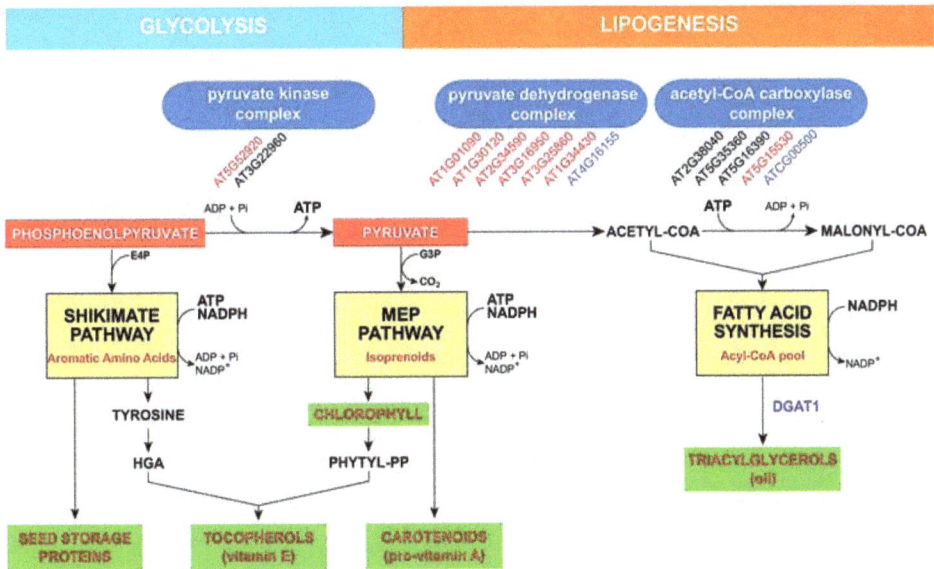

Figure 4. Plastidic intersection of the shikimate pathway, the methyl erythritol phosphate pathway, and lipogenesis. The shikimate pathway, methyl erythritol phosphate (MEP) pathway, and fatty acid synthesis are localized in plastids, and share phosphoenolpyruvate and pyruvate as common biosynthetic precursors. The biosynthesis of triacylglycerols occurs in the endoplasmic reticulum. Arabidopsis Genome Initiative numbers in red indicate that the corresponding gene is downregulated in Arabidopsis *wri1* developing embryos, or upregulated in transgenic plants constitutively expressing the *WRI1* transcription factor. AGI numbers in black indicate that the expression of the corresponding gene is WRI1-independent. AGI numbers in blue indicate that no transcriptomic data was available. Abbreviations: DGAT1, acylCoA:diacylglycerol acyltransferase 1; E4P, erythrose 4-phosphate; G3P, glyceraldehyde 3-phosphate; phytyl-PP, phytyl pyrophosphate.

Targets of the WRI1 transcription factor in plant genomes have been investigated by transcriptomic studies and in vitro enzymatic assays in developing Arabidopsis *wri1* embryos [85–88], in the leaves of transgenic maize lines overexpressing the *ZmWRI1a* gene [89], and in *Nicotiana benthamiana* leaves agroinfiltrated with Arabidopsis, potato, poplar, oat, and nutsedge *WRI1* orthologs [90]. The transcription factor WRI1 regulates the expression of several key biosynthetic genes involved in late glycolysis, fatty acid synthesis, and lipid assembly [85–89]. Thus, WRI1 pushes carbon into fatty acid/lipid metabolism by upregulating key lipid biosynthetic genes. It is interesting to note that the genes encoding almost all of the subunits of the plastidial pyruvate dehydrogenase complex are either downregulated in *wri1* developing embryos or upregulated in transgenic plants overexpressing *WRI1* (Figure 4). Interestingly, tocochromanols are synthetized in chloroplasts as well, and the biosynthetic pathways providing both tocochromanol precursors, i.e., the MEP pathway and the shikimate pathway, use the same biosynthetic precursors, i.e., pyruvate and phosphoenolpyruvate, as those used for fatty acid/lipid synthesis (Figure 4). These suggest that tocochromanol and lipid biosyntheses might compete for these biosynthetic precursors.

Another model explaining the enhancement of tocochromanol metabolism in *wri1* seeds might be considered as well. The transient *WRI1* expression performed in *Nicotiana benthamiana* leaves revealed a novel aspect of WRI1 transcriptome that has not been described neither in transgenic lines constitutively expressing *WRI1* nor in *wri1* seeds. Numerous genes encoding chlorophyll biosynthetic enzymes such as protochlorophyllide oxidoreductase and Mg chelatase, chlorophyll-binding proteins, subunits of the light harvesting complexes, and ATP generation complex subunits were strongly downregulated in tobacco leaves transiently expressing *WRI1* constructs (Supplementary file 6 of Grimberg et al. [90]). The authors concluded that the transient expression of *WRI1* is strongly associated with the downregulation of photosynthesis, including energy and reducing equivalent production, respectively. Interestingly, the endogenous *WRI1* gene is not constitutively expressed in Arabidopsis plants; instead, it is transiently expressed in the developing Arabidopsis embryo [85]. This indicates that the *WRI1* expression in the developing Arabidopsis embryo also might downregulate photosynthesis, and thus the production of ATP, as well as reduce equivalents in embryos. Now, the MEP pathway that produces the building blocks for chloroplastic isoprenoids is strongly dependent on ATP and reducing equivalents such as NADPH (Figure 4). Interestingly, it has been suggested that both ATP and NADPH might be key regulatory components of the MEP pathway [91]. This hypothesis suggests that the enhancement of the tocochromanol metabolism in *wri1* seeds might reflect the higher ATP and NADPH production resulting from the reduced repression of seed photosynthesis.

6. Conclusions and Perspectives

From the first purification of α-tocopherol from wheat germ oil to the latest model of the tocochromanol biosynthetic pathways introducing tocomonoenol biosynthetic genes, our understanding of vitamin E biosynthesis in plants has been considerably enriched [3,21]. In contrast, much less is known about the mechanisms regulating vitamin E biosynthesis in plants. The fact that tocochromanol biosynthesis mobilizes two distinct biosynthetic pathways, the shikimate pathway and the MEP pathway, requires the degradation of two compounds such as L-tyr and chlorophylls, involves a yet-unknown transport system between the cytoplasm and the stroma in plant species in which HGA biosynthesis is localized in the cytoplasm, and is interconnected to fatty acid biosynthesis in plastids, render the understanding of its regulation particularly challenging. Indeed, understanding the regulation of vitamin E biosynthesis will imply that we take up the challenges to understand the regulation of each of these numerous events. Beyond understanding the regulation of vitamin E biosynthesis in plants, it seems important that medical scientists assess the sanitary consequences of the significant vitamin E deficiency independently detected in human populations, including in developed countries. The fundamental role of this vitamin in human reproduction and its benefit in current widespread diseases such as high cholesterol and neurodegenerative pathologies makes it a candidate of choice to improve human health.

Acknowledgments: I am grateful to the University of Fribourg and to the Foundation of Research in Biochemistry and on Vitamins (Epalinges, Switzerland) represented by Claude Bron for supporting the research developed in my laboratory.

Conflicts of Interest: The author declares no conflict of interest.

References

1. Mène-Saffrané, L.; Pellaud, S. Current strategies for vitamin E biofortification of crops. *Curr. Opin. Biotechnol.* **2017**, *44*, 189–197. [CrossRef] [PubMed]

2. Evans, H.M.; Bishop, K.S. On the existence of a hitherto unrecognized dietary factor essential for reproduction. *Science* **1922**, *56*, 650–651. [CrossRef] [PubMed]

3. Evans, H.M.; Emerson, O.H.; Emerson, G.A. The isolation from wheat germ oil of an alcohol, α-tocopherol, having the properties of vitamin E. *J. Biol. Chem.* **1936**, *113*, 319–332. [CrossRef]

4. Galli, F.; Azzi, A.; Birringer, M.; Cook-Mills, J.M.; Eggersdorfer, M.; Frank, J.; Cruciani, G.; Lorkowski, S.; Katar Özer, N. Vitamin E: Emerging aspects and new directions. *Free Radic. Biol. Med.* **2017**, 16–36. [CrossRef] [PubMed]

5. Li, G.X.; Lee, M.J.; Liu, A.B.; Yang, Z.; Lin, Y.; Shih, W.J.; Yang, C.S. δ-tocopherol is more active than α- or γ-tocopherol in inhibiting lung tumorigenesis in vivo. *Cancer Prev. Res.* **2011**, *4*, 404–413. [CrossRef] [PubMed]

6. La Fata, G.; Weber, P.; Mohajeri, M.H. Effects of vitamin E on cognitive performance during ageing and in Alzheimer's disease. *Nutrients* **2014**, *6*, 5453–5472. [CrossRef] [PubMed]

7. Qureshi, A.A.; Burger, W.C.; Peterson, D.M.; Elson, C.E. The structure of an inhibitor of cholesterol biosynthesis isolated from barley. *J. Biol. Chem.* **1986**, *261*, 10544–10550. [PubMed]

8. Qureshi, A.A.; Sami, S.A.; Salser, W.A.; Khan, F.A. Dose-dependent suppression of serum cholesterol by tocotrienol-rich fraction (TRF25) of rice bran in hypercholesterolemic humans. *Atherosclerosis* **2002**, *161*, 199–207. [CrossRef]

9. Maras, J.E.; Bermudez, O.I.; Qiao, N.; Bakun, P.J.; Boody-Alter, E.L.; Tucker, K.L. Intake of α-tocopherol is limited among US adults. *J. Am. Diet. Assoc.* **2004**, *104*, 567–575. [CrossRef] [PubMed]

10. Polito, A.; Intorre, F.; Andriollo-Sanchez, M.; Azzini, E.; Raguzzini, A.; Meunier, N.; Ducros, V.; O'Connor, J.M.; Coudray, C.; Roussel, A.M.; et al. Estimation of intake and status of vitamin A, vitamin E and folate in older European adults: The ZENITH. *Eur. J. Clin. Nutr.* **2005**, *59*, S42–S47. [CrossRef] [PubMed]

11. Kim, Y.N.; Cho, Y.O. Vitamin E status of 20- to 59-year-old adults living in the Seoul metropolitan area of South Korea. *Nutr. Res. Pract.* **2015**, *9*, 192–198. [CrossRef] [PubMed]

12. Shamim, A.A.; Schulze, K.; Merrill, R.D.; Kabir, A.; Christian, P.; Shaikh, S.; Wu, L.; Ali, H.; Labrique, A.B.; Mehra, S.; et al. Tocopherols are associated with risk of miscarriage in rural Bangladesh. *Am. J. Clin. Nutr.* **2015**, *101*, 294–301. [CrossRef] [PubMed]

13. Péter, S.; Friedel, A.; Roos, F.F.; Wyss, A.; Eggersdorfer, M.; Hoffmann, K.; Weber, P. A Systematic review of global alpha-tocopherol status as assessed by nutritional intake levels and blood serum concentrations. *Int. J. Vitam. Nutr. Res.* **2016**, 1–21. [CrossRef] [PubMed]

14. Sussmann, R.A.C.; Angeli, C.B.; Peres, V.J.; Kimura, E.A.; Katzin, A.M. Intraerythrocytic stages of *Plasmodium falciparum* biosynthesize vitamin E. *FEBS Lett.* **2011**, *585*, 3985–3991. [CrossRef] [PubMed]

15. Sussman, R.A.C.; Fotoran, W.L.; Kimura, E.A.; Katzin, A.M. *Plasmodium falciparum* uses vitamin E to avoid oxidative stress. *Parasit. Vectors* **2017**, *10*, 461. [CrossRef] [PubMed]

16. Cassera, M.B.; Gozzo, F.C.; D'Alexandri, F.L.; Merino, E.F.; del Portillo, H.A.; Peres, V.J.; Almeida, I.C.; Eberlin, M.N.; Wunderlich, G.; Wiesner, J.; et al. The methylerythritol phosphate pathway is functionally active in all intraerythrocytic stages of *Plasmodium falciparum*. *J. Biol. Chem.* **2004**, *279*, 51749–51759. [CrossRef] [PubMed]

17. Pellaud, S.; Mène-Saffrané, L. Metabolic origins and transport of vitamin E biosynthetic precursors. *Front. Plant Sci.* **2017**, *8*, 1959. [CrossRef] [PubMed]

18. Collakova, E.; DellaPenna, D. Isolation and functional analysis of homogentisate phytyltransferase from *Synechocystis* sp. PCC 6803 and Arabidopsis. *Plant Physiol.* **2001**, *127*, 1113–1124. [CrossRef] [PubMed]

19. Savidge, B.; Weiss, J.D.; Wong, Y-H.H.; Lassner, M.W.; Mitky, T.A.; Shewmaker, C.K.; Post-Beittenmiller, D.; Valentin, H.E. Isolation and characterization of homogentisate phytyltransferase genes from *Synechocystis* sp. PCC 6803 and Arabidopsis. *Plant Physiol.* **2002**, *129*, 321–332. [CrossRef] [PubMed]

20. Zhang, C.; Cahoon, R.E.; Hunter, S.C.; Chen, M.; Han, J.; Cahoon, E.B. Genetic and biochemical basis for alternative routes of tocotrienol biosynthesis for enhanced vitamin E antioxidant production. *Plant J.* **2013**, *73*, 628–639. [CrossRef] [PubMed]

21. Pellaud, S.; Bory, A.; Chabert, V.; Romanens, J.; Chaisse-Leal, L.; Doan, A.V.; Frey, L.; Gust, A.; Fromm, K.M.; Mène-Saffrané, L. WRINKLED1 and ACYL-COA:DIACYLGLYCEROL ACYLTRANSFERASE1 regulate tocochromanol metabolism in Arabidopsis. *New Phytol.* **2018**, *217*, 1–16. [CrossRef] [PubMed]

22. Yang, W.; Cahoon, R.E.; Hunter, S.C.; Zhang, C.; Han, J.; Borgschulte, T.; Cahoon, E.B. Vitamin E biosynthesis: Functional characterization of the monocot homogentisate geranylgeranyl transferase. *Plant J.* **2011**, *65*, 206–217. [CrossRef] [PubMed]

23. Mène-Saffrané, L.; Jones, A.D.; DellaPenna, D. Plastochromanol-8 and tocopherols are essential lipid-soluble antioxidants during seed desiccation and quiescence in *Arabidopsis*. *Proc. Natl. Acad. Sci. USA* **2010**, *107*, 17815–17820. [CrossRef] [PubMed]

24. Sadre, R.; Gruber, J.; Frentzen, M. Characterization of homogentisate prenyltransferases involved in plastoquinone-9 and tocochromanol biosynthesis. *FEBS Lett.* **2006**, *580*, 5357–5362. [CrossRef] [PubMed]

25. Tian, L.; DellaPenna, D.; Dixon, R.A. The *pds2* mutation is a lesion in the Arabidopsis homogentisate solanesyltransferase gene involved in plastoquinone biosynthesis. *Planta* **2007**, *226*, 1067–1073. [CrossRef] [PubMed]

26. Cheng, Z.; Sattler, S.; Maeda, H.; Sakuragi, Y.; Bryant, D.A.; DellaPenna, D. Highly divergent methyltransferases catalyze a conserved reaction in tocopherol and plastoquinone synthesis in Cyanobacteria and photosynthetic Eukaryotes. *Plant Cell* **2003**, *15*, 2343–2356. [CrossRef] [PubMed]

27. Van Eenennaam, A.L.; Lincoln, K.; Durrett, T.P.; Valentin, H.E.; Shewmaker, C.K.; Thorne, G.M.; Jiang, J.; Baszis, S.R.; Levering, C.K.; Aasen, E.D.; et al. Engineering vitamin E content: From Arabidopsis mutant to soy oil. *Plant Cell* **2003**, *15*, 3007–3019. [CrossRef] [PubMed]

28. Porfirova, S.; Bergmüller, E.; Tropf, S.; Lemke, R.; Dörmann, P. Isolation of an Arabidopsis mutant lacking vitamin E and identification of a cyclase essential for all tocopherol biosynthesis. *Proc. Natl. Acad. Sci. USA* **2002**, *99*, 12495–12500. [CrossRef] [PubMed]

29. Sattler, S.E.; Gilliland, L.U.; Magallanes-Lundback, M.; Pollard, M.; DellaPenna, D. Vitamin E is essential for seed longevity and for preventing lipid peroxidation during germination. *Plant Cell* **2004**, *16*, 1419–1432. [CrossRef] [PubMed]

30. Shintani, D.; DellaPenna, D. Elevating the vitamin E content of plants through metabolic engineering. *Science* **1998**, *282*, 2098–2100. [CrossRef] [PubMed]

31. Bergmüller, E.; Porfirova, S.; Dörmann, P. Characterization of an Arabidopsis mutant deficient in γ-tocopherol methyltransferase. *Plant Mol. Biol.* **2003**, *52*, 1181–1190. [CrossRef] [PubMed]

32. Cahoon, E.B.; Hall, S.E.; Ripp, K.G.; Ganzke, T.S.; Hitz, W.D.; Coughlan, S.J. Metabolic redesign of vitamin E biosynthesis in plants for tocotrienol production and increased antioxidant content. *Nat. Biotechnol.* **2003**, *21*, 1082–1087. [CrossRef] [PubMed]

33. Hosomi, A.; Arita, M.; Sato, Y.; Kiyose, C.; Ueda, T.; Igarashi, O.; Arai, H.; Inoue, K. Affinity of alpha-tocopherol transfer protein as a determinant of the biological activities of vitamin E analogs. *FEBS Lett.* **1997**, *409*, 105–108. [CrossRef]

34. Aeschimann, W.; Staats, S.; Kammer, S.; Olieric, N.; Jeckelmann, J.M.; Fotiadis, D.; Netscher, T.; Rimbach, G.; Cascella, M.; Stocker, A. Self-assembled α-tocopherol transfer protein nanoparticles promote vitamin E delivery across an endothelial barrier. *Sci. Rep.* **2017**, *7*, 4970. [CrossRef] [PubMed]

35. Zbierzak, A.M.; Kanwischer, M.; Wille, C.; Vidi, P.A.; Giavalisco, P.; Lohmann, A.; Briesen, I.; Porfirova, S.; Bréhélin, C.; Kessler, F.; et al. Intersection of the tocopherol and plastoquinol metabolic pathways at the plastoglobule. *Biochem. J.* **2010**, *425*, 389–399. [CrossRef] [PubMed]

36. Tzin, V.; Malitsky, S.; Aharoni, A.; Galili, G. Expression of a bacterial bi-functional chorismate mutase/prephenate dehydratase modulates primary and secondary metabolism associated with aromatic amino acids in Arabidopsis. *Plant J.* **2009**, *60*, 156–167. [CrossRef] [PubMed]

37. Jiang, J.; Jia, H.; Feng, G.; Wang, Z.; Li, J.; Gao, H.; Wang, X. Overexpression of *Medicago sativa TMT* elevates the α-tocopherol content in *Arabidopsis* seeds, alfalfa leaves, and delays dark-induced leaf senescence. *Plant Sci.* **2016**, *249*, 93–104. [CrossRef] [PubMed]

38. Endrigkeit, J.; Wang, X.; Cai, D.; Zhang, C.; Long, Y.; Meng, J.; Jung, C. Genetic mapping, cloning, and functional characterization of the *BnaX.VTE4* gene encoding a γ-tocopherol methyltransferase from oilseed rape. *Theor. Appl. Genet.* **2009**, *119*, 567–575. [CrossRef] [PubMed]

39. Tang, S.; Hass, C.G.; Knapp, S.J. *Ty3/gypsy*-like retrotransposon knockout of a 2-methyl-6-phytyl-1,4-benzoquinone methyltransferase is non-lethal, uncovers a cryptic paralogous mutation, and produces novel tocopherol (vitamin E) profiles in sunflower. *Theor. Appl. Genet.* **2006**, *113*, 783–799. [CrossRef] [PubMed]

40. Lampi, A.M.; Nurmi, T.; Ollilainen, V.; Piironen, V. Tocopherols and tocotrienols in wheat genotypes in the HEALGRAIN diversity screen. *J. Agric. Food Chem.* **2008**, *56*, 9716–9721. [CrossRef] [PubMed]

41. Goffman, F.D.; Velasco, L.; Thies, W. Quantitative determination of tocopherols in single seeds of rapeseed (*Brassica napus* L.). *Lipid* **1999**, *101*, 142–145. [CrossRef]

42. Xu, S.T.; Zhang, D.L.; Cai, Y.; Zhou, Y.; Shah, T.; Ali, F.; Li, Q.; Li, Z.G.; Wang, W.D.; Li, J.S.; et al. Dissecting tocopherols content in maize (*Zea mays* L.), using two segregating populations and high-density single nucleotide polymorphism markers. *BMC Plant Biol.* **2012**, *12*, 201–214. [CrossRef]

43. Lipka, A.E.; Gore, M.A.; Magallanes-Lundback, M.; Mesberg, A.; Lin, H.; Tiede, T.; Chen, C.; Buell, C.R.; Buckler, E.S.; Rocheford, T.; et al. Genome-wide association study and pathway-level analysis of tocochromanol levels in maize grain. *G3 (Bethesda)* **2013**, *3*, 1287–1299. [CrossRef] [PubMed]

44. Ujiie, A.; Yamada, T.; Fujimoto, K.; Endo, Y.; Kitamura, K. Identification of soybean varieties with high α-tocopherol content. *Breed. Sci.* **2005**, *55*, 123–125. [CrossRef]

45. Dwiyanti, M.S.; Yamada, T.; Sato, M.; Abe, J.; Kitamura, K. Genetic variation of γ-tocopherol methyltransferase gene contributes to elevated α-tocopherol content in soybean seeds. *BMC Plant Biol.* **2011**, *11*, 152–168. [CrossRef] [PubMed]

46. Gilliland, L.U.; Magallanes-Lundback, M.; Hemming, C.; Supplee, A.; Koornneef, M.; Bentsink, L.; DellaPenna, D. Genetic basis for natural variation in seed vitamin E levels in *Arabidopsis thaliana*. *Proc. Natl. Acad. Sci. USA* **2006**, *103*, 18834–18841. [CrossRef] [PubMed]

47. Li, H.; Wang, Y.; Han, Y.; Teng, W.; Zhao, X.; Li, Y.; Li, W. Mapping quantitative trait loci (QTLs) underlying seed vitamin E content in soybean with main, epistatic and QTL × environment effects. *Plant Breed.* **2016**, *135*, 208–214. [CrossRef]

48. Shaw, E.J.; Rajcan, I. Molecular mapping of soybean seed tocopherols in the cross 'OAC Bayfield' × 'OAC Shire'. *Plant Breed.* **2017**, *136*, 83–93. [CrossRef]

49. Chander, S.; Guo, Y.Q.; Yang, X.H.; Yan, J.B.; Zhang, Y.R.; Song, T.M.; Li, J.S. Genetic dissection of tocopherol content and composition in maize grain using quantitative trait loci analysis and the candidate gene approach. *Mol. Breed.* **2008**, *22*, 353–365. [CrossRef]

50. Li, Q.; Yang, X.; Xu, S.; Cai, Y.; Zhang, D.; Han, Y.; Li, L.; Zhang, Z.; Gao, S.; Li, J.; et al. Genome-wide association studies identified three independent polymorphisms associated with α-tocopherol content in maize kernels. *PLoS ONE* **2012**, *7*, e36807. [CrossRef] [PubMed]

51. Marwede, V.; Gül, M.K.; Becker, H.C.; Ecke, W. Mapping of QTL controlling tocopherol content in winter oilseed rape. *Plant Breed.* **2005**, *124*, 20–26. [CrossRef]

52. Wang, X.; Zhang, C.; Li, L.; Fritsche, S.; Endrigkeit, J.; Zhang, W.; Long, Y.; Jung, C.; Meng, J. Unraveling the genetic basis of seed tocopherol content and composition in rapeseed (*Brassica napus* L.). *PLoS ONE* **2012**, *7*, e50038. [CrossRef] [PubMed]

53. Oliver, R.E.; Islamovic, E.; Obert, D.E.; Wise, M.L.; Herrin, L.L.; Hang, A.; Harrison, S.A.; Ibrahim, A.; Marshall, J.M.; Miclaus, K.J.; et al. Comparative systems biology reveals allelic variation modulating tocochromanol profiles in barley (*Hordeum vulgare* L.). *PLoS ONE* **2014**, *9*, e96276. [CrossRef] [PubMed]

54. Furuya, T.; Yoshikawa, T.; Kimura, T.; Kaneko, H. Production of tocopherols by cell culture of safflower. *Phytochemistry* **1987**, *26*, 2741–2747. [CrossRef]

55. Caretto, S.; Speth, E.B.; Fachechi, C.; Gala, R.; Zacheo, G.; Giovinazzo, G. Enhancement of vitamin E production in sunflower cell cultures. *Plant Cell Rep.* **2004**, *23*, 174–179. [CrossRef] [PubMed]

56. Karunanandaa, B.; Qi, Q.; Hao, M.; Baszis, S.R.; Jensen, P.K.; Wong, Y.H.H.; Jiang, J.; Venkatramesh, M.; Gruys, K.J.; Moshiri, F.; et al. Metabolically engineered oilseed crops with enhanced seed tocopherol. *Metab. Eng.* **2005**, *7*, 384–400. [CrossRef] [PubMed]

57. Prabhu, P.R.; Hudson, A.O. Identification and partial characterization of an L-tyrosine aminotransferase (TAT) from *Arabidopsis thaliana*. *Biochem. Res. Int.* **2010**, 549572. [CrossRef]

58. Grossmann, K.; Hutzler, J.; Tresch, S.; Christiansen, N.; Looser, R.; Ehrhardt, T. On the mode of action of the herbicides cinmethylin and 5-benzyloxymethyl-1,2-isoxyzolines: Putative inhibitors of plant tyrosine aminotransferase. *Pest Manag. Sci.* **2012**, *68*, 482–492. [CrossRef] [PubMed]

59. Riewe, D.; Koohi, M.; Lisec, J.; Pfeiffer, M.; Lippmann, R.; Schmeichel, J.; Willmitzer, L.; Altmann, T. A tyrosine aminotransferase involved in tocopherol synthesis in Arabidopsis. *Plant J.* **2012**, *71*, 850–859. [CrossRef] [PubMed]

60. Wang, M.; Toda, K.; Maeda, H.A. Biochemical properties and subcellular localization of tyrosine aminotransferases in *Arabidopsis thaliana*. *Phytochemistry* **2016**, *132*, 16–25. [CrossRef] [PubMed]

61. Garcia, I.; Rodgers, M.; Pepin, R.; Hsieh, T.Z.; Matringe, M. Characterization and subcellular compartmentation of recombinant 4-hydroxyphenylpyruvate dioxygenase from Arabidopsis in transgenic tobacco. *Plant Physiol.* **1999**, *119*, 1507–1516. [CrossRef] [PubMed]

62. Falk, J.; Krauss, N.; Dähnhardt, D.; Krupinska, K. The senescence associated gene of barley encoding 4-hydroxyphenylpyruvate dioxygenase is expressed during oxidative stress. *J. Plant. Physiol.* **2002**, *159*, 1245–1253. [CrossRef]

63. Siehl, D.L.; Tao, Y.; Albert, H.; Dong, Y.; Heckert, M.; Madrigal, A.; Lincoln-Cabatu, B.; Lu, J.; Fenwick, T.; Bermudez, E.; et al. Broad 4-hydroxyphenylpyruvate dioxygenase inhibitor herbicide tolerance in soybean with an optimized enzyme and expression cassette. *Plant Physiol.* **2014**, *166*, 1162–1176. [CrossRef] [PubMed]

64. Stacey, M.G.; Cahoon, R.E.; Nguyen, H.T.; Cui, Y.; Sato, S.; Nguyen, C.T.; Phoka, N.; Clark, K.M.; Liang, Y.; Forrester, J.; et al. Identification of homogentisate dioxygenase as a target for vitamin E biofortification in oilseeds. *Plant Physiol.* **2016**, *172*, 1506–1518. [CrossRef] [PubMed]

65. Widhalm, J.R.; Gutensohn, M.; Yoo, H.; Adebesin, F.; Qian, Y.; Guo, L.; Jaini, R.; Lynch, J.H.; McCoy, R.M.; Shreve, J.T.; et al. Identification of a plastidial phenylalanine exporter that influences flux distribution through the phenylalanine biosynthetic network. *Nat. Commun.* **2015**, *6*, 8142–8152. [CrossRef] [PubMed]

66. Rippert, P.; Scimemi, C.; Dubald, M.; Matringe, M. Engineering plant shikimate pathway for production of tocotrienol and improving herbicide resistance. *Plant Physiol.* **2004**, *134*, 92–100. [CrossRef] [PubMed]

67. Graebner, R.C.; Wise, M.; Cuesta-Marcos, A.; Geniza, M.; Blake, T.; Blake, V.C.; Butler, J.; Chao, S.; Hole, D.J.; Horsley, R.; et al. Quantitative trait loci associated with the tocochromanol (vitamin E) pathway in barley. *PLoS ONE* **2015**, *10*, e0133767. [CrossRef] [PubMed]

68. Diepenbrock, C.H.; Kandianis, C.B.; Lipka, A.E.; Magallanes-Lundback, M.; Vaillancourt, B.; Góngora-Castillo, E.; Wallace, J.G.; Cepela, J.; Mesberg, A.; Bradbury, P.J.; et al. Novel loci underlie natural variation in vitamin E levels in maize grain. *Plant Cell* **2017**, *29*, 2374–2392. [CrossRef] [PubMed]

69. Block, A.; Fristedt, R.; Rogers, S.; Kumar, J.; Barnes, B.; Barnes, J.; Elowsky, C.G.; Wamboldt, Y.; Mackenzie, S.A.; Redding, K.; et al. Functional modelling identifies paralogous solanesyl-diphosphate synthases that assemble the side chain of plastoquinone-9 in plastids. *J. Biol. Chem.* **2013**, *288*, 27594–27606. [CrossRef] [PubMed]

70. Hirooka, K.; Bamba, T.; Fukusaki, E.-I.; Kobayashi, A. Cloning and kinetic characterization of Arabidopsis thaliana solanesyl diphosphate synthase. *Biochem. J.* **2003**, *370*, 679–686. [CrossRef] [PubMed]

71. Hirooka, K.; Izumi, Y.; An, C.-I.; Nakazawa, Y.; Fukusaki, E.-I.; Kobayashi, A. Functional analysis of two solanesyl diphosphate synthases from Arabidopsis thaliana. *Biosci. Biotechnol. Biochem.* **2005**, *69*, 592–601. [CrossRef] [PubMed]

72. Eugeni Piller, L.; Besagnia, C.; Ksas, B.; Rumeau, D.; Bréhélin, C.; Glauser, G.; Kessler, F.; Havaux, M. Chloroplast lipid droplet type II NAD(P)H quinone oxidoreductase is essential for prenylquinone metabolism and vitamin K1 accumulation. *Proc. Natl. Acad. Sci. USA* **2011**, *108*, 14354–14359. [CrossRef] [PubMed]

73. Grütter, C.; Alonso, E.; Chougnet, A.; Woggon, W.D. A biomimetic chromanol cyclization leading to α-tocopherol. *Angew. Chem. Int. Ed.* **2006**, *45*, 1126–1130. [CrossRef] [PubMed]

74. Martinis, J.; Glauser, G.; Valimareanu, S.; Kessler, F. A chloroplast ABC1-like kinase regulates vitamin E metabolism in Arabidopsis. *Plant Physiol.* **2013**, *162*, 652–662. [CrossRef] [PubMed]

75. Martinis, J.; Glauser, G.; Valimareanu, S.; Stettler, M.; Zeeman, S.C.; Yamamoto, H.; Shikanai, T.; Kessler, F. ABC1K1/PGR6 kinase: A regulatory link between photosynthetic activity and chloroplast metabolism. *Plant J.* **2014**, *77*, 269–283. [CrossRef] [PubMed]

76. Lundquist, P.K.; Poliakov, A.; Giacomelli, L.; Friso, G.; Appel, M.; McQuinn, R.P.; Krasnoff, S.B.; Rowland, E.; Ponnala, L.; Sun, Q.; et al. Loss of plastoglobule kinases ABC1K1 and ABC1K3 causes conditional degreening, modified prenyl-lipids, and recruitment of the jasmonic acid pathway. *Plant Cell* **2013**, *25*, 1818–1839. [CrossRef] [PubMed]

77. Kim, E.-H.; Lee, Y.; Kim, H.U. Fibrillin 5 is essential for plastoquinone-9 biosynthesis by binding to solanesyl diphosphate synthases in Arabidopsis. *Plant Cell* **2015**, *27*, 2956–2971. [CrossRef] [PubMed]

78. Kim, E.-H.; Lee, D-W.; Lee, K-R.; Jung, S-J.; Jeon, J-S.; Kim, H.U. Conserved function of fibrillin5 in the plastoquinone-9 biosynthetic pathway in Arabidopsis and rice. *Front. Plant Sci.* **2017**, *8*, 1197. [CrossRef] [PubMed]

79. Ischebeck, T.; Zbierzak, A.M.; Kanwischer, M.; Dormann, M. A salvage pathway for phytol metabolism in Arabidopsis. *J. Biol. Chem.* **2006**, *281*, 2470–2477. [CrossRef] [PubMed]

80. Valentin, H.; Lincoln, K.; Moshiri, F.; Jensen, P.K.; Qi, Q.; Venkatesh, T.V.; Karunanandaa, B.; Baszis, S.R.; Norris, S.R.; Savidge, B.; et al. The Arabidopsis *vitamin E pathway gene5-1* mutant reveals a critical role for phytol kinase in seed tocopherol biosynthesis. *Plant Cell* **2006**, *18*, 212–224. [CrossRef] [PubMed]

81. Almeida, J.; da Silva Azevedo, M.; Spicher, L.; Glauser, G.; vom Dorp, K.; Guyer, L.; del Valle Carranza, A.; Asis, R.; Pereira de Souza, A.; Buckeridge, M.; et al. Down-regulation of tomato PHYTOL KINASE strongly impairs tocopherol biosynthesis and affects prenyllipid metabolism in an organ-specific manner. *J. Exp. Bot.* **2016**, *67*, 919–934. [CrossRef] [PubMed]

82. Zhang, C.; Zhang, W.; Ren, G.; Li, D.; Cahoon, R.E.; Chen, M.; Zhou, Y.; Yu, B.; Cahoon, E.B. Chlorophyll synthase under epigenetic surveillance is critical for vitamin E synthesis, and altered expression affects tocopherol levels in Arabidopsis. *Plant Physiol.* **2015**, *168*, 1503–1511. [CrossRef] [PubMed]

83. Vom Dorp, K.; Hölzl, G.; Plohmann, C.; Eisenhut, M.; Abraham, M.; Weber, A.P.M.; Hanson, A.D.; Dörmann, P. Remobilization of phytol from chlorophyll degradation is essential for tocopherol synthesis and growth of Arabidopsis. *Plant Cell* **2015**, *27*, 2846–2859. [CrossRef] [PubMed]

84. Cernac, A.; Benning, C. *WRINKLED1* encodes an AP2/EREB domain protein involved in the control of storage compound biosynthesis in Arabidopsis. *Plant J.* **2004**, *40*, 575–585. [CrossRef] [PubMed]

85. Baud, S.; Santos Mendoza, M.; To, A.; Harscoët, E.; Lepiniec, L.; Dubreucq, B. WRINKLED1 specifies the regulatory action of LEAFY COTYLEDON2 towards fatty acid metabolism during seed maturation in Arabidopsis. *Plant J.* **2007**, *50*, 825–838. [CrossRef] [PubMed]

86. Focks, N.; Benning, C. *wrinkled1*: A novel, low-seed-oil mutant of Arabidopsis with a deficiency in the seed-specific regulation of carbohydrate metabolism. *Plant Physiol.* **1998**, *118*, 91–101. [CrossRef] [PubMed]

87. Ruuska, S.A.; Girke, T.; Benning, C.; Ohlrogge, J.B. Contrapuntal networks of gene expression during Arabidopsis seed filling. *Plant Cell* **2002**, *14*, 1191–1206. [CrossRef] [PubMed]

88. Maeo, K.; Tokuda, T.; Ayame, A.; Mitsui, N.; Kawai, T.; Tsukagoshi, H.; Ishiguro, S.; Nakamura, K. An AP2-type transcription factor, WRINKLED1, of *Arabidopsis thaliana* binds to the AW-box sequence conserved among proximal upstream regions of genes involved in fatty acid synthesis. *Plant J.* **2009**, *60*, 476–487. [CrossRef] [PubMed]

89. Pouvreau, B.; Baud, S.; Vernoud, V.; Morin, V.; Py, C.; Gendrot, G.; Pichon, J.-P.; Rouster, J.; Paul, W.; Rogowsky, P.M. Duplicate maize *Wrinkled1* transcription factors activate target genes involved in seed oil biosynthesis. *Plant Physiol.* **2011**, *156*, 674–686. [CrossRef] [PubMed]

90. Grimberg, Å.; Carlsson, A.S.; Marttila, S.; Bhalerao, R.; Hofvander, P. Transcriptional transitions in *Nicotiana benthamiana* leaves upon induction of oil synthesis by WRINKLED1 homologs from diverse species and tissues. *BMC Plant Biol.* **2015**, *15*, 192. [CrossRef] [PubMed]

91. Banerjee, A.; Sharkey, T.D. Methylerythritol 4-phosphate (MEP) pathway metabolic regulation. *Nat. Prod. Rep.* **2014**, *31*, 1043–1055. [CrossRef] [PubMed]

antioxidants

MDPI

Review

Recent Advances in our Understanding of Tocopherol Biosynthesis in Plants: An Overview of Key Genes, Functions, and Breeding of Vitamin E Improved Crops

Steffi Fritsche [1,*], Xingxing Wang [1,2] and Christian Jung [1]

[1]   Plant Breeding Institute, Christian-Albrechts-University of Kiel, 24118 Kiel, Germany;
     wangxingxing@caas.cn (X.W.); c.jung@plantbreeding.uni-kiel.de (C.J.)
[2]   Institute of Cotton Research of Chinese Academy of Agricultural Sciences, Anyang 455000, China
*   Correspondence: steffi.fritsche@scionresearch.com; Tel.: +64-7343-5448

Received: 31 October 2017; Accepted: 23 November 2017; Published: 1 December 2017

Abstract: Tocopherols, together with tocotrienols and plastochromanols belong to a group of lipophilic compounds also called tocochromanols or vitamin E. Considered to be one of the most powerful antioxidants, tocochromanols are solely synthesized by photosynthetic organisms including plants, algae, and cyanobacteria and, therefore, are an essential component in the human diet. Tocochromanols potent antioxidative properties are due to their ability to interact with polyunsaturated acyl groups and scavenge lipid peroxyl radicals and quench reactive oxygen species (ROS), thus protecting fatty acids from lipid peroxidation. In the plant model species *Arabidopsis thaliana*, the required genes for tocopherol biosynthesis and functional roles of tocopherols were elucidated in mutant and transgenic plants. Recent research efforts have led to new outcomes for the vitamin E biosynthetic and related pathways, and new possible alternatives for the biofortification of important crops have been suggested. Here, we review 30 years of research on tocopherols in model and crop species, with emphasis on the improvement of vitamin E content using transgenic approaches and classical breeding. We will discuss future prospects to further improve the nutritional value of our food.

Keywords: vitamin E; tocopherol; antioxidant; biofortification; crop breeding

---

1. Introduction

Four tocopherols and four tocotrienols form a group of lipid-soluble antioxidants called tocochromanols, commonly known as vitamin E [1,2]. Their basic structure is simple, comprising a polar chromanol ring and a hydrophobic polyprenyl side chain, products of the shikimate and 1-deoxy-D-xylulose 5-phosphate (DOXP) pathways. Tocochromanols with a fully saturated side chain are called tocopherols and those with an unsaturated side chain tocotrienols. The number of methyl groups in the chromanol ring defines the four natural occurring α, β, γ, and δ-tocopherol and tocotrienol subforms [3]. Vitamin E activity involves scavenging peroxyl radicals and quenching reactive oxygen species (ROS). The most active form of vitamin E is α-tocopherol and thus the most needed form of vitamin E in our diet [4]. It is generally found in high concentrations in vegetable oils such as almond, safflower, or canola oil or in other high-fat sources such as nuts, seeds, or grains [5]. Food-grade canola oil, or 00-type rapeseed oil, has a high-quality nutritional composition similar to that of olive oil, and tocopherols are one of the main nutritionally relevant constituents. Despite vitamin E's importance in the human diet, dietary studies show that the recommended daily allowance is often not met, so, improving its quantity and composition has become a target in crop breeding [6].

In this review, we follow the elucidation of the tocopherol biosynthesis pathway genes over the last 30 years of research, with a particular focus on their function and their involvement in plant processes. We will focus on the biosynthesis bottlenecks detected and the transgenic and non-transgenic breeding efforts that have been undertaken in order to improve the antioxidant and nutritional values of important crops.

2. Biosynthesis and Chemical Function

Tocopherol biosynthesis is initiated in the plant's cytoplasm, but, except for this first step, its biosynthesis takes place in the plastids. Here, the necessary enzymes are localized at the inner envelope or in the plastoglobules [7–9]. The biosynthesis starts through the formation of the aromatic head group homogentisic acid (HGA), which is catalyzed by the enzyme *p*-hydroxyphenylpyruvate and is derived from tyrosine degradation [10,11]. The polyprenyl side chain, phytyl diphosphate (phytyl-PP/PDP), is suggested to originate from the DOXP pathway, as well as from the recycling of free phytol derived from the chlorophyll degradation process [12,13]. In *Arabidopsis thaliana* (*A. thaliana*), this step is catalyzed sequentially by two phytol kinases [13,14]. The two substrates, HGA and PDP, are fused together in the following step, mediated by the enzyme homogentisate phytyltransferase (HPT), to 2-methyl-6-phytyl-1,4-benzoquinol (MPBQ). MPBQ, in turn, serves as substrate either for tocopherol cyclase (TC) or MPBQ methyltransferase. MPBQ methyltransferase methylates MPBQ to 2,3-dimethyl-6-phytyl-1,4-benzoquinone (DMPBQ), while tocopherol cyclase transforms both MPBQ and DMPBQ to γ- and δ-tocopherol, respectively. Lastly, the enzyme γ-tocopherol methyltransferase (γ-TMT) catalyzes the conversion of γ- to α-tocopherol and of δ- to β-tocopherol (Figure 1, Table 1).

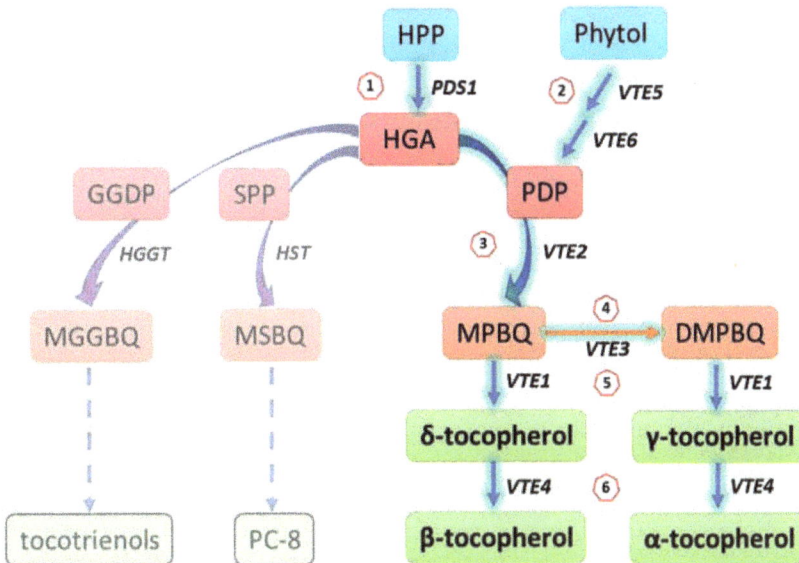

Figure 1. Simplified model of the tocopherol biosynthesis pathway in *A. thaliana* (highlighted branch of the pathway). Compound names are given in the boxes and gene names are indicated in italic. The circled numbers refer to the biosynthesis steps described in detail in Table 1. Adapted from [15]. *HGGT*: homogentisic acid geranylgeranyl transferase; *HST*: homogentisic acid solanesyl transferase; GGDP: geranylgeranyl diphosphate; MGGBQ: 2-methyl-6-geranylgeranyl-1,4-benzoquinol; PC-8: plastochromanol-8; SPP: solanesyl pyrophosphate.

Table 1. Genes and enzymes of the tocopherol biosynthesis in *A. thaliana*.

Biosynthesis Step	Gene	Enzyme	Function	Substrate [1]	Product	At Locus	Reference
1	*PDS1*	HPPD	Head group synthesis	HPP	HGA	AT1G06570	[16–18]
2	*VTE5*	Phytol kinase	Phosphorylation	Phytol + CTP/UTP	Phytyl-P	AT5G04490	[14]
	VTE6	Phytyl -P kinase	Phosphorylation	Phytol-P + CTP	Phytyl-PP(PDP)	AT1G78620	[13]
3	*VTE2*	HPT	Phytylation	HGA + PDP	MPBQ	AT2G18950	[19–23]
4	*VTE3*	MPBQ/ MSBQ MT	Methylation	MPBQ/MSBQ	DMPQ	AT3G63410	[24–26]
5	*VTE1*	TC	Cyclization	MPBQ/DMPQ	γ-tocopherol/ δ-tocopherol	AT4G32770	[27–30]
6	*VTE4*	γ-TMT	Methylation	δ-tocopherol/ γ-tocopherol	α-tocopherol/ β-tocopherol	AT1G64970	[30–33]

[1] Only those compounds related to the tocopherol biosynthesis are mentioned. Genes and their corresponding enzyme names: *PDS1*/HPPD: p-hydroxyphenylpyruvate dioxygenase; *VTE1*/TC: tocopherol cyclase; *VTE2*/HPT: homogentisic acid phytyl transferase; *VTE3*/MT: methyltransferase; *VTE4*/ γ-TMT: γ-tocopherol methyltransferase; *VTE5*/VTE6: phytol kinase. Substrates and products: CTP/UTP: cytidine triphosphate/ uridine triphosphate; DMPBQ: 2,3-dimethyl-6-phytyl-1,4-benzoquinone; HGA: homogentisic acid; HPP: p-hydroxyphenylpyruvate; MPBQ: 2-methyl-6-phytyl-1,4-benzoquinol; MSBQ: 2-methyl-6-solanesyl-1,4-benzoquinol; Phytyl-PP/PDP: phytyl diphosphate; PDP: phytyldiphosphate.

Tocopherols are one of the most valuable antioxidants because of their remarkable chemical mode of action. They interact with polyunsaturated acyl groups and protect polyunsaturated fatty acids from lipid peroxidation by scavenging lipid peroxy radicals and quenching ROS produced e.g., by photosystem II and during membrane lipid peroxidation [34,35]. During this process, tocopherols donate their phenolic hydrogen to lipid-free radicals, thus neutralizing the radical, terminating the autocatalytic lipid peroxidation processes and protecting cell membranes [36,37]. The resulting tocopherol radicals are more stable, are less reactive, and, more importantly, can be reconverted to the corresponding tocopherol by reacting with other antioxidants such as ascorbate or glutathione. This allows each tocopherol to participate in many scavenging reactions before being degraded. Collectively, these properties make tocopherols highly efficient as antioxidants [38]. Furthermore, tocopherols are able to deactivate singlet oxygen ($^1O_2$), which oxidizes, amongst others, membrane lipids, proteins, amino acids, nucleic acids, nucleotides, and carbohydrates [39,40]. The chemical reaction of tocopherols with $^1O_2$ results in the corresponding tocopherol quinones (and other derivatives), some of which have been shown to be potent antioxidants [41–43].

3. Tocopherols in Plants and Mammalians

Tocopherols occur in mammalians, photosynthetic bacteria, fungi, algae, and plants, but only photosynthetic organisms are able to synthesize them [2,3]. The content, composition, and presence of tocopherols varies widely in different plant tissues. They can be found in seeds, fruits, roots, and tubers and are usually present in the green parts of higher plants [15,44]. The most abundant form in leaves is α-tocopherol, whereas the dominating tocopherol form in seeds is γ-tocopherol [2,33]. Nevertheless, in some crops, for instance, sunflower, olive, safflower, wild *Euphorbia*, or grape, α-tocopherol is the main tocopherol form in seeds [45–49]. Total tocopherol content and composition strongly changes under conditions of plant oxidative stress including high light, salinity, drought, and low temperatures [3,31,32,50,51]. Both plant growth and development affect the levels of tocopherol content and composition, which changes for example during senescence, chloroplast to chromoplast conversion, fruit ripening, and seed development [51–55].

Since 1922, when the term vitamin E was introduced by Evans and Bishop [56], the role of vitamin E has been extensively studied in mammalian systems and the beneficial effects demonstrated. Their antioxidant and radical scavenging mechanisms have been described in a variety of studies [57–60]. In these studies, tocopherols and tocotrienols are termed vitamin E as they usually occur as a mixture in food. Because vitamin E cannot be synthesized by humans, it is an essential component of our nutrition. A sufficient uptake of vitamin E can help to prevent neurological disorders and chronic diseases, especially those believed to have an oxidative stress component such as atherosclerosis, cataracts, and cancer [60–69]. In recent years, studies focused also on the role of other vitamin E forms such as γ-tocopherol, δ-tocopherol, and γ-tocotrienol as well as on the non-antioxidant molecular mechanisms to explain the regulatory effects of vitamin E on signal transduction and gene expression in mammals [70–72].

According to the German Society for Nutrition (Deutsche Gesellschaft für Ernährung e.V.), the recommended daily amount of vitamin E for an adult between 25 and 51 years is 14 mg/day. Nuts, seeds, and vegetable oils are among the best sources for the dietary uptake but also green leafy vegetables, fruits, and cereals. Soybean, rapeseed, and corn oil are the best dietary sources for γ-tocopherol, whereas the most biologically active form of vitamin E, α-tocopherol, can be found in wheat germ oil, sunflower, and olive oils [34,73]. Of the naturally occurring α-forms, the stereoisomers RRR-α-tocopherols have the highest biological activity. They can be stored and transported in the body due to specific selection by the hepatic α-tocopherol transfer protein (α-TTP) [34]. Thus, the increase in vitamin E content and the enrichment of α-tocopherol in vegetable oils is of particular interest in crop breeding.

4. Studying the Plant Tocopherol Biosynthesis Key Genes

In leaves, tocopherols are located in chloroplast membranes, plastoglobules, and thylakoid membranes and are thus in close proximity to the photosynthetic apparatus. For this reason, it was believed that tocopherols were the most important antioxidants in plants for preventing damage to the apparatus from photosynthesis-derived ROS. However, recent studies indicate that tocopherols only partially contribute to a considerable variety of plant antioxidants such as carotenoids, ascorbates, glutathiones, and flavonoids, which also contribute to photo-protection [74–76]. Nevertheless, tocopherols are crucial for the inhibition of non-enzymatic lipid peroxidation during stress conditions, supporting the germination of seedlings and are needed for the protection of seed storage lipids [15,32,77]. They are involved in a plethora of others functions such as the activation of plant defense responses, intracellular signaling, transcript regulation, and function in membrane stability [32,77–79].

Step 1: Homogentisic Acid Synthesis

The first committed step of the tocopherol biosynthesis is initiated by the gene *PDS1*, known to encode the enzyme p-hydroxyphenylpyruvate dioxygenase (HPPD) and was initially detected in carrot cells [10]. The corresponding *A. thaliana pds1* mutant was devoid of tocopherols and plastoquinones and had a lethal photobleached phenotype, thus showing that *PDS1* is a key gene of the tocopherol biosynthesis pathway [16,17]. In *Escherichia coli* (*E. coli*) expression studies, the protein HPPD catalyzed the accumulation of HGA. The subsequent oxidative polymerization of HGA to an ochronotic pigment resulted in a brownish coloration of the growth medium [10,80–82].

Constitutive expression of *PDS1* in *A. thaliana* plants led to elevated tocopherol concentrations of up to 43% in leaves and 28% in seeds, without a change in tocopherol composition [18]. In other studies, overexpression of the gene in tobacco leaves or in *A. thaliana* seeds resulted only in slightly increased tocopherol concentrations [83–85]. It was reasoned that HGA not only serves as a substrate for tocochromanols synthesis but also for other secondary plant metabolites e.g., plastoquinones [17]. Higher tocopherol concentrations were achieved when *PDS1* was co-expressed with downstream genes within the tocopherol biosynthetic pathway, the tocopherol content in seeds of *A. thaliana*, rapeseed, and corn kernels increased up to 1.8-, 2.0- and 3-fold, respectively [86–88].

The expression of *PDS1* is also linked to fruit ripening as well as leaf senescence in rice and barley [54,81,89]. Both ripening and leaf aging are correlated with an increase in tocopherol concentration, which is thought to be induced by ethylene treatment and oxidative stress [54,81]. This was underlined by sequence motifs for abscisic acid- and ethylene-specific response elements that were found in the promoter region of *PDS1*, indicating that these compounds can affect tocopherol biosynthesis.

Step 2: Polyprenyl Side Chain Synthesis

Uptake experiments of cotyledon-derived soybean suspension cultures with different tocopherol precursors demonstrated that PDP, together with HGA, has a high impact on tocopherol synthesis [86]. PDP, the polyprenyl side chain of tocopherols, can originate from two sources: direct reduction of geranylgeranyl diphosphate (GGDP) and free phytol, a product of chlorophyll degradation [14,90–92]. The latter was detected in experiments where the addition of free phytol in safflower cell cultures resulted in an 18-fold increase in total tocopherol content [93]. Additionally, feeding experiments of *A. thaliana* seedlings with labeled PDP demonstrated the integration of free phytol into tocopherols [92]. Further evidence was the detection of an *A. thaliana* vte5-1 mutant plant, which had a 20% reduced seed tocopherol concentration compared to the wild type and increased amounts of free phytol. The corresponding VTE5 protein had a high homology to a dolichol kinase and was able to catalyze the phosphorylation of free phytol into PDP [14].

Interestingly, tomato plants that have *VTE5*-like knockdown genes and are deficient in leaf and fruit tocopherol displayed multifaceted impacted metabolisms. While the chlorophyll content was not affected, the knockdown resulted in an alteration of the prenyl lipid profile in fruits and in an increase in the fatty acid phytyl ester synthesis in leaves, leading to changes in photosynthesis and sugar partitioning, which in turn affected tomato fruit quality [94].

The long-missing second enzyme of the phosphorylation pathway, phytyl phosphate kinase (VTE6) has recently been discovered using a phylogenetic approach [13]. The transgenic constitutive overexpression of the gene resulted in a 2-fold increase in PDP that led to a higher γ-tocopherol accumulation in seeds. The corresponding Arabidopsis mutant plants were tocopherol-deficient in leaves, and impaired in seed longevity and plant growth [13]. Intriguingly, the authors speculated that the impaired growth of the *vte6* mutants might be an effect of phytol accumulation rather than tocopherol shortage, resulting in a disruption of chlorophyll and galactolipid synthesis in chloroplast membranes. Recent studies also showed that tocopherol levels are indirectly regulated through the contribution of chlorophyll synthases by reducing PDP precursors, thus providing new possible alternative routes for improving tocopherol content [91,95].

Step 3: MPBQ (2-Methyl-6-Phytyl-1,4-Benzoquinol) Synthesis

The condensation of HGA and PDP to MPBQ is catalyzed by the enzyme HPT, which is encoded by the gene *VTE2* [96,97]. *A. thaliana* plants lacking *VTE2* are completely deficient in all tocopherol derivatives and all pathway precursors, indicating that this is a limiting step in tocopherol biosynthesis [98]. Transgenic approaches downregulating *VTE2* mRNA levels by antisense RNA led to a 10-fold reduction of tocopherol content in *A. thaliana* seeds. In contrast, overexpression of *VTE2* resulted in up to a 2-fold increased tocopherol concentration in seeds and up to a 4.4-fold increase in leaves [20,22]. Interestingly, in rapeseed, only minor increases were attained by seed-specific expression of the *VTE2* gene from the bacteria Synechocystis sp. PCC 6803, suggesting species-specific differences in tocopherol biosynthesis regulation [86]. The constitutive expression of the lettuce homolog (*LsHPT*) caused an increase up to 4-fold of the α- and 2.6-fold of γ-tocopherol content in lettuce leaves in comparison to the non-transformed plants [99].

The lack of tocopherols in *A. thaliana* vte2 mutants made their functional role particularly apparent in those plants. Seedlings exhibited severe defects during germination and early seedling growth due to high concentrations of non-enzymatic lipid peroxides and hydroxy fatty acids, indicating that tocopherols are crucial for the prevention of lipid oxidation during these important developmental stages. The seedlings were disturbed in root growth and in cotyledon expansion and showed a slowed storage lipid catabolism [21]. Furthermore, the authors showed that seed viability was heavily impacted in *vte2* mutants compared to the wild type, probably due to the non-enzymatic oxidation of storage lipids. The accumulation of lipid peroxidation products in *vte2* mutants was correlated with an increased expression of many defense-related genes, showing another main role of tocopherols in the regulation of plant defense responses by modulating the levels of lipid peroxidation products [21]. The developmental and molecular phenotypes of *vte2* are abolished if tri-unsaturated fatty acids are suppressed as shown in the quadruple mutant *fad3-2fad7-2fad8vte2-1*, proving that there is a *vte2*-specific phenotype [100]. The presence of plastochromanol-8 (PC-8), a homolog of γ-tocotrienol with a C40 prenyl side chain, suppresses the phenotypes associated with lower plant fitness and seed longevity in *vte2* plants [101]. PC-8 has been found in vegetable oils and shown to display antioxidant activity, comparable to that of tocopherols [102,103].

Under normal growth conditions, as well as high light stress, adult *vte2* plants are indistinguishable from wild-type plants, confirming that tocopherols are substitutable by other antioxidants at this developmental stage [21,27,32]. However, if the same plants are subjected to low temperature, severe changes in gene transcription, as well as biochemical and physiological phenotypes, can be observed. The plants are compromised in transfer cell development and photoassimilate export, resulting in growth impairment [32,78,104]. Maeda et al., found that the specific underlying mechanisms need to be further investigated but speculate that plastid-localized tocopherols may have a role in the endoplasmic reticulum (ER) membrane lipid biosynthesis causing these effects [104].

Step 4: DMPBQ (2,3-dimethyl-6-phytyl-1,4-benzoquinone) Synthesis

The enzyme that methylates MPBQ to DMPBQ was first identified in *Synechocystis* sp. PCC 6803 and shares substrate specificities with the corresponding protein in *A. thaliana*, MPBQ methyltransferase [25,26]. The enzyme is encoded by the gene *VTE3* and without the catalysis of the prenylquinol DMPBQ, α- and γ-tocopherol could not be produced. The enzyme is also involved in plastoquinone synthesis and is able to use 2-methyl-6-solanesyl-1,4-benzoquinol (MSBQ) for the formation of plastoquinol-9 (PQ-9).

Several *A. thaliana vte3* mutants were isolated that had different phenotypes according to their origin and the severity of the mutation. Plants with a mutated *vet3-1* allele, derived from ethyl methanesulfonate (EMS) mutagenesis, were found to have a partial loss-of-function of the enzyme. As a result, an altered tocopherol composition along with reduced MPBQ/MSBQ methyltransferase activity was observed. In these plants, both δ- and β-tocopherol concentrations in leaves were increased at the expense of α-tocopherol, whereas γ-tocopherol was replaced by higher δ-tocopherol levels, but PQ amounts were not affected in seeds [25,26]. In contrast, plants with the non-functional *vte3-2* allele or Ds-tagged *A. thaliana* plants, which have a complete gene disruption, were deficient in α- and γ-tocopherol, as well as PQ. These mutants exhibited a pale green phenotype, abnormal chloroplasts and did not survive beyond the seedling stage [24,25].

The overexpression of *VTE3* in *A. thaliana* and soybean did not have a substantial impact on the tocopherol amounts but severely changed the tocopherol composition [26,75]. In seeds of transgenic soybean, increased γ- and α-tocopherol concentrations with simultaneous strong reductions in δ- and β-tocopherols concentrations were observed [26]. The seed-specific co-expression of *A. thaliana VTE3* and *VTE4* (γ-TMT) in soybean seeds shifted the tocopherol composition in favor of α-tocopherol (>90%). This is especially of interest for the genetic engineering of crop plants, in which seeds are naturally enriched in γ-tocopherol (e.g., rapeseed). The nutritional value of those plants could be enhanced by changing the composition towards α-tocopherol.

Step 5: Synthesis of γ-Tocopherol and δ-Tocopherol

Similar to the *vte2* mutant plants, *vte1* mutant plants lack all tocopherol forms due to the absence of a functional tocopherol cyclase (TC) enzyme. They accumulate the intermediate product DMPBQ, the un-cyclized precursor of γ-tocopherol, instead [27]. The expression of *AtVTE1* cDNA in *E. coli* efficiently converted DMPBQ into γ-tocopherol, demonstrating the cyclase activity of the enzyme [4,55,105]. Seed-specific expression of the *VTE1* gene from *A. thaliana* and maize (*Zea mays*) in transgenic rapeseed (*Brassica napus*) increased concentrations of all tocopherol forms in the seed oil of T_1 and T_2 generation plants. However, using different *B. napus VTE1* alleles, only a 1.5-fold increase in seed α-tocopherol was produced [82,105].

Interestingly, expression of the endogenous *VTE1* gene in *A. thaliana* plants under a constitutive promoter resulted in a 7-fold increase of the leaf tocopherol concentration and an extreme shift from α- to γ-tocopherol [28]. The reason for this shift was recently unraveled by Zbierzak et al. (2010). The authors showed that the co-elution of PC-8 with γ-tocopherol during High-performance liquid chromatography (HPLC) separation masked the result of increased γ-tocopherol concentrations [106]. PC-8 shares the same head group with γ-tocopherol and is derived from a cyclization step of PQ-9, the redox component of photosystem II. PQ-9 accumulates in *A. thaliana vte1* mutants, whereas PC-8 accumulates with tocopherols in *VTE1*-overexpressing plants, showing that TC is also able to cycle PQ-9 into PC-8 [105,106]. Accordingly, *AtVTE1* overexpression in rapeseed induces a 2.4-fold increase in PC-8 levels [105]. Nevertheless, the proportion of PC-8 with respect to total tocochromanols only ranged between 5 and 10%, suggesting that the majority of the effects observed in *vte1* mutants are indeed caused due to the absence of tocopherols [106].

The *A. thaliana vte1* mutant is not inhibited in its development, despite being deficient in tocopherols and $^1O_2$ accumulation in chloroplasts [107]. Germination, growth, chlorophyll content, and photosynthetic quantum yield were similar to the wild type and only slightly different in response to photo-oxidative stress and/or low temperatures [21,27,108]. The absence of these phenotypes is explained by the accumulation of the redox-active DMPBQ, whose quinol form can have similar antioxidant properties to tocopherol [21]. Furthermore, the absence of tocopherols in *vte1* mutants resulted in an increase of ascorbate and glutathione, which belong to the most abundant antioxidants in higher plants. This effect was reversed when *VTE1* was overexpressed: their concentrations were reduced to 40–60% relative to wild-type plants [28,75]. It is known that the ascorbate–glutathione cycle is linked to the regeneration of tocopheroxyl radicals, the oxidized forms of tocopherol and the result of the reaction of tocopherols with lipid peroxyl radicals [36]. Hence, ascorbate deficiency in *A. thaliana vte1* mutants caused increased lipid peroxidation in water-stressed plants [109]. Furthermore, the overexpression of *VTE1* led to enhanced tolerance against salt stress in rice and drought stress in tobacco plants, providing evidence that crop species with increased tolerance to environmental stresses as well as high tocopherol content could be developed [110,111].

Intriguingly, another analysis of *A. thaliana vte1* mutants revealed a different role of tocopherols and indicated that α-tocopherol may have an impact on cellular signaling by altering plant hormone levels [112,113]. In these mutants, an age-dependent increase of jasmonic acid levels was detected. The levels were up to 2.4-fold higher compared to the wild type and accompanied by reduced plant growth and increased anthocyanin production. It is suggested that the reaction is indirectly triggered by tocopherols by controlling the extent of lipid peroxidation and therefore the accumulation of lipid hydroperoxides, which, in turn, are used for jasmonic acid synthesis [114]. The role of jasmonic acid, especially during senescence, is associated with the downregulation of housekeeping genes, such as photosynthetic genes, and with the upregulation of defense genes related to biotic and abiotic stress [115,116]. These findings further support the hypothesis that tocopherols may, albeit indirectly, contribute to gene expression regulation [78].

The analysis of the maize gene *sucrose export defective1* (*SXD1*) led to the suggestion that tocopherols might play a role in regulating carbon translocation. The maize *sxd1* mutant has a *vte1* mutant resembling phenotype that is characterized, amongst others, by the leaf-specific accumulation of starch and anthocyanin, and the deposition of callose in phloem parenchyma transfer cells. These characteristics are coincident with a loss of the symplastic transport, as well as a typical increase of the soluble sugar content due to the sucrose export deficiency [117–119]. RNAi inhibition of *StSXD1* expression in potato also induces defects in carbohydrate transport [120], showing that the functions of tocopherols are not assigned to a specific photosynthesis type and may be conserved in monocot and dicot plant species [55].

Step 6: Synthesis of α-Tocopherol and β-Tocopherol

The final step of the tocopherol synthesis is catalyzed by the gene *VTE4* and encodes a key enzyme (γ-TMT) that methylates both δ- and γ-tocopherol to β- and α-tocopherol, respectively (Figure 1). Correspondingly, α-tocopherol was absent in leaves of *vte4 A. thaliana* mutants, whereas high levels of γ-tocopherol accumulated [31]. Development and growth of the mutant plants was similar to that of the wild type, with only slight differences during oxidative stress, e.g., in chlorophyll content and photosynthetic quantum yield. The accumulation of γ-tocopherol had no impact on the amount of fatty acids or on lipid hydrolysis, indicating that γ-tocopherol is able to functionally substitute α-tocopherol [31,32].

VTE4-silenced tobacco plants had a decreased tolerance against salt-induced stress but an increased tolerance towards sorbitol stress and methyl viologen, an inductor of photo-oxidative stress. These findings suggest specific functions for different tocopherol forms with respect to stress, e.g., γ-tocopherol in the desiccation tolerance of seeds [50]. This hypothesis is supported by the fact that γ-tocopherol is the naturally predominant form in most seeds. Sattler et al. (2004) could show that a loss of γ-tocopherol increases the oxidation of polyunsaturated fatty acids and thereby diminishes seed longevity in *A. thaliana* [21]. Intriguingly, only in mature leaves of salt-stressed *vte4* mutant plants transcription levels of jasmonic acid- and ethylene-signaling genes were downregulated, but not in *vte1* mutants, or in wild-type plants. This suggests that γ-tocopherol, rather than α-tocopherol, plays a role in the regulation of these genes during osmotic stress [121].

The exclusive overexpression of the *VTE4* gene in leaves changed both tocopherol composition and content, resulting in elevated α-tocopherol values and up to a 30% increase in total tocopherol content [75]. Seed-specific expression of the gene nearly converted the entire amount of γ-tocopherol into α-tocopherol in soybean and *A. thaliana* [26,33]. Because most plant species accumulate mainly γ-tocopherol in seeds (>95%), γ-TMT activity might be a limiting bottleneck for α-tocopherol synthesis in seeds. This is consistent with the isolation of low α-tocopherol sunflower mutants, the natural wild type of which normally exhibit up to 90% α-tocopherol in kernels [122]. In these sunflower mutants, two *VTE4* paralogues are reduced or disrupted in their expression, leading to the accumulation of γ-tocopherol (>90%). This confirms that *VTE4* expression in sunflower seeds directly regulates the amount of α-tocopherol.

The fact that γ-TMT activity is limiting, becomes particularly clear when plants are exposed to abiotic stress. High levels of γ- and δ-tocopherol levels were observed in stressed leaves of *VTE2* overexpressing plants but not when *VTE2* and *VTE4* were simultaneously expressed [123]. Moreover, the concomitant overexpression of *VTE2* and *VTE4* was additive in *A. thaliana* leaves and seeds. The total tocopherol content increased up to 12-fold, while, at the same time, all of the γ-tocopherol was converted to α-tocopherol [20,75]. Thus, *VTE4* is suggested to be a key gene for metabolic engineering of enhanced α-tocopherol levels in crop plants. Similarly, using a transgenic approach, Endrigkeit et al., showed that a *B. napus VTE4* homolog was able to increase the α-tocopherol content 50-fold in transgenic Arabidopsis seeds, providing a promising candidate for the marker-assisted selection of α-tocopherol content in rapeseed [124].

5. Breeding for Higher Vitamin E Content

Improving vitamin E content in staple crops has been a major aim of crop breeding due to its benefits for health and oil quality [125]. Vitamin E content and composition vary widely in different crops, such as rapeseed, maize, rice, soybean, and barley [126–130]. These variations provide an incentive for breeding varieties that have superior vitamin E content. Quantitative genetic approaches, which map quantitative trait loci (QTLs) onto linkage maps or which detect associations between markers and phenotypes, are powerful methods to dissect complex traits in crops [131]. To date, dozens of QTL and association mapping studies related to vitamin E have been carried out in major crops including rapeseed, maize, soybean, rice, barley, and tomato [95,126,127,129,130,132–134]. As an example, Marwede et al., detected eight QTL-related to tocopherol content and composition in a doubled haploid rapeseed population [135]. However, Wang et al. discovered 33 QTLs and 61 associated loci for rapeseed tocopherol content and composition in a joint QTL, candidate gene, and association mapping study [132]. In another candidate-gene based association analysis, alleles of genes encoding the key enzymes of the core biosynthetic pathway associated with tocopherol content and composition in rapeseed were identified [126]. Intriguingly, the found QTLs explained only a small part of the phenotypic variation in tocopherol content and composition in rapeseed (5–30%). Diepenbrock et al. found 52 QTLs associated with vitamin E content using a 5000 line U.S. nested association mapping (NAM) panel [95]. Surprisingly, of the 14 resolved to individual genes, six novel genes affecting tocochromanols in plants were identified. These included unexpectedly two chlorophyll biosynthetic enzymes being major determinants of tocopherol content in non-photosynthetic maize grains, allowing new insights into tocopherol biosynthesis regulation [95]. Nevertheless, these QTL and association mapping analyses provide the foundation for improving vitamin E content in crops but also indicate that the genetic mechanism of vitamin E biosynthesis in crops remains incomplete and needs more comprehensive research. Some crop species are also facing breeding bottlenecks due to their low genetic inheritance in elite germplasm and have to use introgression from wild species, mutagenesis, or biotech methods to enhance different forms and levels of tocopherol [135–138].

Moreover, Quadrana et al., analyzed *VTE3* alleles and detected that their differential expression is associated with differences in methylation of a short-interspersed nuclear element (SINE) retrotransposon located in the promoter region that is responsible for the regulation of vitamin E content in tomatoes [139]. In another study, it was found that a chlorophyll synthase was able to induce the production of small interfering RNAs, accompanied with an increase in tocopherol levels [91]. These observations indicate that the tocopherol biosynthetic pathway might be regulated epigenetically and could also partially explain the low phenotypic variation found in QTL studies.

The vitamin E biosynthetic pathway has been well elucidated in model species, and select genes have been transformed and overexpressed individually or collectively in various plants to improve vitamin E content and composition [26,87,125,128,140–149]. For oil crops, the enhancement of γ-tocopherol is important in order to prevent oil peroxidation and thus to improve oil quality [150]. Since α-tocopherol is considered to have the highest nutritional value for humans and livestock, breeding crops with high α-tocopherol concentrations and good nutritional value is a target. For example, Yusuf et al., was able to elevate α-tocopherol concentrations 6-fold by overexpressing Arabidopsis *VTE4* in *B. juncea* plants [151]. In another study, α-tocopherol content was increased up to 11 times in soybean by overexpressing *VTE4* [152,153]. Interestingly, by transgenically expressing the *VTE1* gene, rice and tobacco plants with an increased tolerance against salt and drought stress, respectively, and a simultaneously enhanced tocopherol content were developed [110,111]. These findings provide prospects for breeding vitamin E biofortified crops that are even tolerant against environmental stresses through biotechnology methods.

In contrast to attempts to improve α-tocopherol content, it is relatively difficult to significantly increase the total vitamin E content in crops. The most successful attempts at increasing total vitamin E content have occurred in rapeseed and soybean, with a nearly 2.4-fold and 15-fold enhancement, respectively, of total vitamin E content after co-overexpression of several key genes for vitamin E biosynthesis [86,87]. Collectively, these observations indicate that the tocopherol biosynthetic pathway and its regulation are complex. Nevertheless, natural variations and transgenic strategies can be used in vitamin E biofortification breeding.

6. Conclusions and Challenges

Present studies using model plant species demonstrate that vitamin E biosynthesis genes are well understood. The use of mutant plants has helped with the discovery and analysis of the roles of tocopherols in plant processes beyond their antioxidant function. However, open questions remain about the specific functions of the different tocopherol forms as well as their concrete roles in signaling and defense responses. Additionally, plastoquinones and tocotrienols, which are known to be derived from the same head group as tocopherols, have gained attention in recent years but most of their roles still remain elusive [17,55,102,154,155].

Recently, new genes and epigenetic mechanisms have been discovered through investigations of the chlorophyll degradation pathway, transgenic and phylogenetic approaches, or by conducting joint linkage/genome-wide association studies [13,91,95]. Moreover, even in crops with hexaploid genomes, such as oat, the genes of the tocopherol pathway have been identified via deep sequencing and by using the high conservation of the tocopherol biosynthesis genes between plant species [156].

Natural variations of vitamin E content have been investigated and demonstrated wide variations in/among different crops, which suggest a tremendous genetic potential for breeding vitamin E improved crops. On this basis, several QTL/association mapping studies have been carried out and have detected a range of genetic loci associated with vitamin E content and composition, but of which many detected loci cannot pinpoint candidate genes [95,129,130,132,157–159]. Furthermore, epigenetics may be involved in regulating the vitamin E biosynthesis and can partially explain the low phenotypic variation of the detected genetic loci, indicating that vitamin E biosynthesis in crops may be more complex than that in model plants [91,139]. Transgenic approaches have also been used for vitamin E improvement in crops, and an increased α-tocopherol content has been achieved [20]. Elevating vitamin E content, even by stacking candidate genes, is still a challenge, but tocopherol enhancing alleles of the pathway genes have been identified, for example, in *B. napus*, and can now be used for marker-assisted breeding [86,126].

To date, many crops have been sequenced including vitamin E-rich staple crops, such as rapeseed, soybean, maize, and rice [160–163]. These sequences will be used in the process of cloning new genes from QTL and enable us to fully understand the genetic basis of vitamin E variations and biosynthesis. Thanks to the rapid development of biotechnology, many methods for efficient genome-editing have become available [164]. For example, CRISPR/Cas9 (Clustered Regularly Interspaced Short Palindromic Repeats/ CRISPR-associated protein 9), which is one of the most popular genome-editing methods at present, can efficiently and precisely edit gene/genes in mammals and plants. A genome-editing approach based on the cloned genes from QTL in crops and CRISPR/Cas9 can be a promising alternative strategy in breeding vitamin E biofortification crops [165].

Acknowledgments: We are grateful for Kim McGrouter and Michelle Harnett from Scion, Rotorua, as well as the two anonymous reviewers for critically reading the manuscript and suggesting substantial improvements.

Conflicts of Interest: The authors declare no conflict of interest.

References

1. DellaPenna, D.; Pogson, B.J. Vitamin synthesis in plants: Tocopherols and carotenoids. *Annu. Rev. Plant Biol.* **2006**, *57*, 711–738. [CrossRef] [PubMed]
2. Grusak, M.A.; DellaPenna, D. Improving the nutrient composition of plants to enhance human nutrition and health. *Annu. Rev. Plant Biol.* **1999**, *50*, 133–161. [CrossRef] [PubMed]
3. Munné-Bosch, S.; Alegre, L. The function of tocopherols and tocotrienols in plants. *Crit. Rev. Plant Sci.* **2002**, *21*, 31–57. [CrossRef]
4. DellaPenna, D.; Last, R.L. Progress in the dissection and manipulation of plant vitamin E biosynthesis. *Physiol. Plant.* **2006**, *126*, 356–368. [CrossRef]
5. DellaPenna, D. A decade of progress in understanding vitamin E synthesis in plants. *J. Plant Physiol.* **2005**, *162*, 729–737. [CrossRef] [PubMed]

6. Péter, S.; Friedel, A.; Roos, F.F.; Wyss, A.; Eggersdorfer, M.; Hoffmann, K.; Weber, P. A systematic review of global alpha-tocopherol status as assessed by nutritional intake levels and blood serum concentrations. *Int. J. Vitam. Nutr. Res.* **2016**, *14*, 1–21. [CrossRef] [PubMed]

7. Soll, J.; Douce, R.; Schultz, G. Site of biosynthesis of alpha-tocopherol in spinach chloroplasts. *FEBS Lett.* **1980**, *112*, 243–246. [CrossRef]

8. Vidi, P.A.; Kanwischer, M.; Baginsky, S.; Austin, J.R.; Csucs, G.; Dormann, P.; Kessler, F.; Brehelin, C. Tocopherol cyclase (VTE1) localization and vitamin E accumulation in chloroplast plastoglobule lipoprotein particles. *J. Biol. Chem.* **2006**, *281*, 11225–11234. [CrossRef] [PubMed]

9. Soll, J.; Schultz, G.; Joyard, J.; Douce, R.; Block, M.A. Localization and synthesis of prenylquinones in isolated outer and inner envelope membranes from spinach chloroplasts. *Arch. Biochem. Biophys.* **1985**, *238*, 290–299. [CrossRef]

10. Garcia, I.; Rodgers, M.; Lenne, C.; Rolland, A.; Sailland, A.; Matringe, M. Subcellular localization and purification of a *p*-hydroxyphenylpyruvate dioxygenase from cultured carrot cells and characterization of the corresponding cDNA. *Biochem. J.* **1997**, *325 Pt 3*, 761–769. [CrossRef] [PubMed]

11. Riewe, D.; Koohi, M.; Lisec, J.; Pfeiffer, M.; Lippmann, R.; Schmeichel, J.; Willmitzer, L.; Altmann, T. A tyrosine aminotransferase involved in tocopherol synthesis in *Arabidopsis*. *Plant J.* **2012**, *71*, 850–859. [CrossRef] [PubMed]

12. Hortensteiner, S. Chlorophyll degradation during senescence. *Annu. Rev. Plant Biol.* **2006**, *57*, 55–77. [CrossRef] [PubMed]

13. Vom Dorp, K.; Hölzl, G.; Plohmann, C.; Eisenhut, M.; Abraham, M.; Weber, A.P.; Hanson, A.D.; Dörmann, P. Remobilization of phytol from chlorophyll degradation is essential for tocopherol synthesis and growth of *Arabidopsis*. *Plant Cell* **2015**, *27*, 2846–2859. [CrossRef] [PubMed]

14. Valentin, H.E.; Lincoln, K.; Moshiri, F.; Jensen, P.K.; Qi, Q.; Venkatesh, T.V.; Karunanandaa, B.; Baszis, S.R.; Norris, S.R.; Savidge, B.; et al. The Arabidopsis vitamin E pathway gene5-1 mutant reveals a critical role for phytol kinase in seed tocopherol biosynthesis. *Plant Cell* **2006**, *18*, 212–224. [CrossRef] [PubMed]

15. Mene-Saffrane, L.; DellaPenna, D. Biosynthesis, regulation and functions of tocochromanols in plants. *Plant Physiol. Biochem.* **2010**, *48*, 301–309. [CrossRef] [PubMed]

16. Norris, S.R.; Shen, X.; DellaPenna, D. Complementation of the Arabidopsis pds1 mutation with the gene encoding *p*-hydroxyphenylpyruvate dioxygenase. *Plant Physiol.* **1998**, *117*, 1317–1323. [CrossRef] [PubMed]

17. Norris, S.R.; Barrette, T.R.; DellaPenna, D. Genetic dissection of carotenoid synthesis in Arabidopsis defines plastoquinone as an essential component of phytoene desaturation. *Plant Cell* **1995**, *7*, 2139–2149. [CrossRef] [PubMed]

18. Tsegaye, Y.; Shintani, D.K.; DellaPenna, D. Overexpression of the enzyme *p*-hydroxyphenolpyruvate dioxygenase in *Arabidopsis* and its relation to tocopherol biosynthesis. *Plant Physiol. Biochem.* **2002**, *40*, 913–920. [CrossRef]

19. Schwab, R.; Palatnik, J.F.; Riester, M.; Schommer, C.; Schmid, M.; Weigel, D. Specific effects of microRNAs on the plant transcriptome. *Dev. Cell* **2005**, *8*, 517–527. [CrossRef] [PubMed]

20. Collakova, E.; DellaPenna, D. Homogentisate phytyltransferase activity is limiting for tocopherol biosynthesis in *Arabidopsis*. *Plant Physiol.* **2003**, *131*, 632–642. [CrossRef] [PubMed]

21. Sattler, S.E.; Gilliland, L.U.; Magallanes-Lundback, M.; Pollard, M.; DellaPenna, D. Vitamin E is essential for seed longevity and for preventing lipid peroxidation during germination. *Plant Cell* **2004**, *16*, 1419–1432. [CrossRef] [PubMed]

22. Savidge, B.; Weiss, J.D.; Wong, Y.H.H.; Lassner, M.W.; Mitsky, T.A.; Shewmaker, C.K.; Post-Beittenmiller, D.; Valentin, H.E. Isolation and characterization of homogentisate phytyltransferase genes from *Synechocystis* sp. PCC 6803 and *Arabidopsis*. *Plant Physiol.* **2002**, *129*, 321–332. [CrossRef] [PubMed]

23. Venkatesh, T.V.; Karunanandaa, B.; Free, D.L.; Rottnek, J.M.; Baszis, S.R.; Valentin, H.E. Identification and characterization of an *Arabidopsis* homogentisate phytyltransferase paralog. *Planta* **2006**, *223*, 1–11. [CrossRef] [PubMed]

24. Motohashi, R.; Ito, T.; Kobayashi, M.; Taji, T.; Nagata, N.; Asami, T.; Yoshida, S.; Yamaguchi-Shinozaki, K.; Shinozaki, K. Functional analysis of the 37 kDa inner envelope membrane polypeptide in chloroplast biogenesis using a Ds-tagged *Arabidopsis* pale-green mutant. *Plant J.* **2003**, *34*, 719–731. [CrossRef] [PubMed]

25. Cheng, Z.; Sattler, S.; Maeda, H.; Sakuragi, Y.; Bryant, D.A.; DellaPenna, D. Highly divergent methyltransferases catalyze a conserved reaction in tocopherol and plastoquinone synthesis in cyanobacteria and photosynthetic eukaryotes. *Plant Cell* **2003**, *15*, 2343–2356. [CrossRef] [PubMed]

26. Van Eenennaam, A.L.; Lincoln, K.; Durrett, T.P.; Valentin, H.E.; Shewmaker, C.K.; Thorne, G.M.; Jiang, J.; Baszis, S.R.; Levering, C.K.; Aasen, E.D.; et al. Engineering vitamin E content: From *Arabidopsis* mutant to soy oil. *Plant Cell* **2003**, *15*, 3007–3019. [CrossRef] [PubMed]

27. Porfirova, S.; Bergmüller, E.; Tropf, S.; Lemke, R.; Dörmann, P. Isolation of an *Arabidopsis* mutant lacking vitamin E and identification of a cyclase essential for all tocopherol biosynthesis. *Proc. Natl. Acad. Sci. USA* **2002**, *99*, 12495–12500. [CrossRef] [PubMed]

28. Kanwischer, M.; Porfirova, S.; Bergmuller, E.; Dormann, P. Alterations in tocopherol cyclase activity in transgenic and mutant plants of *Arabidopsis* affect tocopherol content, tocopherol composition, and oxidative stress. *Plant Physiol.* **2005**, *137*, 713–723. [CrossRef] [PubMed]

29. Sattler, S.E.; Cahoon, E.B.; Coughlan, S.J.; DellaPenna, D. Characterization of tocopherol cyclases from higher plants and cyanobacteria. Evolutionary implications for tocopherol synthesis and function. *Plant Physiol.* **2003**, *132*, 2184–2195. [CrossRef] [PubMed]

30. Semchuk, N.M.; Lushchak, O.V.; Falk, J.; Krupinska, K.; Lushchak, V.I. Inactivation of genes, encoding tocopherol biosynthetic pathway enzymes, results in oxidative stress in outdoor grown *Arabidopsis thaliana*. *Plant Physiol. Biochem.* **2009**, *47*, 384–390. [CrossRef] [PubMed]

31. Bergmüller, E.; Porfirova, S.; Dörmann, P. Characterization of an *Arabidopsis* mutant deficient in γ-tocopherolmethyltransferase. *Plant Mol. Biol.* **2003**, *52*, 1181–1190. [CrossRef] [PubMed]

32. Maeda, H.; Song, W.; Sage, T.L.; DellaPenna, D. Tocopherols play a crucial role in low-temperature adaptation and Phloem loading in *Arabidopsis*. *Plant Cell* **2006**, *18*, 2710–2732. [CrossRef] [PubMed]

33. Shintani, D.; DellaPenna, D. Elevating the vitamin E content of plants through metabolic engineering. *Science* **1998**, *282*, 2098–2100. [CrossRef] [PubMed]

34. Brigelius-Flohe, R.; Traber, M.G. Vitamin E: Function and metabolism. *FASEB J.* **1999**, *13*, 1145–1155. [PubMed]

35. Krieger-Liszkay, A.; Fufezan, C.; Trebst, A. Singlet oxygen production in photosystem II and related protection mechanism. *Photosynth. Res.* **2008**, *98*, 551–564. [CrossRef] [PubMed]

36. Kamal-Eldin, A.; Appelqvist, L.A. The chemistry and antioxidant properties of tocopherols and tocotrienols. *Lipids* **1996**, *31*, 671–701. [CrossRef] [PubMed]

37. Liebler, D.C. The role of metabolism in the antioxidant function of vitamin E. *Crit. Rev. Toxicol.* **1993**, *23*, 147–169. [CrossRef] [PubMed]

38. Fukuzawa, K.; Tokumura, A.; Ouchi, S.; Tsukatani, H. Antioxidant activities of tocopherols on Fe^{2+}-ascorbate-induced lipid peroxidation in lecithin liposomes. *Lipids* **1982**, *17*, 511–513. [CrossRef] [PubMed]

39. Halliwell, B.; Gutteridge, J.M. *Free Radicals in Biology and Medicine*; Oxford University Press: New York, NY, USA, 2015.

40. Straight, R.C.; Spikes, J.D. Photosensitized oxidation of biomolecules. In *Singlet O_2*; Frimer, A.A., Ed.; CRC Press: Boca Raton, FL, USA, 1985; Volume 4, pp. 91–143.

41. Lass, A.; Sohal, R.S. Electron transport-linked ubiquinone-dependent recycling of alpha-tocopherol inhibits autooxidation of mitochondrial membranes. *Arch. Biochem. Biophys.* **1998**, *352*, 229–236. [CrossRef] [PubMed]

42. Siegel, D.; Bolton, E.M.; Burr, J.A.; Liebler, D.C.; Ross, D. The reduction of α-tocopherolquinone by human NAD(P)H: Quinone oxidoreductase: The role of α-tocopherolhydroquinone as a cellular antioxidant. *Mol. Pharmacol.* **1997**, *52*, 300–305. [PubMed]

43. Wu, J.H.; Croft, K.D. Vitamin E metabolism. *Mol. Aspects Med.* **2007**, *28*, 437–452. [CrossRef] [PubMed]

44. Horvath, G.; Wessjohann, L.; Bigirimana, J.; Jansen, M.; Guisez, Y.; Caubergs, R.; Horemans, N. Differential distribution of tocopherols and tocotrienols in photosynthetic and non-photosynthetic tissues. *Phytochemistry* **2006**, *67*, 1185–1195. [CrossRef] [PubMed]

45. Velasco, L.; Fernández-Martínez, J.; Garcia-Ruiz, R.; Domínguez, J. Genetic and environmental variation for tocopherol content and composition in sunflower commercial hybrids. *J. Agric. Sci.* **2002**, *139*, 425–429. [CrossRef]

46. Bruni, R.; Muzzoli, M.; Ballero, M.; Loi, M.C.; Fantin, G.; Poli, F.; Sacchetti, G. Tocopherols, fatty acids and sterols in seeds of four Sardinian wild Euphorbia species. *Fitoterapia* **2004**, *75*, 50–61. [CrossRef] [PubMed]

47. Horvath, G.; Wessjohann, L.; Bigirimana, J.; Monica, H.; Jansen, M.; Guisez, Y.; Caubergs, R.; Horemans, N. Accumulation of tocopherols and tocotrienols during seed development of grape (*Vitis vinifera* L. cv. Albert Lavallee). *Plant Physiol. Biochem.* **2006**, *44*, 724–731. [CrossRef] [PubMed]

48. Cayuela, J.A.; García, J.F. Sorting olive oil based on alpha-tocopherol and total tocopherol content using near-infra-red spectroscopy (NIRS) analysis. *J. Food Eng.* **2017**, *202*, 79–88. [CrossRef]

49. Gotor, A.A.; Farkas, E.; Berger, M.; Labalette, F.; Centis, S.; Daydé, J.; Calmon, A. Determination of tocopherols and phytosterols in sunflower seeds by NIR spectrometry. *Eur. J. Lipid Sci. Technol.* **2007**, *109*, 525–530. [CrossRef]

50. Abbasi, A.R.; Hajirezaei, M.; Hofius, D.; Sonnewald, U.; Voll, L.M. Specific roles of alpha- and gamma-tocopherol in abiotic stress responses of transgenic tobacco. *Plant Physiol.* **2007**, *143*, 1720–1738. [CrossRef] [PubMed]

51. Arango, Y.; Heise, K.-P. Localization of α-tocopherol synthesis in chromoplast envelope membranes of *Capsicum annuum* L. fruits. *J. Exp. Bot.* **1998**, *49*, 1259–1262. [CrossRef]

52. Abushita, A.A.; Hebshi, E.A.; Daood, H.G.; Biacs, P.A. Determination of antioxidant vitamins in tomatoes. *Food Chem.* **1997**, *60*, 207–212. [CrossRef]

53. Arrom, L.; Munné-Bosch, S. Tocopherol composition in flower organs of *Lilium* and its variations during natural and artificial senescence. *Plant Sci.* **2010**, *179*, 289–295. [CrossRef]

54. Falk, J.; Krauß, N.; Dähnhardt, D.; Krupinska, K. The senescence associated gene of barley encoding 4-hydroxyphenylpyruvate dioxygenase is expressed during oxidative stress. *J. Plant Physiol.* **2002**, *159*, 1245–1253. [CrossRef]

55. Falk, J.; Munne-Bosch, S. Tocochromanol functions in plants: Antioxidation and beyond. *J. Exp. Bot.* **2010**, *61*, 1549–1566. [CrossRef] [PubMed]

56. Evans, H.M.; Bishop, K.S. On the Existence of a Hitherto Unrecognized Dietary Factor Essential for Reproduction. *Science* **1922**, *56*, 650–651. [CrossRef] [PubMed]

57. Azzi, A.; Gysin, R.; Kempna, P.; Munteanu, A.; Negis, Y.; Villacorta, L.; Visarius, T.; Zingg, J.M. Vitamin E mediates cell signaling and regulation of gene expression. *Ann. N. Y. Acad. Sci.* **2004**, *1031*, 86–95. [CrossRef] [PubMed]

58. Azzi, A.; Ricciarelli, R.; Zingg, J.M. Non-antioxidant molecular functions of α-tocopherol (vitamin E). *FEBS Lett.* **2002**, *519*, 8–10. [CrossRef]

59. Zingg, J.M.; Azzi, A. Non-antioxidant activities of vitamin E. *Curr. Med. Chem.* **2004**, *11*, 1113–1133. [CrossRef] [PubMed]

60. Schneider, C. Chemistry and biology of vitamin E. *Mol. Nutr. Food Res.* **2005**, *49*, 7–30. [CrossRef] [PubMed]

61. Kushi, L.H.; Folsom, A.R.; Prineas, R.J.; Mink, P.J.; Wu, Y.; Bostick, R.M. Dietary antioxidant vitamins and death from coronary heart disease in postmenopausal women. *N. Engl. J. Med.* **1996**, *334*, 1156–1162. [CrossRef] [PubMed]

62. Weinstein, S.J.; Wright, M.E.; Lawson, K.A.; Snyder, K.; Mannisto, S.; Taylor, P.R.; Virtamo, J.; Albanes, D. Serum and dietary vitamin E in relation to prostate cancer risk. *Cancer Epidemiol. Biomark. Prev.* **2007**, *16*, 1253–1259. [CrossRef] [PubMed]

63. Bramley, P.M.; Elmadfa, I.; Kafatos, A.; Kelly, F.J.; Manios, Y.; Roxborough, H.E.; Schuch, W.; Sheehy, P.J.A.; Wagner, K.H. Vitamin E. *J. Sci. Food Agric.* **2000**, *80*, 913–938. [CrossRef]

64. Öhrvall, M.; Vessby, B.; Sundlöf, G. Gamma, but not alpha, tocopherol levels in serum are reduced in coronary heart disease patients. *J. Intern. Med.* **1996**, *239*, 111–117. [CrossRef] [PubMed]

65. Schuelke, M.; Mayatepek, E.; Inter, M.; Becker, M.; Pfeiffer, E.; Speer, A.; Hubner, C.; Finckh, B. Treatment of ataxia in isolated vitamin E deficiency caused by alpha-tocopherol transfer protein deficiency. *J. Pediatr.* **1999**, *134*, 240–244. [CrossRef]

66. Witztum, J.L.; Steinberg, D. Role of oxidized low density lipoprotein in atherogenesis. *J. Clin. Investig.* **1991**, *88*, 1785–1792. [CrossRef] [PubMed]

67. Traber, M.G.; Frei, B.; Beckman, J.S. Vitamin E revisited: Do new data validate benefits for chronic disease prevention? *Curr. Opin. Lipidol.* **2008**, *19*, 30–38. [CrossRef] [PubMed]

68. Niki, E. Do free radicals play causal role in atherosclerosis? Low density lipoprotein oxidation and vitamin E revisited. *J. Clin. Biochem. Nutr.* **2010**, *48*, 3–7. [CrossRef] [PubMed]

69. Colombo, M.L. An update on vitamin E, tocopherol and tocotrienol-perspectives. *Molecules* **2010**, *15*, 2103–2113. [CrossRef] [PubMed]

70. Azzi, A. Many tocopherols, one vitamin E. *Mol. Asp. Med.* **2017**. [CrossRef] [PubMed]

71. Azzi, A.; Meydani, S.N.; Meydani, M.; Zingg, J.M. The rise, the fall and the renaissance of vitamin E. *Arch. Biochem. Biophys.* **2016**, *595*, 100–108. [CrossRef] [PubMed]

72. Jiang, Q. Natural forms of vitamin E: Metabolism, antioxidant, and anti-inflammatory activities and their role in disease prevention and therapy. *Free Radic. Biol. Med.* **2014**, *72*, 76–90. [CrossRef] [PubMed]

73. Dietrich, M.; Traber, M.G.; Jacques, P.F.; Cross, C.E.; Hu, Y.; Block, G. Does γ-tocopherol play a role in the primary prevention of heart disease and cancer? A review. *J. Am. Coll. Nutr.* **2006**, *25*, 292–299. [CrossRef] [PubMed]

74. Kloz, M.; Pillai, S.; Kodis, G.; Gust, D.; Moore, T.A.; Moore, A.L.; van Grondelle, R.; Kennis, J.T. Carotenoid photoprotection in artificial photosynthetic antennas. *J. Am. Chem. Soc.* **2011**, *133*, 7007–7015. [CrossRef] [PubMed]

75. Li, Y.; Zhou, Y.; Wang, Z.A.; Sun, X.F.; Tang, K.X. Engineering tocopherol biosynthetic pathway in *Arabidopsis* leaves and its effect on antioxidant metabolism. *Plant Sci.* **2010**, *178*, 312–320. [CrossRef]

76. Asensi-Fabado, M.A.; Munne-Bosch, S. Vitamins in plants: Occurrence, biosynthesis and antioxidant function. *Trends Plant Sci.* **2010**, *15*, 582–592. [CrossRef] [PubMed]

77. Azzi, A. Molecular mechanism of α-tocopherol action. *Free Radic. Biol. Med.* **2007**, *43*, 16–21. [CrossRef] [PubMed]

78. Sattler, S.E.; Mene-Saffrane, L.; Farmer, E.E.; Krischke, M.; Mueller, M.J.; DellaPenna, D. Nonenzymatic lipid peroxidation reprograms gene expression and activates defense markers in *Arabidopsis* tocopherol-deficient mutants. *Plant Cell* **2006**, *18*, 3706–3720. [CrossRef] [PubMed]

79. Hyun, T.K.; Kumar, K.; Rao, K.P.; Sinha, A.K.; Roitsch, T. Role of α-tocopherol in cellular signaling: α-tocopherol inhibits stress-induced mitogen-activated protein kinase activation. *Plant Biotechnol. Rep.* **2011**, *5*, 19–25. [CrossRef]

80. Denoya, C.D.; Skinner, D.D.; Morgenstern, M.R. A *Streptomyces avermitilis* gene encoding a 4-hydroxyphenylpyruvic acid dioxygenase-like protein that directs the production of homogentisic acid and an ochronotic pigment in *Escherichia coli. J. Bacteriol.* **1994**, *176*, 5312–5319. [CrossRef] [PubMed]

81. Singh, R.K.; Ali, S.A.; Nath, P.; Sane, V.A. Activation of ethylene-responsive p-hydroxyphenylpyruvate dioxygenase leads to increased tocopherol levels during ripening in mango. *J. Exp. Bot.* **2011**, *62*, 3375–3385. [CrossRef] [PubMed]

82. Fritsche, S.; Wang, X.; Nichelmann, L.; Suppanz, I.; Hadenfeldt, S.; Endrigkeit, J.; Meng, J.; Jung, C. Genetic and functional analysis of tocopherol biosynthesis pathway genes from rapeseed (*Brassica napus* L.). *Plant Breed.* **2014**, *133*, 470–479. [CrossRef]

83. Falk, J.; Andersen, G.; Kernebeck, B.; Krupinska, K. Constitutive overexpression of barley 4-hydroxy phenylpyruvate dioxygenase in tobacco results in elevation of the vitamin E content in seeds but not in leaves. *FEBS Lett.* **2003**, *540*, 35–40. [CrossRef]

84. Rippert, P.; Scimemi, C.; Dubald, M.; Matringe, M. Engineering plant shikimate pathway for production of tocotrienol and improving herbicide resistance. *Plant Physiol.* **2004**, *134*, 92–100. [CrossRef] [PubMed]

85. Li, Y.; Wang, Z.; Sun, X.; Tang, K. Current opinions on the functions of tocopherol based on the genetic manipulation of tocopherol biosynthesis in plants. *J. Integr. Plant Biol.* **2008**, *50*, 1057–1069. [CrossRef] [PubMed]

86. Karunanandaa, B.; Qi, Q.; Hao, M.; Baszis, S.R.; Jensen, P.K.; Wong, Y.H.; Jiang, J.; Venkatramesh, M.; Gruys, K.J.; Moshiri, F.; et al. Metabolically engineered oilseed crops with enhanced seed tocopherol. *Metab. Eng.* **2005**, *7*, 384–400. [CrossRef] [PubMed]

87. Raclaru, M.; Gruber, J.; Kumar, R.; Sadre, R.; Lühs, W.; Zarhloul, M.; Friedt, W.; Frentzen, M.; Weier, D. Increase of the tocochromanol content in transgenic *Brassica napus* seeds by overexpression of key enzymes involved in prenylquinone biosynthesis. *Mol. Breed.* **2006**, *18*, 93–107. [CrossRef]

88. Naqvi, S.; Farre, G.; Zhu, C.; Sandmann, G.; Capell, T.; Christou, P. Simultaneous expression of *Arabidopsis* rho-hydroxyphenylpyruvate dioxygenase and MPBQ methyltransferase in transgenic corn kernels triples the tocopherol content. *Transgenic Res.* **2011**, *20*, 177–181. [CrossRef] [PubMed]

89. Kleber-Janke, T.; Krupinska, K. Isolation of cDNA clones for genes showing enhanced expression in barley leaves during dark-induced senescence as well as during senescence under field conditions. *Planta* **1997**, *203*, 332–340. [CrossRef] [PubMed]

90. Keller, Y.; Bouvier, F.; D'Harlingue, A.; Camara, B. Metabolic compartmentation of plastid prenyllipid biosynthesis—Evidence for the involvement of a multifunctional geranylgeranyl reductase. *Eur. J. Biochem.* **1998**, *251*, 413–417. [CrossRef] [PubMed]

91. Zhang, C.; Zhang, W.; Ren, G.; Li, D.; Cahoon, R.E.; Chen, M.; Zhou, Y.; Yu, B.; Cahoon, E.B. Chlorophyll synthase under epigenetic surveillance is critical for vitamin E synthesis, and altered expression affects tocopherol levels in *Arabidopsis*. *Plant Physiol.* **2015**, *168*, 1503–1511. [CrossRef] [PubMed]

92. Ischebeck, T.; Zbierzak, A.M.; Kanwischer, M.; Dormann, P. A salvage pathway for phytol metabolism in *Arabidopsis*. *J. Biol. Chem.* **2006**, *281*, 2470–2477. [CrossRef] [PubMed]

93. Furuya, T.; Yoshikawa, T.; Kimura, T.; Kaneko, H. Production of tocopherols by cell culture of safflower. *Phytochemistry* **1987**, *26*, 2741–2747. [CrossRef]

94. Almeida, J.; Azevedo Mda, S.; Spicher, L.; Glauser, G.; vom Dorp, K.; Guyer, L.; del Valle Carranza, A.; Asis, R.; de Souza, A.P.; Buckeridge, M.; et al. Down-regulation of tomato PHYTOL KINASE strongly impairs tocopherol biosynthesis and affects prenyllipid metabolism in an organ-specific manner. *J. Exp. Bot.* **2016**, *67*, 919–934. [CrossRef] [PubMed]

95. Diepenbrock, C.H.; Kandianis, C.B.; Lipka, A.E.; Magallanes-Lundback, M.; Vaillancourt, B.; Gongora-Castillo, E.; Wallace, J.G.; Cepela, J.; Mesberg, A.; Bradbury, P.; et al. Novel Loci Underlie Natural Variation in Vitamin E Levels in Maize Grain. *Plant Cell* **2017**, *29*. [CrossRef] [PubMed]

96. Collakova, E.; DellaPenna, D. Isolation and functional analysis of homogentisate phytyltransferase from *Synechocystis* sp. PCC 6803 and *Arabidopsis*. *Plant Physiol.* **2001**, *127*, 1113–1124. [CrossRef] [PubMed]

97. Schledz, M.; Seidler, A.; Beyer, P.; Neuhaus, G. A novel phytyltransferase from *Synechocystis* sp. PCC 6803 involved in tocopherol biosynthesis. *FEBS Lett.* **2001**, *499*, 15–20. [CrossRef]

98. Sattler, S.E.; Cheng, Z.; DellaPenna, D. From *Arabidopsis* to agriculture: Engineering improved Vitamin E content in soybean. *Trends Plant Sci.* **2004**, *9*, 365–367. [CrossRef] [PubMed]

99. Ren, W.; Zhao, L.; Wang, Y.; Cui, L.; Tang, Y.; Sun, X.; Tang, K. Overexpression of homogentisate phytyltransferase in lettuce results in increased content of vitamin E. *AJB* **2011**, *10*, 14046–14051.

100. Mene-Saffrane, L.; Davoine, C.; Stolz, S.; Majcherczyk, P.; Farmer, E.E. Genetic removal of tri-unsaturated fatty acids suppresses developmental and molecular phenotypes of an *Arabidopsis* tocopherol-deficient mutant. Whole-body mapping of malondialdehyde pools in a complex eukaryote. *J. Biol. Chem.* **2007**, *282*, 35749–35756. [CrossRef] [PubMed]

101. Mene-Saffrane, L.; Jones, A.D.; DellaPenna, D. Plastochromanol-8 and tocopherols are essential lipid-soluble antioxidants during seed desiccation and quiescence in *Arabidopsis*. *Proc. Natl. Acad. Sci. USA* **2010**, *107*, 17815–17820. [CrossRef] [PubMed]

102. Ksas, B.; Becuwe, N.; Chevalier, A.; Havaux, M. Plant tolerance to excess light energy and photooxidative damage relies on plastoquinone biosynthesis. *Sci. Rep.* **2015**, *5*, 10919. [CrossRef] [PubMed]

103. Olejnik, D.; Gogolewski, M.; Nogala-Kałucka, M. Isolation and some properties of plastochromanol-8. *Mol. Nutr. Food Res.* **1997**, *41*, 101–104. [CrossRef]

104. Maeda, H.; Song, W.; Sage, T.; Dellapenna, D. Role of callose synthases in transfer cell wall development in tocopherol deficient *Arabidopsis* mutants. *Front. Plant Sci.* **2014**, *5*, 46. [CrossRef] [PubMed]

105. Kumar, R.; Raclaru, M.; Schusseler, T.; Gruber, J.; Sadre, R.; Lühs, W.; Zarhloul, K.M.; Friedt, W.; Enders, D.; Frentzen, M.; et al. Characterisation of plant tocopherol cyclases and their overexpression in transgenic *Brassica napus* seeds. *FEBS Lett.* **2005**, *579*, 1357–1364. [CrossRef] [PubMed]

106. Zbierzak, A.M.; Kanwischer, M.; Wille, C.; Vidi, P.A.; Giavalisco, P.; Lohmann, A.; Briesen, I.; Porfirova, S.; Brehelin, C.; Kessler, F.; et al. Intersection of the tocopherol and plastoquinol metabolic pathways at the plastoglobule. *Biochem. J.* **2010**, *425*, 389–399. [CrossRef] [PubMed]

107. Rastogi, A.; Yadav, D.K.; Szymanska, R.; Kruk, J.; Sedlarova, M.; Pospisil, P. Singlet oxygen scavenging activity of tocopherol and plastochromanol in *Arabidopsis thaliana*: Relevance to photooxidative stress. *Plant Cell Environ.* **2014**, *37*, 392–401. [CrossRef] [PubMed]

108. Havaux, M.; Eymery, F.; Porfirova, S.; Rey, P.; Dormann, P. Vitamin E protects against photoinhibition and photooxidative stress in *Arabidopsis thaliana*. *Plant Cell* **2005**, *17*, 3451–3469. [CrossRef] [PubMed]

109. Munné-Bosch, S.; Alegre, L. Interplay between ascorbic acid and lipophilic antioxidant defences in chloroplasts of water-stressed *Arabidopsis* plants. *FEBS Lett.* **2002**, *524*, 145–148. [CrossRef]

110. Liu, X.; Hua, X.; Guo, J.; Qi, D.; Wang, L.; Liu, Z.; Jin, Z.; Chen, S.; Liu, G. Enhanced tolerance to drought stress in transgenic tobacco plants overexpressing VTE1 for increased tocopherol production from *Arabidopsis thaliana*. *Biotechnol. Lett.* **2008**, *30*, 1275–1280. [CrossRef] [PubMed]

111. Ouyang, S.; He, S.; Liu, P.; Zhang, W.; Zhang, J.; Chen, S. The role of tocopherol cyclase in salt stress tolerance of rice (*Oryza sativa*). *Sci. China Life Sci.* **2011**, *54*, 181–188. [CrossRef] [PubMed]

112. Munne-Bosch, S. Linking tocopherols with cellular signaling in plants. *New Phytol.* **2005**, *166*, 363–366. [CrossRef] [PubMed]

113. Munné-Bosch, S.; Weiler, E.W.; Alegre, L.; Müller, M.; Düchting, P.; Falk, J. α-Tocopherol may influence cellular signaling by modulating jasmonic acid levels in plants. *Planta* **2007**, *225*, 681–691. [CrossRef] [PubMed]

114. Schaller, F. Enzymes of the biosynthesis of octadecanoid-derived signalling molecules. *J. Exp. Bot.* **2001**, *52*, 11–23. [CrossRef] [PubMed]

115. Creelman, R.A.; Mullet, J.E. Biosynthesis and Action of Jasmonates in Plants. *Annu. Rev. Plant Physiol. Plant Mol. Biol.* **1997**, *48*, 355–381. [CrossRef] [PubMed]

116. Wasternack, C. Jasmonates: An update on biosynthesis, signal transduction and action in plant stress response, growth and development. *Ann. Bot.* **2007**, *100*, 681–697. [CrossRef] [PubMed]

117. Botha, C.; Cross, R.; Van Bel, A.; Peter, C. Phloem loading in the sucrose-export-defective (SXD-1) mutant maize is limited by callose deposition at plasmodesmata in bundle sheath—Vascular parenchyma interface. *Protoplasma* **2000**, *214*, 65–72. [CrossRef]

118. Provencher, L.M.; Miao, L.; Sinha, N.; Lucas, W.J. Sucrose export defective1 encodes a novel protein implicated in chloroplast-to-nucleus signaling. *Plant Cell* **2001**, *13*, 1127–1141. [CrossRef] [PubMed]

119. Russin, W.A.; Evert, R.F.; Vanderveer, P.J.; Sharkey, T.D.; Briggs, S.P. Modification of a Specific Class of Plasmodesmata and Loss of Sucrose Export Ability in the sucrose export defective1 Maize Mutant. *Plant Cell* **1996**, *8*, 645–658. [CrossRef] [PubMed]

120. Hofius, D.; Hajirezaei, M.R.; Geiger, M.; Tschiersch, H.; Melzer, M.; Sonnewald, U. RNAi-mediated tocopherol deficiency impairs photoassimilate export in transgenic potato plants. *Plant Physiol.* **2004**, *135*, 1256–1268. [CrossRef] [PubMed]

121. Cela, J.; Chang, C.; Munné-Bosch, S. Accumulation of γ-rather than α-tocopherol alters ethylene signaling gene expression in the *vte4* mutant of *Arabidopsis thaliana*. *Plant Cell Physiol.* **2011**, *52*, 1389–1400. [CrossRef] [PubMed]

122. Hass, C.G.; Tang, S.; Leonard, S.; Traber, M.G.; Miller, J.F.; Knapp, S.J. Three non-allelic epistatically interacting methyltransferase mutations produce novel tocopherol (vitamin E) profiles in sunflower. *Theor. Appl. Genet.* **2006**, *113*, 767–782. [CrossRef] [PubMed]

123. Collakova, E.; DellaPenna, D. The role of homogentisate phytyltransferase and other tocopherol pathway enzymes in the regulation of tocopherol synthesis during abiotic stress. *Plant Physiol.* **2003**, *133*, 930–940. [CrossRef] [PubMed]

124. Endrigkeit, J.; Wang, X.; Cai, D.; Zhang, C.; Long, Y.; Meng, J.; Jung, C. Genetic mapping, cloning, and functional characterization of the *BnaX.VTE4* gene encoding a gamma-tocopherol methyltransferase from oilseed rape. *Theor. Appl. Genet.* **2009**, *119*, 567–575. [CrossRef] [PubMed]

125. Hunter, S.C.; Cahoon, E.B. Enhancing vitamin E in oilseeds: Unraveling tocopherol and tocotrienol biosynthesis. *Lipids* **2007**, *42*, 97–108. [CrossRef] [PubMed]

126. Fritsche, S.; Wang, X.; Li, J.; Stich, B.; Kopisch-Obuch, F.J.; Endrigkeit, J.; Leckband, G.; Dreyer, F.; Friedt, W.; Meng, J.; et al. A candidate gene-based association study of tocopherol content and composition in rapeseed (*Brassica napus*). *Front Plant Sci.* **2012**, *3*, 129. [CrossRef] [PubMed]

127. Muzhingi, T.; Palacios-Rojas, N.; Miranda, A.; Cabrera, M.L.; Yeum, K.J.; Tang, G. Genetic variation of carotenoids, vitamin E and phenolic compounds in Provitamin A biofortified maize. *J. Sci. Food Agric.* **2017**, *97*, 793–801. [CrossRef] [PubMed]

128. Wang, X.Q.; Yoon, M.Y.; He, Q.; Kim, T.S.; Tong, W.; Choi, B.W.; Lee, Y.S.; Park, Y.J. Natural variations in OsgammaTMT contribute to diversity of the alpha-tocopherol content in rice. *Mol. Genet. Genomics* **2015**, *290*, 2121–2135. [CrossRef] [PubMed]

129. Shaw, E.J.; Rajcan, I.; Morris, B. Molecular mapping of soybean seed tocopherols in the cross 'OAC Bayfield' × 'OAC Shire'. *Plant Breed.* **2017**, *136*, 83–93. [CrossRef]

130. Graebner, R.C.; Wise, M.; Cuesta-Marcos, A.; Geniza, M.; Blake, T.; Blake, V.C.; Butler, J.; Chao, S.; Hole, D.J.; Horsley, R.; et al. Quantitative Trait Loci Associated with the Tocochromanol (Vitamin E) Pathway in Barley. *PLoS ONE* **2015**, *10*, e0133767. [CrossRef] [PubMed]

131. Mauricio, R. Mapping quantitative trait loci in plants: Uses and caveats for evolutionary biology. *Nat. Rev. Genet.* **2001**, *2*, 370–381. [CrossRef] [PubMed]

132. Wang, X.; Zhang, C.; Li, L.; Fritsche, S.; Endrigkeit, J.; Zhang, W.; Long, Y.; Jung, C.; Meng, J. Unraveling the genetic basis of seed tocopherol content and composition in rapeseed (*Brassica napus* L.). *PLoS ONE* **2012**, *7*, e50038. [CrossRef] [PubMed]

133. Lipka, A.E.; Gore, M.A.; Magallanes-Lundback, M.; Mesberg, A.; Lin, H.; Tiede, T.; Chen, C.; Buell, C.R.; Buckler, E.S.; Rocheford, T.; et al. Genome-wide association study and pathway-level analysis of tocochromanol levels in maize grain. *G3 (Bethesda)* **2013**, *3*, 1287–1299. [CrossRef] [PubMed]

134. Luby, C.H.; Maeda, H.A.; Goldman, I.L. Genetic and phenological variation of tocochromanol (vitamin E) content in wild (*Daucus carota* L. var. carota) and domesticated carrot (*D. carota* L. var. sativa). *Hortic. Res.* **2014**, *1*, 14015. [CrossRef] [PubMed]

135. Rani, R.; Sheoran, R.; Sharma, B. Perspectives of breeding for altering sunflower oil quality to obtain novel oils. *Int. J. Curr. Microbiol. App. Sci.* **2017**, *6*, 949–962. [CrossRef]

136. Rauf, S.; Jamil, N.; Tariq, S.A.; Khan, M.; Kausar, M.; Kaya, Y. Progress in modification of sunflower oil to expand its industrial value. *J. Sci. Food Agric.* **2017**, *97*, 1997–2006. [CrossRef] [PubMed]

137. Seyis, F.; Snowdon, R.; Luhs, W.; Friedt, W. Molecular characterization of novel resynthesized rapeseed (*Brassica napus*) lines and analysis of their genetic diversity in comparison with spring rapeseed cultivars. *Plant Breed.* **2003**, *122*, 473–478. [CrossRef]

138. Hasan, M.; Seyis, F.; Badani, A.; Pons-Kühnemann, J.; Friedt, W.; Lühs, W.; Snowdon, R.J. Analysis of genetic diversity in the *Brassica napus* L. gene pool using SSR markers. *Genet. Resour. Crop Evol.* **2006**, *53*, 793–802. [CrossRef]

139. Quadrana, L.; Almeida, J.; Asis, R.; Duffy, T.; Dominguez, P.G.; Bermudez, L.; Conti, G.; Correa da Silva, J.V.; Peralta, I.E.; Colot, V.; et al. Natural occurring epialleles determine vitamin E accumulation in tomato fruits. *Nat. Commun.* **2014**, *5*, 3027. [CrossRef] [PubMed]

140. Farre, G.; Sudhakar, D.; Naqvi, S.; Sandmann, G.; Christou, P.; Capell, T.; Zhu, C. Transgenic rice grains expressing a heterologous rho-hydroxyphenylpyruvate dioxygenase shift tocopherol synthesis from the gamma to the alpha isoform without increasing absolute tocopherol levels. *Transgenic Res.* **2012**, *21*, 1093–1097. [CrossRef] [PubMed]

141. Chaudhary, N.; Khurana, P. Cloning, functional characterisation and transgenic manipulation of vitamin E biosynthesis genes of wheat. *Funct. Plant Biol.* **2013**, *40*, 1129–1136. [CrossRef]

142. Espinoza, A.; San Martin, A.; Lopez-Climent, M.; Ruiz-Lara, S.; Gomez-Cadenas, A.; Casaretto, J.A. Engineered drought-induced biosynthesis of alpha-tocopherol alleviates stress-induced leaf damage in tobacco. *J. Plant Physiol.* **2013**, *170*, 1285–1294. [CrossRef] [PubMed]

143. Chen, D.; Chen, H.; Zhang, L.; Shi, X.; Chen, X. Tocopherol-deficient rice plants display increased sensitivity to photooxidative stress. *Planta* **2014**, *239*, 1351–1362. [CrossRef] [PubMed]

144. Jin, S.; Daniell, H. Expression of gamma-tocopherol methyltransferase in chloroplasts results in massive proliferation of the inner envelope membrane and decreases susceptibility to salt and metal-induced oxidative stresses by reducing reactive oxygen species. *Plant Biotechnol. J.* **2014**, *12*, 1274–1285. [CrossRef] [PubMed]

145. Tanaka, H.; Yabuta, Y.; Tamoi, M.; Tanabe, N.; Shigeoka, S. Generation of transgenic tobacco plants with enhanced tocotrienol levels through the ectopic expression of rice homogentisate geranylgeranyl transferase. *Plant Biotechnol. J.* **2015**, *32*, 233–238. [CrossRef]

146. Che, P.; Zhao, Z.Y.; Glassman, K.; Dolde, D.; Hu, T.X.; Jones, T.J.; Obukosia, S.; Wambugu, F.; Albertsen, M.C. Elevated vitamin E content improves all-trans beta-carotene accumulation and stability in biofortified sorghum. *Proc. Natl. Acad. Sci. USA* **2016**, *113*, 11040–11045. [CrossRef] [PubMed]

147. Levac, D.; Cázares, P.; Yu, F.; De Luca, V. A picrinine N-methyltransferase belongs to a new family of γ-tocopherol-like methyltransferases found in medicinal plants that make biologically active monoterpenoid indole alkaloids. *Plant Physiol.* **2016**, *170*, 1935–1944. [CrossRef] [PubMed]

148. Fleta-Soriano, E.; Munne-Bosch, S. Enhanced plastochromanol-8 accumulation during reiterated drought in maize (*Zea mays* L.). *Plant Physiol. Biochem.* **2017**, *112*, 283–289. [CrossRef] [PubMed]

149. Liao, P.; Chen, X.; Wang, M.; Bach, T.J.; Chye, M.L. Improved fruit α-tocopherol, carotenoid, squalene and phytosterol contents through manipulation of Brassica juncea 3-HYDROXY-3-METHYLGLUTARYL-COA SYNTHASE1 in transgenic tomato. *Plant Biotechnol. J.* **2017**. [CrossRef] [PubMed]

150. Demurin, Y.; Skoric, D.; Karlovic, D. Genetic variability of tocopherol composition in sunflower seeds as a basis of breeding improved oil quality. *Plant Breed.* **1996**, 33–36. [CrossRef]

151. Yusuf, M.A.; Sarin, N.B. Antioxidant value addition in human diets: Genetic transformation of Brassica juncea with gamma-TMT gene for increased alpha-tocopherol content. *Transgenic Res.* **2007**, *16*, 109–113. [CrossRef] [PubMed]

152. Vinutha, T.; Maheswari, C.; Bansal, N.; Prashat, G.R.; Krishnan, V.; Kumari, S.; Dahuja, A.; Sachdev, A.; Rai, R. Expression analysis of γ-tocopherol methyl transferase genes and α-tocopherol content in developing seeds of soybean [*Glycine max* (L.) Merr.]. *IJBB* **2015**, *52*, 267–273.

153. Tavva, V.S.; Kim, Y.H.; Kagan, I.A.; Dinkins, R.D.; Kim, K.H.; Collins, G.B. Increased alpha-tocopherol content in soybean seed overexpressing the *Perilla frutescens* gamma-tocopherol *methyltransferase* gene. *Plant Cell. Rep.* **2007**, *26*, 61–70. [CrossRef] [PubMed]

154. Siles, L.; Cela, J.; Munne-Bosch, S. Vitamin E analyses in seeds reveal a dominant presence of tocotrienols over tocopherols in the Arecaceae family. *Phytochemistry* **2013**, *95*, 207–214. [CrossRef] [PubMed]

155. Matringe, M.; Ksas, B.; Rey, P.; Havaux, M. Tocotrienols, the unsaturated forms of vitamin E, can function as antioxidants and lipid protectors in tobacco leaves. *Plant Physiol.* **2008**, *147*, 764–778. [CrossRef] [PubMed]

156. Gutierrez-Gonzalez, J.J.; Garvin, D.F. Subgenome-specific assembly of vitamin E biosynthesis genes and expression patterns during seed development provide insight into the evolution of oat genome. *Plant Biotechnol. J.* **2016**, *14*, 2147–2157. [CrossRef] [PubMed]

157. Marwede, V.; Gul, M.K.; Becker, H.C.; Ecke, W. Mapping of QTL controlling tocopherol content in winter oilseed rape. *Plant Breed.* **2005**, *124*, 20–26. [CrossRef]

158. Li, H.Y.; Liu, H.C.; Han, Y.P.; Wu, X.X.; Teng, W.L.; Liu, G.F.; Li, W.B. Identification of QTL underlying vitamin E contents in soybean seed among multiple environments. *Theor. Appl. Genet.* **2010**, *120*, 1405–1413. [CrossRef] [PubMed]

159. Liu, H.; Cao, G.; Wu, D.; Jiang, Z.; Han, Y.; Li, W.; Morris, B. Quantitative trait loci underlying soybean seed tocopherol content with main additive, epistatic and QTL × environment effects. *Plant Breed.* **2017**. [CrossRef]

160. Ding, Y.; Li, H.; Chen, L.L.; Xie, K. Recent Advances in Genome Editing Using CRISPR/Cas9. *Front. Plant Sci.* **2016**, *7*, 703. [CrossRef] [PubMed]

161. Doudna, J.A.; Gersbach, C.A. Genome editing: The end of the beginning. *Genome Biol.* **2015**, *16*, 292. [CrossRef] [PubMed]

162. Feng, Z.; Zhang, B.; Ding, W.; Liu, X.; Yang, D.L.; Wei, P.; Cao, F.; Zhu, S.; Zhang, F.; Mao, Y.; et al. Efficient genome editing in plants using a CRISPR/Cas system. *Cell Res.* **2013**, *23*, 1229–1232. [CrossRef] [PubMed]

163. Gao, W.; Long, L.; Tian, X.; Xu, F.; Liu, J.; Singh, P.K.; Botella, J.R.; Song, C. Genome Editing in Cotton with the CRISPR/Cas9 System. *Front. Plant Sci.* **2017**, *8*, 1364. [CrossRef] [PubMed]

164. Cardi, T.; Neal Stewart, C., Jr. Progress of targeted genome modification approaches in higher plants. *Plant Cell Rep.* **2016**, *35*, 1401–1416. [CrossRef] [PubMed]

165. Jung, C.; Capistrano-Gossmann, G.; Braatz, J.; Sashidhar, N.; Melzer, S. Recent developments in genome editing and applications in plant breeding. *Plant Breed.* **2017**. [CrossRef]

antioxidants

MDPI

Review

Genotypic and Environmental Effects on Tocopherol Content in Almond

Ossama Kodad [1], Rafel Socias i Company [2],* and José M. Alonso [2]

[1] Département Arboriculture-Viticuture, École Nationale d'Agriculture de Meknès,
Meknès BP S/40, Morocco; osama.kodad@yahoo.es

[2] Unidad de Hortofruticutura, Centro de Investigación y Tecnología Agroalimentaria de Aragón (CITA),
Av. Montañana 930, 50059 Zaragoza, Spain; jmalonsos@aragon.es

* Correspondence: rsocias@cita-aragon.es; Tel.: +34-976-716-310

Received: 17 November 2017; Accepted: 3 January 2018; Published: 5 January 2018

Abstract: Almond is the most important nut species worldwide and almond kernels show the highest levels of tocopherols among all nuts. In almond, tocopherols not only play a substantial role as a healthy food for human consumption, but also in protecting lipids against oxidation and, thus, lengthening the storage time of almond kernels. The main tocopherol homologues detected in almond in decreasing content and biological importance are α-, γ-, δ-, and β-tocopherol. Tocopherol concentration in almond depends on the genotype and the environment, such as the climatic conditions of the year and the growing management of the orchard. The range of variability for the different tocopherol homologues is of 335–657 mg/kg of almond oil for α-, 2–50 for γ-, and 0.1–22 for β-tocopherol. Drought and heat have been the most important stresses affecting tocopherol content in almond, with increased levels at higher temperatures and in water deficit conditions. The right cultivar and the most appropriate growing conditions may be selected to obtain crops with effective kernel storage and for the most beneficial effects of almond consumption for human nutrition and health.

Keywords: almond; *Prunus amygdalus*; tocopherols; genotype; climate

---

1. Introduction

Almond is the most important tree nut crop in terms of commercial production [1]. An adaptation to harsh climates combined with an ability to develop a deep and extensive root system has allowed almond to exploit a wide range of ecological niches. Almond is well-adapted to the Mediterranean climate, characterized by mild winters and dry, hot summers. This adaptation has led to early bloom and rapid early shoot growth because of the low chilling requirements of almond. Almond also shows high tolerance to summer drought and heat. Almond has traditionally been the earliest temperate fruit tree crop to bloom, which limited growing to areas relatively free from spring frosts before the release of late-blooming cultivars by different breeding programs, since frosts at bloom or early fruit development can reduce, and even completely nullify, the crop. Since almond is naturally self-incompatible, it often requires cross-pollination, which further acts to promote genetic variability and adaptability to diverse environments [2].

The edible part of the almond nut is the kernel, considered an important food crop with a high nutritional and medicinal value. The oldest and most extensive medical system that first recorded the health uses for almonds derives from ancient Greeks and then the Persians, and later in traditional Chinese medicine and Indian Ayurvedic medicine [3]. From medieval times to the 18th century, almond nuts were a source of substitute "milk" [4], and also it was used as thickener before starch was "discovered" [3]. Almond consumption has almost doubled in the last 20 years [5], a fact that highlights how this consumption has evolved from a convenient snack food and component of a high number of confectioneries, to an important food which is increasingly recognized as essential for maintaining and increasing human health. Recent nutritional and medical studies have associated the regular consumption of almonds with a wide range of health benefits, including protection from cancer [6,7], obesity [8–10], diabetes [11,12], and heart diseases [13–17]. Almond kernels may be consumed in many different ways, blanched or unblanched, raw or combined, and/or mixed with other nuts. They can also be transformed to produce marzipan, nougat, almond milk, and almond flour, incorporated in many pastries and ice creams. Almonds may also be combined in other products and gastronomic specialties [18]. The high nutritive value of almond kernels is mainly due to their high lipid content. This lipid fraction, even constituting an important source of calories, does not contribute to cholesterol formation because of their high level of unsaturated fatty acids, mainly mono-unsaturated fatty acids [19]. Although almond kernels are high in energy, humans compensate their effect with their high satiety value [20]. The absorption of energy from almond kernels is rather inefficient, having been suggested that their chronic consumption may raise resting energy expenditure [21]. Acute and longer-term almond ingestion may help in regulating body weight [17], modulating fluctuations of blood glucose [22], total low density/high density lipoprotein cholesterol ratio, and triglycerides [23].

Almond kernel quality must be high in order to fulfill not only the industry requirements, but also to be attractive for the consumers [24]. Until recently only the kernel physical traits were considered when trying to establish almond quality [18], but the kernel chemical composition appears to be essential when establishing the best raw material for the different industrial applications and the high diversity of almond confectioneries [24]. In view of the high lipid fraction in the almond kernel, the quality of the almond oil is considered as the most important feature in the evaluation of almond quality. Different parameters related to the lipid fraction have been suggested for quality evaluation of the almond kernels, such as the amount of oil content in the kernel (fat percentage over the kernel dry weight), the percentage of oleic acid of the total fatty acids, the ratio of the percentages of oleic/linoleic acids (O/L), and, especially, the tocopherol concentration in the almond oil [24,25].

Early reviews on the composition of the almond nuts [18,26] did not include tocopherols as an important component of almond kernels. More recently tocopherols were already included in several reviews [27,28], but most of them were primarily descriptive without attempts to assess quality, particularly as it relates not only to industrial requirements, but also to breeding goals and approaches. Extensive variability in the chemical composition has been demonstrated among cultivars; additionally the importance of differences in geographical origins, as well as climatic and growing conditions have also been demonstrated. Despite this extensive information, little is known concerning the genetic control and inheritance of biochemical components of almond quality. For all these reasons, this review summarizes the current knowledge of the almond kernel tocopherol composition and factors affecting its variability.

2. Tocopherol Variability in Almond

Almost all studies carried out on vitamin content in almond are limited to those having an antioxidant effect, mainly tocopherols. The main tocopherol homologues detected in almond in decreasing importance are α-, γ-, δ-, and β-tocopherol. The main biochemical function of tocopherols is considered to be the protection of polyunsaturated fatty acids against peroxidation [29]. They also have protective roles in human health since recent data indicate that they have hypo-cholesterolemic, anti-cancer, and neuroprotective properties [30]. Tocopherol concentrations have been determined in many vegetable oils, having been correlated with their antioxidant activity, since this activity may depend on the ratio of % total tocopherols/% polyunsaturated fatty acids [31]. Tocopherol concentration plays an important role in protecting almond lipids against oxidation, thus increasing the possibilities of lengthening kernel storage [31–33].

The range of variability of the different tocopherol homologues has been reported for different almond cultivars and genotypes from different countries [27,34–44], being summarized in Table 1. The most biologically active form of vitamin E is α-tocopherol, being preferentially utilized in the human body over the other homologues [45], ranging from 200 mg/kg oil in some Australian cultivars [44] to 656.7 mg/kg oil in some local accessions from Morocco [37]. At high temperatures γ-tocopherol has been reported to be much more effective as an inhibitor of polymerization and protection against oxidation than α-tocopherol [46,47]. The range of variability of this homologue was from 2.4 mg/kg oil in some local Moroccan almond seedlings [37] to 50.2 mg/kg oil in some almond selections from the CITA breeding program [35,48]. For δ-tocopherol, the range of variability is reduced as compared to the other homologues and ranged from 0.1 mg/kg oil in some local Moroccan almond seedlings [37] to 22.0 mg/kg oil in some Spanish almond cultivars [39].

Table 1. Tocopherol homologue concentration in almond kernel and kernel oil.

Tocopherol Homologue	Range of Variability		Origin	Reference
	mg/kg Kernel	mg/kg Oil		
α-tocopherol	85–190		Spain	[42]
		335.3–551.7		[39]
		309–656.7	Morocco	[37]
	180–320		California	[38]
	250–840		Italy	[32]
		350–471		[43]
		370–571	Argentina	[40]
		200	Australia	[44]
β-tocopherol	50–80		Italy	[32]
	1.2		Australia	[44]
γ-tocopherol		6.1–50.2	Spain	[34,39]
	1.4–8.4			[42]
		75	Italy	[31]
		2.4–13.5	Morocco	[37]
	5.7		Australia	[44]
δ-tocopherol	0.2–1.6		Spain	[42]
		0.2–22		[34,39]
		0.1–0.3	Morocco	[37]

3. Environmental Effects

Tocopherol levels increase in response to a variety of abiotic stresses, considered as evidence of its protective role [49]. Tocopherol concentration in almond oil depends on the genotype and the climatic conditions of the year [36,38–40,42], as well as on the environmental conditions of the growing region [36,38]. Drought and heat have been the most important stresses studied until now on the expression of the chemical compounds in fruit trees, including almond. The climatic conditions of the year, mainly temperature, affect the concentration of the different tocopherol homologues in several nut crops [50], indicating that these components depend on the temperature and the occurrence of drought during fruit or nut growth.

The effect of drought stress on the oil percentage in almond kernels and on its composition is not clear since the results of different studies undertaken in this field are ambiguous. However, studies in other species reported that water stress appears to promote tocopherol synthesis [51–53]. In almond there was no obvious relationship between almond tocopherol content and the degree of water deficiency [54], since little variation in kernel oil tocopherols was observed under moderate water stress, except for the more severe deficit of 70% SDI (sustained deficit irrigation) which for all components, except γ-tocopherol, had higher values than the control. In this study, the minor tocopherol homologues appeared to be slightly more responsive to deficit irrigation than the main homologue, α-tocopherol, but their small proportion had little impact on the amount of the total tocopherol content [54]. In the "Nonpareil" almond cultivar, the highest accumulation rate of α-tocopherol takes place during the period from 74 to 94 days after anthesis [44], indicating that the critical period to add nutrient and water to ensure normal accumulation of these chemical compounds is during this period. Tocopherol concentration has been reported to be affected by drought stress in olive oil [55], but not in *Brassica napus* [56], by temperature in soybean [57], and by a combination of both in shea butter (*Vitellaria paradoxa* C.F. Gaertn.) [50]. In olive oil, the total tocopherols and α-tocopherol were highly influenced by the crop-year rainfalls, with the highest concentration from olives harvested during the driest year [58]. The increase in tocopherol content might contribute to the prevention of plant oxidative damage in drought conditions [59,60].

Higher tocopherol concentrations were found in almond kernels harvested in years with higher temperatures suggesting that this environmental factor could affect tocopherol synthesis during kernel development [34]. A study on the effect of two contrasting environments, with drought and heat conditions in Morocco, and irrigated and cold conditions in Spain, concluded that the tocopherol content, mainly of α-tocopherol, increased under the warmer climate conditions of Morocco [37]. More recently, higher α-tocopherol concentrations (~646 µg/g in average) were found when the almond kernel development mostly coincided with spring and summer months with warmer mean temperature in a study in arid Northwestern Argentina [40]. Similar results were reported for α-tocopherol in some almond genotypes grown under hot and dry conditions in Afghanistan [61]. Temperature during seed development in sunflower has shown to affect oil yield and tocopherol concentration since high temperatures may have a negative effect in oil synthesis, but not in tocopherol concentration [62,63]. In soybean seed, it has been reported that the increasing temperature from 23 °C to 28 °C can significantly increase tocopherol levels [64]. Thus, the temperature could be considered an important parameter playing a great role in tocopherol accumulation in different species, including almond.

Concerning the effect of solar radiation, in "Nonpareil" almond kernels the concentration of α-tocopherol was increased after a mild solar UV radiation supplement using the white weed mat, which may have altered metabolic pathways and stimulated α-tocopherol accumulation in almond lipids [44]. In sunflower, it has been reported that an increase in intercepted solar radiation per plant increased the amount of tocopherol per grain [65].

4. Genetic Effects

The evaluation of tocopherol concentration in the oil of different almond cultivars and genotypes showed high variability of the different tocopherol homologues, with a great effect of the environment [28]. However, it has been reported that the cultivars "Atocha", "Desmayo Rojo", "Desmayo Largueta" from Spain, and "Ferraduel" from France showed stable and similar year to year values for α-tocopherol; whereas "Atocha", "Ferraduel", and "Fournat de Brézenaud", and "Yaltinskij" from Ukraine were also stable for γ-tocopherol content in almond kernels produced under contrasting climatic conditions, with drought and heat conditions in Morocco, and irrigated and cold conditions in Spain [36]. These results confirmed that the stability of each tocopherol homologue depends on the specific characteristics of the genotype [34,35]. The estimation of the heritability of the different tocopherols isomers were estimated for the first time in almond [35], reporting that the content of γ-tocopherol showed high heritability estimates, with $h^2 = 60.0\%$, whereas α-tocopherol showed lower heritability ($h^2 = 20.5\%$). These results confirm that the tocopherol content in almond oil kernel is under polygenic control, as previously suggested [28].

The biosynthetic pathway of vitamin E in plants was biochemically elucidated several years ago, and all enzymes in this pathway were localized to the inner chloroplast envelope [66–68]. Today, the availability of complete genome sequences, in particular from *Arabidopsis* and *Synechocystis* sp. PCC6803, all biosynthetic genes in tocopherol biosynthesis have been identified and cloned to date [69,70]. A mutation in the gene VTE5 (PCT) of *Arabidopsis* lead to the discovery of its function, since it is encoding a protein with phytol kinase activity, directly involved in the biosynthetic pathway of tocopherol [70,71]. Tocopherol QTL analysis found that up to 65% of the markers were co-located in certain genomic regions of maize, including the candidate genes PDS1 (*p*-hydroxyphenylpyruvate dioxygenase; HPPD) and VTE4 (γ-tocopherol methyltransferase; γ-TMT), thus showing that a single QTL may affect more than one tocopherol homologue [71]. After exploring mutant inbred lines in sunflower, three loci (*m* = *Tph*1, *g* = *Tph*2, and *d*) were shown to disrupt synthesis in α-tocopherol production; additionally, the loci losing function in these mutations enhanced synthesis of other tocopherols [72]. In almond no studies have been conducted to elucidate the pathway biosynthesis of tocopherol isomers and the genes involved in this process, but, recently, five different QTLs believed to control the tocopherol concentration in the almond kernel oil have been identified [73]. More studies are required to understand the biosynthetic pathway of these biochemicals components to elucidate the gene involved in these pathways. This information will be of great interest for the breeders to improve tocopherol content in the future, since the almond kernel is used more and more in different industrial processes using high temperatures in their application [24].

5. Conclusions

The present information on the different effects on tocopherol content in almond is scarce. Only descriptive results have been published, not allowing a critical comparison among them, since some results are given as tocopherol content in the almond kernel and others in the kernel oil, not always stating the oil concentration of the almond kernel. Although genetic and environmental effects affecting tocopherol content in almond have been described, no interaction between them has ever been established. The present scenario of climatic change and of the shift of almond growing to warmer regions [1] may have a positive effect on the tocopherol content in almond kernels due to the effect of higher temperatures increasing this content. Consequently, the right cultivar and the most appropriate growing conditions may be selected in order to have almond crops not only allowing a better effective kernel storage, but also holding the most beneficial effects for human nutrition and health with their consumption. The objective of selecting for high tocopherol content in a breeding progeny is easily attainable because of its high heritability, whenever the adequate parents are chosen. As a consequence, these components could be considered as selection criteria in almond breeding programs as already suggested [24]. At present the possibilities of selecting cultivars, growing regions, and orchard practices offer an optimistic outlook for increasing tocopherol content in almond.

Acknowledgments: This work was supported by the grant RTA2014-00062-00-00 of the Spanish INIA, and the activity of the Consolidated Research Group A12 of Aragon.

Conflicts of Interest: The authors declare no conflict of interest.

References

1. Gradziel, T.M.; Curtis, R.; Socias i Company, R. Production and growing regions. In *Almonds: Botany, Production and Uses*; Socias i Company, R., Gradizel, T.M., Eds.; CABI: Wallingford, UK, 2017; pp. 70–86.
2. Socias i Company, R.; Felipe, A.J. Almond: A diverse germplasm. *HortScience* **1992**, *27*, 717–718.
3. Albala, K. Almonds along the Silk Road: The exchange and adaptation of ideas from West to East. *Petits Propos Culin.* **2009**, *88*, 17–32.
4. Mori, A.M.; Considine, R.V.; Mattes, R.D. Acute and second-meal effects of almond form in impaired glucose tolerant adults: A randomized crossover trial. *Nutr. Metab.* **2011**, *8*, 6. [CrossRef] [PubMed]
5. Ryan, N.T. World almond market. In *Almonds: Botany, Production and Uses*; Socias i Company, R., Gradizel, T.M., Eds.; CABI: Wallingford, UK, 2017; pp. 449–459.
6. Davis, P.A.; Iwasashi, C.K. Whole almonds and almond fractions reduce aberrant crypt foci in a rat model of colon carcinogenesis. *Cancer Lett.* **2001**, *165*, 27–33. [CrossRef]
7. Davis, P.A.; Law, S.; Wong, J. Colonic interposition after esophagectomy for cancer. *Arch. Surg.* **2003**, *138*, 303–308. [CrossRef] [PubMed]

8. Ren, Y.; Waldron, K.W.; Pacy, J.F.; Ellis, P.R. Chemical and histochemical characterization of cell wall polysaccharides in almond seeds in relation to lipid bioavailability. In *Biologically Active Phytochemicals in Food*; Pfannhauser, W., Fenwick, G.R., Khokhar, S., Eds.; Royal Society of Chemistry: Cambridge, UK, 2001; pp. 448–452.

9. Fraser, G.E.; Bennett, H.W.; Jaceldo, K.B.; Sabaté, J. Effect on body weight of a free 76 kilojoule (320 calorie) daily supplement of almonds for six months. *J. Am. Coll. Nutr.* **2002**, *21*, 275–283. [CrossRef] [PubMed]

10. Kendall, C.W.; Jenkins, D.J.; Marchie, A.; Ren, Y.; Ellis, P.R.; Lapsley, K.G. Energy availability from almonds: Implications for weight loss and cardiovascular health. A randomized controlled dose-response trial. *FASEB J.* **2003**, *17*, A339.

11. Lovejoy, J.C.; Most, M.M.; Lefevre, M.; Greenway, F.L.; Rood, J.C. Effect of diets enriched in almonds on insulin action and serum lipids in adults with normal glucose tolerance or type 2 diabetes. *Am. J. Clin. Nutr.* **2002**, *76*, 1000–1006. [PubMed]

12. Scott, L.W.; Balasubramanyam, A.; Kimball, K.T.; Ahrens, A.K.; Fordis, C.M.; Ballantyne, C.M. Long-term, randomonized clinical trial of two diets in the metabolic syndrome and type 2 diabetes. *Diabetes Care* **2003**, *26*, 2481–2482. [CrossRef] [PubMed]

13. Spiller, G.A.; Jenkins, D.J.A.; Cragen, L.N.; Gates, J.E.; Bosello, O.; Berra, K.; Rudd, C.; Stevenson, M.; Superko, R. Effect of a diet high in monounsaturated fat from almonds on plasma-cholesterol and lipoproteins. *J. Am. Coll. Nutr.* **1992**, *11*, 126–130. [PubMed]

14. Fulgoni, V.L.; Abbey, M.; Davis, P.; Jenkins, D.; Lovejoy, J.; Most, M.; Sabaté, J.; Spiller, G. Almonds lower blood cholesterol and LDL-cholesterol but not HDL-cholesterol in human subjects: Results of a meta-analysis. *FASEB J.* **2002**, *16*, A981–A982.

15. Hyson, D.A.; Schneeman, B.O.; Davis, P.A. Almonds and almond oil have similar effects on plasma lipids and LDL oxidation in healthy men and women. *J. Nutr.* **2002**, *132*, 703–707. [PubMed]

16. Jenkins, D.J.A.; Kendall, C.W.C.; Marchie, A.; Parker, T.L.; Connelly, P.W.; Qian, W.; Haight, J.S.; Faulkner, D.; Vidgen, E.; Lapsley, K.G.; et al. Dose response of almonds on coronary heart disease risk factors—Blood lipids, oxidized LDL, lipoprotein(a), homocysteine and pulmonary nitric oxide: A randomized controlled cross-over trial. *Circulation* **2002**, *106*, 1327–1332. [CrossRef] [PubMed]

17. Sabaté, J.; Haddad, E.; Tanzman, J.S.; Jambazian, P.; Rajaram, S. Serum lipid response to the graduated enrichment of a Step I diet with almonds: A randomized feeding trial. *Am. J. Clin. Nutr.* **2003**, *77*, 1379–1384. [PubMed]

18. Schirra, M. Postharvest technology and utilization of almonds. *Hortic. Rev.* **1997**, *20*, 267–292.

19. Sabaté, J.; Hook, D.G. Almonds, walnuts, and serum lipids. In *Lipids in Human Nutrition*; Spiller, G.A., Ed.; CRC Press: Boca Raton, FL, USA, 1996; pp. 137–144.

20. Hollis, J.; Mattes, R. Effect of chronic consumption of almonds on body weight in healthy humans. *Br. J. Nutr.* **2007**, *98*, 651–656. [CrossRef] [PubMed]

21. Cassady, B.A.; Hollis, J.H.; Fulford, A.D.; Considine, R.V.; Mattes, R.D. Mastication of almonds: Effects of lipid bioaccessibility, appetite, and hormone response. *Am. J. Clin. Nutr.* **2009**, *89*, 794–800. [CrossRef] [PubMed]

22. Jenkins, D.J.; Kendall, C.W.; Josse, A.R.; Salvatore, S.; Brighenti, F.; Augustin, L.S.; Ellis, P.R.; Vidgen, E.; Rao, A.V. Almonds decrease postprandial glycemia, insulinemia, and oxidative damage in healthy individuals. *J. Nutr.* **2006**, *136*, 2987–2992. [PubMed]

23. Foster, G.D.; Shantz, K.L.; Vander Veur, S.S.; Oliver, T.L.; Lent, M.R.; Virus, A.; Szapary, P.; Rader, D.J.; Zemel, B.S.; Gilden-Tsai, A. A randomized trial of the effects of an almond-enriched, hypocaloric diet in the treatment of obesity. *Am. J. Clin. Nutr.* **2012**, *96*, 249–254. [CrossRef] [PubMed]

24. Socias i Company, R.; Kodad, O.; Alonso, J.M.; Gradziel, T.M. Almond quality: A breeding perspective. *Hortic. Rev.* **2008**, *34*, 197–238.

25. Kodad, O.; Socias i Company, R. Variability of oil content and of major fatty acid composition in almond (*Prunus amygdalus* Batsch) and its relationshipwith kernel quality. *J. Agric. Food Chem.* **2008**, *56*, 4096–4101. [CrossRef] [PubMed]

26. Saura Calixto, F.; Cañellas, J.; Soler, L. *La Almendra: Composición, Variedades, Desarrollo y Maduración*; INIA: Madrid, Spain, 1988.

27. Yada, S.; Lapsley, K.; Huang, G. A review of composition studies of cultivated almonds: Macronutrients and micronutrients. *J. Food Compos. Anal.* **2011**, *24*, 469–480. [CrossRef]

28. Kodad, O. Chemical composition of almond nuts. In *Almonds: Botany, Production and Uses*; Socias i Company, R., Gradizel, T.M., Eds.; CABI: Wallingford, UK, 2017; pp. 428–449.

29. Kamal-Eldin, A.; Appelqvist, L.A. The chemistry and antioxidant properties of tocopherols and tocotrienols. *Lipids* **1996**, *31*, 671–701. [CrossRef] [PubMed]

30. Sen, C.K.; Khanna, S.; Roy, S. Tocotrienols in health and disease: The other half of the natural vitamin E family. *Mol. Asp. Med.* **2007**, *28*, 692–728. [CrossRef] [PubMed]

31. Senesi, E.; Rizzolo, A.; Colombo, C.; Testoni, A. Influence of pre-processing storage conditions on peeled almond quality. *Ital. J. Food Sci.* **1996**, *2*, 115–125.

32. Zacheo, G.; Cappello, M.S.; Gallo, A.; Santino, A.; Cappello, A.R. Changes associated with postharvest ageing in almond seeds. *Lebensm. Wiss. Technol.* **2000**, *33*, 415–423. [CrossRef]

33. García-Pascual, P.; Mateos, M.; Carbonell, V.; Salazar, D.M. Influence of storage conditions on the quality of shelled and roasted almonds. *Biosyst. Eng.* **2003**, *84*, 201–209. [CrossRef]

34. Kodad, O.; Socias i Company, R.; Prats, M.S.; López Ortiz, M.C. Variability in tocopherol concentrations in almond oil and its use as a selection criterion in almond breeding. *J. Hortic. Sci. Biotechnol.* **2006**, *81*, 501–507. [CrossRef]

35. Font i Forcada, C.; Kodad, O.; Juan, T.; Estopañán, G.; Socias i Company, R. Genetic variability and pollen effect on the transmission of the chemical components of the almond kernel. *Span. J. Agric. Res.* **2011**, *9*, 781–789. [CrossRef]

36. Kodad, O.; Estopañán, G.; Juan, T.; Mamouni, A.; Socias i Company, R. Tocopherol concentration in almond oil: Genetic variation and environmental effect under warm conditions. *J. Agric. Food Chem.* **2011**, *59*, 6137–6141. [CrossRef] [PubMed]

37. Kodad, O.; Estopañán, G.; Juan, T.; Socias i Company, R. Protein content and oil composition of almond from Moroccan seedlings: Genetic diversity, oil quality and geographical origin. *J. Am. Oil Chem. Soc.* **2013**, *90*, 243–252. [CrossRef]

38. Yada, S.; Huang, G.; Lapsley, K. Natural variability in the nutrient composition of California-grown almonds. *J. Food Compos. Anal.* **2013**, *30*, 80–85. [CrossRef]

39. Kodad, O.; Estopañán, G.; Juan, T.; Socias i Company, R. Tocopherol concentration in almond oil from Moroccan seedlings: Geographical origin and post-harvest implications. *J. Food Compos. Anal.* **2014**, *33*, 161–165. [CrossRef]

40. Maestri, D.; Martínez, M.; Bodoira, R.; Rossi, Y.; Oviedo, A.; Pierantozzi, P.; Torres, M. Variability in almond oil chemical traits from traditional cultivars and native genetic resources from Argentina. *Food Chem.* **2015**, *170*, 55–61. [CrossRef] [PubMed]

41. Kodad, O.; Alonso, J.M.; Espiau, M.T.; Estopañán, G.; Juan, T.; Socias i Company, R. Chemometric characterization of almond germplasm: Compositional aspects involved in quality and breeding. *J. Am. Soc. Hortic. Sci.* **2011**, *136*, 273–281.

42. López-Ortiz, C.M.; Prats-Moya, S.; Beltrán Sanahuja, A.; Maestre-Pérez, S.E.; Grané-Teruel, N.; Martín-Carratalá, M.L. Comparative study of tocopherol homologue content in four almond oil cultivars during two consecutive years. *J. Food Compos. Anal.* **2008**, *21*, 144–151. [CrossRef]

43. Rizzolo, A.; Baldo, C.; Polesello, A. Application of high-performance liquid chromatography to the analysis of niacin and biotin in Italian almond cultivars. *J. Chromatogr.* **1991**, *553*, 187–192. [CrossRef]

44. Zhu, Y. Almond (*Prunus dulcis* (Mill.) D.A. Webb) Fatty Acids and Tocopherols under Different Conditions. Ph.D. Thesis, University of Adelaide, Adelaide, Australia, March 2014.

45. Brigelius-Flohé, R.; Kelly, F.J.; Salonen, J.T.; Neuzil, J.; Zingg, J.M.; Azzi, A. The European perspective on vitamin E: Current knowledge and future research. *Am. J. Clin. Nutr.* **2002**, *76*, 703–716. [PubMed]

46. Lampi, A.M.; Kamal-Eldin, A. Effect of α-and γ-tocopherols on thermal polymerization of purified high-oleic sunflower triacylglycerols. *J. Am. Oil Chem. Soc.* **1998**, *75*, 1699–1703. [CrossRef]

47. Warner, K.; Neff, W.E.; Eller, E.J. Enhancing quality and oxidative stability of aged fried food with gamma-tocopherol. *J. Agric. Food Chem.* **2003**, *51*, 623–627. [CrossRef] [PubMed]

48. Kodad, O. Criterios de Selección y de Evaluación de Nuevas Obtenciones Autocompatibles en un Programa de Mejora Genética del Almendro. Ph.D. Thesis, University of Lleida, Lleida, Spain, 2006.

49. Munné-Bosch, S.; Alegre, L. The function of tocopherols and tocotrienols in plants. *Crit. Rev. Plant Sci.* **2002**, *21*, 31–57. [CrossRef]

50. Maranz, S.; Wiesman, Z. Influence of climate on the tocopherol content of sheabutter. *J. Agric. Food Chem.* **2004**, *52*, 2934–2937. [CrossRef] [PubMed]

51. Ali, Q.; Ashraf, M.; Anwar, F. Physico-chemical attributes of seed oil from drought stressed sunflower (*Helianthus annus* L.) plants. *Grasas Aceites* **2009**, *60*, 475–481.

52. Ali, Q.; Ashraf, M.; Anwar, F. Seed composition and seed oil antioxidant activity of maize under water stress. *J. Am. Oil Chem. Soc.* **2010**, *87*, 1179–1187. [CrossRef]

53. Britz, S.J.; Kremer, D.F. Warm temperatures or drought during seed maturation increase free α-tocopherol in seed of soybean (*Glycine max* [L.] Merr.). *J. Agric. Food Chem.* **2002**, *50*, 6058–6063. [CrossRef] [PubMed]

54. Zhu, Y.; Taylor, C.; Sommer, K.; Wilkinson, K.; Wirthensohn, M. Influence of deficit irrigation strategies on fatty acid and tocopherol concentration of almond (*Prunus dulcis*). *Food Chem.* **2015**, *173*, 821–826. [CrossRef] [PubMed]

55. Romero, M.P.; Tovar, M.J.; Ramos, T.; Motilva, M.J. Effect of crop season on the composition of virgin olive oil with protected designation of "Les Garrigues". *J. Am. Oil Chem. Soc.* **2003**, *80*, 423–430. [CrossRef]

56. Richards, A.; Wijesundera, C.; Salisbury, P. Genotype and growing environment effects on the tocopherols and fatty acids of *Brassica napus* and *B. juncea*. *J. Am. Oil Chem. Soc.* **2008**, *87*, 469–481. [CrossRef]

57. Almonor, G.O.; Fenner, G.P.; Wilson, R.F. Temperature effects on tocopherols composition in soybeans with genetically improved oil quality. *J. Am. Oil Chem. Soc.* **1998**, *75*, 591–596. [CrossRef]

58. Beltrán, G.; Jiménez, A.; del Rio, C.; Sánchez, S.; Martínez, L.; Uceda, M.; Aguilera, M.P. Variability of vitamin E in virgin olive oil by agronomical and genetic factors. *J. Food Compos. Anal.* **2010**, *23*, 633–639. [CrossRef]

59. Munné-Bosch, S.; Schwarz, K.; Alegre, L. Enhanced formation of alpha-tocopherol and highly oxidized abietane diterpenes in water-stressed rosemary plants. *Plant Physiol.* **1999**, *121*, 1047–1052. [CrossRef] [PubMed]

60. Munné-Bosch, S.; Peñuelas, J. Drought-induced oxidative stress in strawberry tree (*Arbutus unedo* L.) growing in Mediterranean field conditions. *Plant Sci.* **2004**, *166*, 1105–1110. [CrossRef]

61. Zamany, A.J.; Samadi, G.R.; Kim, D.H.; Keum, Y.S.; Saini, R.K. Comparative study of tocopherol contents and fatty acids composition in twenty almond cultivars of Afghanistan. *J. Am. Oil Chem. Soc.* **2017**, *6*, 805–817. [CrossRef]

62. Baydar, H.; Erbaf, S. Influence of seed development and seed position on oil, fatty acids and total tocopherol contents in sunflower (*Helianthus annuus* L.). *Turk. J. Agric. For.* **2005**, *29*, 179–186.

63. Ayerdi-Gotor, A.; Berger, M.; Labalette, F.; Centis, S.; Daydé, J.; Calmon, A. Variabilité des teneurs et compositions des composés mineurs dans l'huile de tournesol au cours du développement du capitule. Partie I—Tocophérols. *OCL* **2006**, *13*, 206–212. [CrossRef]

64. Britz, S.J.; Kremer, D.F.; Kenworthy, W.J. Tocopherols in soybean seeds: Genetic variation and environmental effects in field-grown crops. *J. Am. Oil Chem. Soc.* **2008**, *85*, 931–936. [CrossRef]

65. Izquierdo, N.G.; Nolasco, S.; Mateo, C.; Santos, D.; Aguirrezábal, L.A.N. Relationship between oil tocopherol concentration and oil weight per grain in several crop species. *Crop Pasture Sci.* **2011**, *62*, 1088–1097. [CrossRef]

66. Soll, J.; Schultz, G. Comparison of geranylgenariol and phytyl substituted methyl-quinols in the tocopherol synthesis of spinach chloroplasts. *Biochem. Biophys. Res. Commun.* **1979**, *91*, 715–720. [CrossRef]

67. Soll, J.; Kemmerling, M.; Schultz, G. Tocopherol and plastoquinone synthesis in spinach chloroplast subfractions. *Arch. Biochem. Biophys.* **1980**, *204*, 544–550. [CrossRef]

68. Soll, J.; Schultz, G. Phytol synthesis from geranylgeraniol in spinach chloroplasts. *Biochem. Biophys. Res. Commun.* **1981**, *99*, 907–912. [CrossRef]

69. Herbers, K. Vitamin production in transgenic plants. *J. Plant Physiol.* **2003**, *160*, 821–829. [CrossRef] [PubMed]

70. Gilliland, L.U.; Magallanes-Lundback, M.; Hemming, C.; Supplee, A.; Koornneef, M.; Leo, B.; Della Penna, D. Genetic basis for natural variation in seed vitamin E levels in *Arabidopsis thaliana*. *Proc. Natl. Acad. Sci. USA* **2006**, *49*, 18834–18841. [CrossRef] [PubMed]

71. Chander, S.; Guo, Y.G.; Yang, X.H.; Yan, J.B.; Zhang, Y.R.; Song, T.M.; Li, J.S. Genetic dissection of tocopherol concentration and composition in maize grain using quantitative trait loci analysis and the candidate gene approach. *Mol. Breed.* **2008**, *22*, 353–365. [CrossRef]

72. Tang, S.; Hass, C.G.; Knapp, S.J. *Ty3/gypsy*-like retrotransposon knockout of a 2-methyl-6-phytyl-1, 4-benzoquinone methyltransferase is non-lethal, uncovers a cryptic paralogous mutation, and produces novel tocopherol (vitamin E) profiles in sunflower. *Theor. Appl. Genet.* **2006**, *113*, 783–799. [CrossRef] [PubMed]

73. Font i Forcada, C.; Fernández i Martí, A.; Socias i Company, R. Mapping quantitative trait loci for kernel composition in almond. *BMC Genet.* **2012**, *13*, 47. [CrossRef] [PubMed]

antioxidants

MDPI

Review

Vitamin E Bioavailability: Mechanisms of Intestinal Absorption in the Spotlight

Emmanuelle Reboul ⓘ

Aix Marseille University, INRA, INSERM, C2VN, 13005 Marseille, France; Emmanuelle.Reboul@univ-amu.fr;
Tel.: +33-049-1324-278

Received: 31 October 2017; Accepted: 16 November 2017; Published: 22 November 2017

Abstract: Vitamin E is an essential fat-soluble micronutrient whose effects on human health can be attributed to both antioxidant and non-antioxidant properties. A growing number of studies aim to promote vitamin E bioavailability in foods. It is thus of major interest to gain deeper insight into the mechanisms of vitamin E absorption, which remain only partly understood. It was long assumed that vitamin E was absorbed by passive diffusion, but recent data has shown that this process is actually far more complex than previously thought. This review describes the fate of vitamin E in the human gastrointestinal lumen during digestion and focuses on the proteins involved in the intestinal membrane and cellular transport of vitamin E across the enterocyte. Special attention is also given to the factors modulating both vitamin E micellarization and absorption. Although these latest results significantly improve our understanding of vitamin E intestinal absorption, further studies are still needed to decipher the molecular mechanisms driving this multifaceted process.

Keywords: tocopherol; intestine; mixed micelles; dietary lipids; food matrix; membrane transporters; uptake; enterocytes; chylomicrons; HDL; fat-soluble vitamins

1. Introduction

Tocochromanols, a subset of isoprenoids better known as vitamin E, include four tocopherols and four tocotrienols (Figure 1). These lipophilic antioxidants are synthesized by plants and other photosynthetic organisms only [1]. The base of the molecule of tocopherol is a hydroxychromane nucleus upon which a phytyl saturated chain of 16 carbon atoms is fixed. Three of these carbons are asymmetric, which entails the possibility of the existence of eight stereoisomers. The different tocopherols are distinguished from each other by the number and the position of the methyl groups attached to the nucleus. RRR-α-tocopherol is the most common in nature and the tocopherol with the highest biological activity. In biological tests for vitamin evaluation (fetal resorption tests), β and γ-tocopherol display a reduced vitamin activity (from 15 to 30%), and δ-tocopherol is almost inactive. Tocotrienols are distinguished from tocopherols by the presence of three double bonds on the side chain. Only α and β-tocotrienol appear to have a significant vitamin activity [2].

The hydroxychromane nucleus of vitamin E can react with peroxyl radicals, generating a hydroperoxide (which can be inactivated by specific enzymes) and a tocopheryl radical (which can be regenerated by vitamin C or coenzyme Q10). This property places vitamin E at the forefront of anti-radical defense systems [2]. However, the beneficial effects of vitamin E in human health may also be due to the ability of its phosphorylated metabolite to modulate signal transduction and gene expression in numerous conditions, including inflammation and immune system disorders [3].

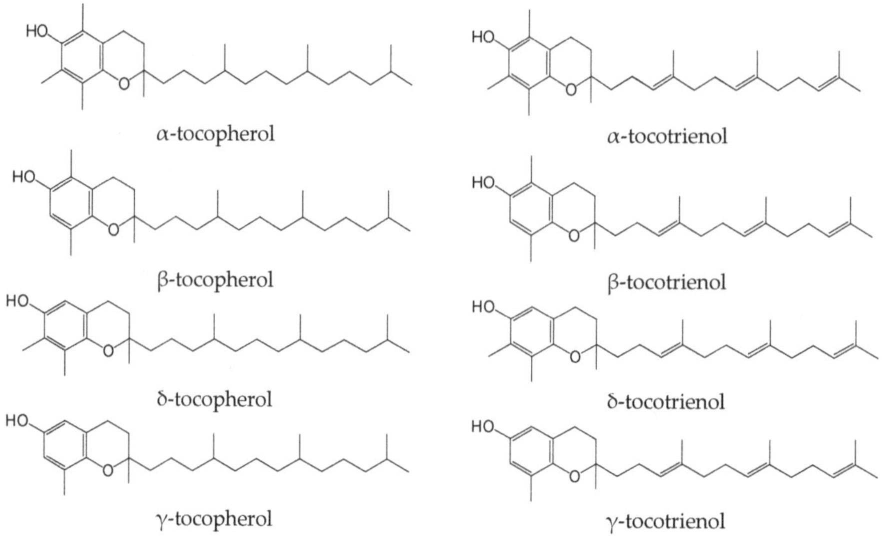

Figure 1. Vitamin E vitamers.

The main sources of vitamin E are vegetable oils and seeds. It can also be found in smaller quantities in some fruits and vegetables (Table 1). In France and Europe in general, α-tocopherol is the most consumed vitamin E vitamer [2], while in the US, it is γ-tocopherol [4].

Table 1. Vitamin E food content [2,5]. Average values in brackets.

Foods	Vitamin E Content (mg/100 g)
Sunflower oil	0.1–90 (58.3)
Sunflower seeds	0.01–57.6 (42.3)
Other vegetal oils	0.1–30
Almonds	0.01–24 (14.6)
Butter	1.5–2.3 (2.11)
Fatty fish	0.9–2
Fruits and vegetables (spinach, tomatoes, etc.)	0.8–2

In Europe, the Recommended Dietary Allowance (RDA) for vitamin E is 11 mg per day (as an α-tocopherol-equivalent) for women and 13 mg per day for men [6], while the RDA is 15 mg for all adults in the US [7]. Although there are no real vitamin E deficiencies in Western countries, three surveys carried out in France (Burgundy, ESVITAF, and Val-de-Marne surveys) showed that more than 30% of French people consumed less than 8 mg of vitamin E a day [8], which was recently confirmed in other European countries [9] as well as in the US, where over 90% of the population do not consume the estimated average requirements [10]. As recently shown, the low dietary intake of vitamin E may be worsened by the low stability of vitamin E in vegetable oils [11].

2. Vitamin E Digestion Process

2.1. The Fate of Vitamin E in the Gastrointestinal Tract

The first phase of the digestion–absorption process is the dissolution of vitamin E in the lipid phase of the meal. This phase is then emulsified into lipid droplets at both gastric and duodenal levels. No metabolism of vitamin E (i.e., degradation or absorption) appears to exist in the stomach.

In addition, the size of the droplets does not seem to have any effect on the efficiency of the subsequent absorption of the vitamin E in healthy humans [12].

In the duodenum, vitamin E is incorporated, along with lipid digestion products, in mixed micelles, structures that are theoretically essential for its absorption by the enterocyte. Indeed, mixed micelles can solubilize hydrophobic components and diffuse into the unstirred water layer (glycocalix) to approach the brush border membrane of the enterocytes.

2.2. Factors Affecting Vitamin E Transfer to Mixed Micelles

Numerous factors can affect vitamin E bioaccessibility (i.e., the fraction of vitamin E recovered in the mixed micelles compared to the initial amount of vitamin E provided by the meal) and thus in turn vitamin E bioavailability. The main factor is the food matrix in which vitamin E is embedded. For instance, it was shown that vitamin E bioaccessibility was low in apples but almost total in bananas, bread, or lettuce [13], and that vitamin E from durum wheat pasta was more bioaccessible than from pasta containing 10% eggs [14]. Unfortunately, it was not possible to identify the biochemical parameters of green leafy vegetables (cell-wall content, pectin, tannin, ...) governing α-tocopherol bioaccessibility [15]. However, as for other lipid micronutrients, matrix disruption can enhance vitamin E transfer to mixed micelles [16], while thermal or high pressure treatments have either no or negative effects [17].

Another important factor determining vitamin E bioaccessibility is the amount of fat provided in the meal, as fat likely facilitates vitamin E extraction from its food matrix, stimulates biliary secretion, and promotes micelle formation. It was first shown that various fats and oils, as well as long-chain triacylglycerols, did not significantly enhance vitamin E bioaccessibility. This was in contrast to more hydrophobic microconstituents such as β-carotene [18]. However, further data consistently showed that tocopherol acetate bioaccessibility was higher in long-chain rather than medium-chain triglyceride emulsions, probably due to a greater solubilization capacity of mixed micelles formed from long chain fatty acids and an enhanced conversion into tocopherol [19–21]. Vitamin E bioaccessibility was also increased by the presence of phospholipids [22].

2.3. Vitamin E Ester Hydrolysis

It is acknowledged that only the free forms of vitamin E are absorbed by the intestinal mucosa, suggesting that the esterified forms are hydrolyzed beforehand. This hydrolysis is probably carried out by cholesteryl ester hydrolase, also known as bile salt-dependent lipase [23]. Conversely to what we observed for retinyl esters [24], neither pancreatic lipase nor pancreatic lipase-related protein 2 were able to hydrolyze tocopheryl esters [25]. Surprisingly, it has been recently reported that α-tocopherol acetate absorption was equivalent to that of free α-tocopherol in the absence of both digestive enzymes and bile salts in healthy subjects [26]. This result signifies that enzymes originating from the enterocytes, such as endoplasmic reticulum esterases [27], are able to realize this hydrolysis in a very efficient manner.

3. Vitamin E Absorption Mechanisms by the Enterocyte

3.1. Apical Transport at the Brush Border Level

When approaching the brush border membrane, mixed micelles are supposed to dissociate due to the existing pH gradient. The released constituents can then be captured by different more or less specific systems to be absorbed by the enterocyte. For more than 30 years, due to the first results obtained in rat intestinal everted sacs [28,29], vitamin E absorption has been considered to occur by passive diffusion through enterocyte apical membrane. However, we showed for the first time in 2006 that α- and γ-tocopherol absorption was mediated, at least partly, by scavenger receptor class B type I (SR-BI) [30]. It was also shown that NPC1 like intracellular cholesterol transporter 1 (NPC1L1) was involved in α-tocopherol [31,32] and γ-tocotrienol [33] absorption. We finally highlighted the additional role of CD36 molecule (CD36) in the tocopherol absorption process [34].

These three membrane proteins have primarily been described as cholesterol transporters in the intestine [35,36]. However, they do display a relatively broad substrate specificity. Besides cholesterol, SRBI can mediate carotenoid [37,38], vitamin D [39], and K [40] transport; NPC1L1 is involved in phytosterols [41], vitamin D [39], K [42] and lutein [43] uptake; and CD36 is involved in very long chain fatty acid [44], vitamin D [39], K [40] and carotenoid [37] absorption. It is thus not surprising that they are involved in vitamin E uptake as well. Both SR-BI [45] and NPC1L1 [46] were showed to traffic in clathrin-coated lipid vesicles after a lipid load. The fact that these transporters seem to selectively mediate the transport of some molecules present in mixed micelles is an argument in favor of a direct interaction with their ligands. Interestingly, it was recently demonstrated that α-tocopherol competed with cholesterol to bind to the NPC1L1-N terminal domain, promoting NPC1L1 endocytosis [47].

CD36 [48] and SR-BI [49] have recently been described as intestinal lipid sensors, and they appear to be key modulators of chylomicron secretion [50,51]. These roles suggest that their impact on vitamin E transport may actually be a consequence of their role in other lipid absorption process. Indeed, by promoting lipid fluxes through the enterocyte, these receptors would create a driving force for absorption of minor lipids such as micronutrients, due to the lipid gradient.

We observed that the presence of tocopherol in structures mimicking mixed micelles (i.e., containing a biliary salt and at least oleic acid) was necessary for transporter-dependent absorption in Caco-2 cells [32]. This is in agreement with another study in which we showed that both SR-BI and CD36 extracellular loops were able to bind postprandial mixed micelles in a more efficient manner than interprandial micelles [52]. However, this does not indicate whether this interaction is the first step of either a sensing or an absorptive process—or if the two phenomena are coexisting.

Finally, it is worth mentioning that we observed that vitamin E could be effluxed back to the lumen after being absorbed in Caco-2 cells. This efflux was SR-BI-dependent and was increased in the presence of acceptors, e.g., mixed micelles that did not contain vitamin E [30]. The in vivo relevance of such observation should be further investigated.

3.2. Vitamin E Trafficking across the Enterocyte

Once absorbed, the fate of vitamin E across the enterocyte has been poorly described. Being hydrophobic, vitamin E likely localizes into organelle membranes, cytosolic lipid droplets, or traffic bound to binding proteins. Subcellular localization revealed that vitamin E could accumulate in microsomal membranes, i.e., the endoplasmic reticulum, Golgi, lysosomal and peroxisomal membranes [53]. However, it should be noted that in this study, vitamin E was delivered to cells with Tween 40, which may have influenced its targeting within the cells compared to a delivery in its physiological vehicles (i.e., mixed micelles).

Targeting to microsomal membranes may occur either via clathrin-coated vesicles as mentioned above, or thanks to the intervention of cytosolic carriers. A tocopherol-associated protein (TAP) expressed in the intestine has been shown to bind vitamin E in human tissues [54], but it actually displays a weak affinity towards tocopherols [55]. Sec14p-like proteins TAP1, 2 and 3, also expressed in the human intestine, are probably better candidates as they improved tocopherol transport to mitochondria as efficiently as the α-tocopherol transport protein (α-TTP) [56]. Additional research is needed to definitely confirm their role in the enterocyte.

3.3. Basolateral Secretion to the Lymph or to the Blood Circulation

Most of the vitamin E is incorporated into chylomicrons in its free form at the Golgi apparatus level before being released to the lymph.

In mice, it has been showed that in addition to this apolipoprotein-B (apoB)-dependent route, an non-apoB pathway could exist [57]. This non-apoB route involves ATB-binding cassette A1 (ABCA1) that allows the secretion of vitamin E via intestinal High Density Lipoproteins (HDL) [58], and maybe the ATP binding cassette sub-family G member 1 ABCG1 [59,60]. However, this pathway seems to remain minor in humans. Indeed, mutations in microsomal triglyceride transfer protein (MTP) or in secretion-associated Ras related GTPase 1B (SAR1B), lead to abetalipoproteinemia and chylomicron retention diseases, respectively [61]. These pathologies, characterized by a lack of chylomicrons, are associated with a massive impairment of vitamin E absorption that is not balanced by another pathway [62,63].

3.4. Vitamin E Absorption Site in the Intestine

It has long been assumed that vitamin E, as with other lipids and lipid micronutrients, is absorbed in the upper half of the small intestine [64]. However, recent work from our laboratory highlighted that vitamin E absorption was in fact mainly located in the distal part of mouse small intestine, i.e., in the distal jejunum and the ileum [32,65]. This data seems conflicting with the fact that identified vitamin E intestinal transporters, i.e., scavenger receptors and NPC1L1, have been described as mainly expressed in the duodenum and the jejunum, respectively [36,66,67]. However, this can be partly explained by the subcellular localization of these proteins. For instance, SR-BI is mostly expressed at the apical side of the duodenal enterocytes, but it is present on the basolateral surface of the distal intestine [68]. Besides, a postmortem study in humans showed that the expression of CD36, NPC1L1, and ABCA1 was highly variable and displayed a bell-shape pattern, with the highest levels in the ileum [69].

3.5. Factors Modulating Vitamin E Absorption by the Intestinal Cell

As for vitamin E transfer to mixed micelles, numerous factors can influence vitamin E transport across the intestinal cell, which likely explains the important variations observed regarding vitamin E absorption efficiency. Indeed, different studies report efficiency in ranges of 10–95% [70–72]. However when deuterium-labeled vitamin E was used to assess absorption, this range was reduced to 10–33% [73].

The intestine does not seem to specifically discriminate between vitamin E stereoisomers [74], or between α- and γ-tocopherol [30,75]. However, a study found a higher absorption of α-tocopherol compared to γ- and δ-tocopherol in lymph-cannulated rats [76], which is consistent with the existence of a ω-hydroxylase that preferentially metabolized these two last vitamers in 3′ and 5′ carboxychromanol metabolites that can be excreted in the urine [77].

The food matrix can also influence specifically this step of vitamin E absorption. The presence of fibers did not modify vitamin E absorption in humans [78,79]. Conversely, lipids can be classified as effectors of vitamin E absorption as they can promote chylomicron formation. It is interesting to note that a minimal quantity of fat of 3 g was required for an optimal tocopherol absorption, and that increasing further this amount did not led to a better vitamin E bioavailability [80]. This was partly confirmed in another trial where α-tocopherol-acetate was almost negligible when ingested with 2.7 g fat [81]. However, these data are conflicting with other studies showing that the higher the amount of fat, the better vitamin E absorption [73,82], or conversely that dairy fat from whole milk does not increase vitamin E absorption compared to low-fat milk [83]. Mono and polyunsaturated fatty acid seem to promote vitamin E absorption compared to saturated fatty acids in cockerels [84] and in Caco-2 cells [85]. Conversely, phosphatidylcholine decreased α-tocopherol absorption efficiency in rats [86,87], an effect that was reversed by the presence of lysophosphatidylcholine [86]. Authors suggested that vitamin E was associated with phospholipids, leading to a low uptake. As we showed that the presence of phosphatidylcholine in mixed micelles was associated with a decreased binding of mixed micelles on scavenger receptor extracellular loops [52], we suggest that neutral phospholipids can also impact vitamin E absorption by modifying micellar interaction with the membrane proteins responsible for its uptake.

We showed that α-tocopherol E could compete for absorption with other lipid micronutrients such as γ-tocopherol and carotenoids [30,88], as well as vitamin A, D, and K [65] in Caco-2 cells. Except for vitamin A, these competitions are presumably due to common uptake pathways involving cholesterol transporters. Vitamin A uptake mechanisms are still unknown. However, it has been hypothesized that vitamin E was protecting vitamin A against oxidation in the intestine, leading to vitamin E degradation and reduced absorption in chickens [89]. Polyphenols such as naringenin could also reduce vitamin E uptake in Caco-2 cells [88]. The underlying mechanisms still need to be resolved, but we can suggest that polyphenols can impair (micro) nutrient absorption by interfering with membrane protein functioning, as previously shown with digestive enzymes [90]. Although this is still debated [91], a study showed that phytosterols (2.2 g per day during 1 week) could inhibit vitamin E absorption in normocholesterolemic subjects [92].

Finally, it is noteworthy that genetic factors including polymorphisms in genes coding for vitamin E and lipid intestinal metabolism such as SR-BI, CD36, ABCA1, ABCG1, or apoB have been associated with a modulation of vitamin E bioavailability in humans [93].

4. Conclusions

Overall, this review highlights the fact that the molecular mechanisms of both intraluminal fate and intestinal absorption of vitamin E are only partly understood to date. The discovery of vitamin E intestinal transporters with broad substrate specificity has raised many questions with respect to the potential interactions with dietary lipids during the vitamin E absorption process. Besides, it is likely that other proteins involved in vitamin E absorption still remain to be identified. The modulation of the activity of these proteins, including the existence of functional polymorphisms in their encoding genes and the regulation of their expression levels by epigenetic to post-translational factors, may explain much of the observed large interindividual variation in postprandial responses to vitamin E. Further dedicated investigations are needed to address these presumptions in order to offer adequate vitamin E-tailored recommendations to individuals.

Conflicts of Interest: The author declares no conflict of interest.

References

1. DellaPenna, D. A decade of progress in understanding vitamin E synthesis in plants. *J. Plant Physiol.* **2005**, *162*, 729–737. [CrossRef] [PubMed]
2. Martin, A. *Apports Nutritionnels Conseillés Pour la Population Française*, 3rd ed.; Tec & Doc. Lavoisier: Paris, France, 2001; p. 605.
3. Azzi, A.; Meydani, S.N.; Meydani, M.; Zingg, J.M. The rise, the fall and the renaissance of vitamin E. *Arch. Biochem. Biophys.* **2016**, *595*, 100–108. [CrossRef] [PubMed]
4. Jiang, Q.; Christen, S.; Shigenaga, M.K.; Ames, B.N. gamma-Tocopherol, the major form of vitamin E in the US diet, deserves more attention. *Am. J. Clin. Nutr.* **2001**, *74*, 714–722. [PubMed]
5. Table de Composition Nutritionnelle des Aliments Ciqual. 2016. Available online: https://www.anses.fr/fr/content/ciqual-la-table-de-composition-nutritionnelle-des-aliments (accessed on 31 October 2017).
6. EFSA Panel on Dietetic Products, Nutrition, and Allergies (NDA). Dietary Reference Values for vitamin E as α-tocopherol. *EFSA J.* **2015**, *13*. [CrossRef]
7. Medicine, I.O. *Dietary Reference Intakes for Vitamin C, Vitamin E, Selenium, and Carotenoids*; The National Academies Press: Washington, DC, USA, 2000; p. 529.
8. De Carvalho, M.J.; Guilland, J.C.; Moreau, D.; Boggio, V.; Fuchs, F. Vitamin status of healthy subjects in Burgundy (France). *Ann. Nutr. Metab.* **1996**, *40*, 24–51. [CrossRef] [PubMed]
9. Troesch, B.; Hoeft, B.; McBurney, M.; Eggersdorfer, M.; Weber, P. Dietary surveys indicate vitamin intakes below recommendations are common in representative western countries. *Br. J. Nutr.* **2012**, *108*, 692–698. [CrossRef] [PubMed]
10. Traber, M.G. Vitamin E inadequacy in humans: Causes and consequences. *Adv. Nutr.* **2014**, *5*, 503–514. [CrossRef] [PubMed]
11. Pignitter, M.; Stolze, K.; Gartner, S.; Dumhart, B.; Stoll, C.; Steiger, G.; Kraemer, K.; Somoza, V. Cold fluorescent light as major inducer of lipid oxidation in soybean oil stored at household conditions for eight weeks. *J. Agric. Food Chem.* **2014**, *62*, 2297–2305. [CrossRef] [PubMed]

12. Borel, P.; Pasquier, B.; Armand, M.; Tyssandier, V.; Grolier, P.; Alexandre-Gouabau, M.C.; Andre, M.; Senft, M.; Peyrot, J.; Jaussan, V.; et al. Processing of vitamin A and E in the human gastrointestinal tract. *Am. J. Physiol. Gastrointest. Liver Physiol.* **2001**, *280*, G95–G103. [PubMed]
13. Reboul, E.; Richelle, M.; Perrot, E.; Desmoulins-Malezet, C.; Pirisi, V.; Borel, P. Bioaccessibility of carotenoids and vitamin E from their main dietary sources. *J. Agric. Food Chem.* **2006**, *54*, 8749–8755. [CrossRef] [PubMed]
14. Werner, S.; Bohm, V. Bioaccessibility of carotenoids and vitamin e from pasta: Evaluation of an in vitro digestion model. *J. Agric. Food Chem.* **2011**, *59*, 1163–1170. [CrossRef] [PubMed]
15. Sriwichai, W.; Berger, J.; Picq, C.; Avallone, S. Determining Factors of Lipophilic Micronutrient Bioaccessibility in Several Leafy Vegetables. *J. Agric. Food Chem.* **2016**, *64*, 1695–1701. [CrossRef] [PubMed]
16. Mandalari, G.; Faulks, R.M.; Rich, G.T.; Lo Turco, V.; Picout, D.R.; Lo Curto, R.B.; Bisignano, G.; Dugo, P.; Dugo, G.; Waldron, K.W.; et al. Release of protein, lipid, and vitamin E from almond seeds during digestion. *J. Agric. Food Chem.* **2008**, *56*, 3409–3416. [CrossRef] [PubMed]
17. Cilla, A.; Alegria, A.; de Ancos, B.; Sanchez-Moreno, C.; Cano, M.P.; Plaza, L.; Clemente, G.; Lagarda, M.J.; Barbera, R. Bioaccessibility of tocopherols, carotenoids, and ascorbic acid from milk- and soy-based fruit beverages: Influence of food matrix and processing. *J. Agric. Food Chem.* **2012**, *60*, 7282–7290. [CrossRef] [PubMed]
18. Nagao, A.; Kotake-Nara, E.; Hase, M. Effects of fats and oils on the bioaccessibility of carotenoids and vitamin E in vegetables. *Biosci. Biotechnol. Biochem.* **2013**, *77*, 1055–1060. [CrossRef] [PubMed]
19. Yang, Y.; McClements, D.J. Vitamin E bioaccessibility: Influence of carrier oil type on digestion and release of emulsified alpha-tocopherol acetate. *Food Chem.* **2013**, *141*, 473–481. [CrossRef] [PubMed]
20. Yang, Y.; Decker, E.A.; Xiao, H.; McClements, D.J. Enhancing vitamin E bioaccessibility: Factors impacting solubilization and hydrolysis of alpha-tocopherol acetate encapsulated in emulsion-based delivery systems. *Food Funct.* **2015**, *6*, 84–97. [CrossRef] [PubMed]
21. Yang, Y.; Xiao, H.; McClements, D.J. Impact of Lipid Phase on the Bioavailability of Vitamin E in Emulsion-Based Delivery Systems: Relative Importance of Bioaccessibility, Absorption, and Transformation. *J. Agric. Food Chem.* **2017**, *65*, 3946–3955. [CrossRef] [PubMed]
22. Yang, Y.; McClements, D.J. Vitamin E and vitamin E acetate solubilization in mixed micelles: Physicochemical basis of bioaccessibility. *J. Colloid Interface Sci.* **2013**, *405*, 312–321. [CrossRef] [PubMed]
23. Lombardo, D.; Guy, O. Studies on the substrate specificity of a carboxyl ester hydrolase from human pancreatic juice. II. Action on cholesterol esters and lipid-soluble vitamin esters. *Biochim. Biophys. Acta* **1980**, *611*, 147–155. [CrossRef]
24. Reboul, E.; Berton, A.; Moussa, M.; Kreuzer, C.; Crenon, I.; Borel, P. Pancreatic lipase and pancreatic lipase-related protein 2, but not pancreatic lipase-related protein 1, hydrolyze retinyl palmitate in physiological conditions. *Biochim. Biophys. Acta* **2006**, *1761*, 4–10. [CrossRef] [PubMed]
25. Desmarchelier, C.; Tourniaire, F.; Preveraud, D.P.; Samson-Kremser, C.; Crenon, I.; Rosilio, V.; Borel, P. The distribution and relative hydrolysis of tocopheryl acetate in the different matrices coexisting in the lumen of the small intestine during digestion could explain its low bioavailability. *Mol. Nutr. Food Res.* **2013**, *57*, 1237–1245. [CrossRef] [PubMed]
26. Nagy, K.; Ramos, L.; Courtet-Compondu, M.C.; Braga-Lagache, S.; Redeuil, K.; Lobo, B.; Azpiroz, F.; Malagelada, J.R.; Beaumont, M.; Moulin, J.; et al. Double-balloon jejunal perfusion to compare absorption of vitamin E and vitamin E acetate in healthy volunteers under maldigestion conditions. *Eur. J. Clin. Nutr.* **2013**, *67*, 202–206. [CrossRef] [PubMed]
27. Mathias, P.M.; Harries, J.T.; Peters, T.J.; Muller, D.P. Studies on the in vivo absorption of micellar solutions of tocopherol and tocopheryl acetate in the rat: Demonstration and partial characterization of a mucosal esterase localized to the endoplasmic reticulum of the enterocyte. *J. Lipid Res.* **1981**, *22*, 829–837. [PubMed]
28. Muralidhara, K.S.; Hollander, D. Intestinal absorption of alpha-tocopherol in the unanesthetized rat. The influence of luminal constituents on the absorptive process. *J. Lab. Clin. Med.* **1977**, *90*, 85–91. [PubMed]
29. Hollander, D.; Rim, E.; Muralidhara, K.S. Mechanism and site of small intestinal absorption of alpha-tocopherol in the rat. *Gastroenterology* **1975**, *68*, 1492–1499. [PubMed]
30. Reboul, E.; Klein, A.; Bietrix, F.; Gleize, B.; Malezet-Desmoulins, C.; Schneider, M.; Margotat, A.; Lagrost, L.; Collet, X.; Borel, P. Scavenger receptor class B type I (SR-BI) is involved in vitamin E transport across the enterocyte. *J. Biol. Chem.* **2006**, *281*, 4739–4745. [CrossRef] [PubMed]

31. Narushima, K.; Takada, T.; Yamanashi, Y.; Suzuki, H. Niemann-pick C1-like 1 mediates alpha-tocopherol transport. *Mol. Pharmacol.* **2008**, *74*, 42–49. [CrossRef] [PubMed]

32. Reboul, E.; Soayfane, Z.; Goncalves, A.; Cantiello, M.; Bott, R.; Nauze, M.; Terce, F.; Collet, X.; Comera, C. Respective contributions of intestinal Niemann-Pick C1-like 1 and scavenger receptor class B type I to cholesterol and tocopherol uptake: In vivo v. in vitro studies. *Br. J. Nutr.* **2012**, *107*, 1296–1304. [CrossRef] [PubMed]

33. Abuasal, B.; Sylvester, P.W.; Kaddoumi, A. Intestinal absorption of gamma-tocotrienol is mediated by Niemann-Pick C1-like 1: In situ rat intestinal perfusion studies. *Drug Metab. Dispos.* **2010**, *38*, 939–945. [CrossRef] [PubMed]

34. Goncalves, A.; Roi, S.; Nowicki, M.; Niot, I.; Reboul, E. Cluster-determinant 36 (CD36) impacts on vitamin E postprandial response. *Mol. Nutr. Food Res.* **2014**, *58*, 2297–2306. [CrossRef] [PubMed]

35. Werder, M.; Han, C.H.; Wehrli, E.; Bimmler, D.; Schulthess, G.; Hauser, H. Role of scavenger receptors SR-BI and CD36 in selective sterol uptake in the small intestine. *Biochemistry* **2001**, *40*, 11643–11650. [CrossRef] [PubMed]

36. Altmann, S.W.; Davis, H.R., Jr.; Zhu, L.J.; Yao, X.; Hoos, L.M.; Tetzloff, G.; Iyer, S.P.; Maguire, M.; Golovko, A.; Zeng, M.; et al. Niemann-Pick C1 Like 1 protein is critical for intestinal cholesterol absorption. *Science* **2004**, *303*, 1201–1204. [CrossRef] [PubMed]

37. Borel, P.; Lietz, G.; Goncalves, A.; Szabo de Edelenyi, F.; Lecompte, S.; Curtis, P.; Goumidi, L.; Caslake, M.J.; Miles, E.A.; Packard, C.; et al. CD36 and SR-BI Are Involved in Cellular Uptake of Provitamin A Carotenoids by Caco-2 and HEK Cells, and Some of Their Genetic Variants Are Associated with Plasma Concentrations of These Micronutrients in Humans. *J. Nutr.* **2013**, *143*, 448–456. [CrossRef] [PubMed]

38. Reboul, E.; Abou, L.; Mikail, C.; Ghiringhelli, O.; Andre, M.; Portugal, H.; Jourdheuil-Rahmani, D.; Amiot, M.J.; Lairon, D.; Borel, P. Lutein transport by Caco-2 TC-7 cells occurs partly by a facilitated process involving the scavenger receptor class B type I (SR-BI). *Biochem. J.* **2005**, *387*, 455–461. [CrossRef] [PubMed]

39. Reboul, E.; Goncalves, A.; Comera, C.; Bott, R.; Nowicki, M.; Landrier, J.F.; Jourdheuil-Rahmani, D.; Dufour, C.; Collet, X.; Borel, P. Vitamin D intestinal absorption is not a simple passive diffusion: Evidences for involvement of cholesterol transporters. *Mol. Nutr. Food Res.* **2011**, *55*, 691–702. [CrossRef] [PubMed]

40. Goncalves, A.; Margier, M.; Roi, S.; Collet, X.; Niot, I.; Goupy, P.; Caris-Veyrat, C.; Reboul, E. Intestinal scavenger receptors are involved in vitamin K1 absorption. *J. Biol. Chem.* **2014**, *289*, 30743–30752. [CrossRef] [PubMed]

41. Davis, H.R., Jr.; Zhu, L.J.; Hoos, L.M.; Tetzloff, G.; Maguire, M.; Liu, J.; Yao, X.; Iyer, S.P.; Lam, M.H.; Lund, E.G.; et al. Niemann-Pick C1 Like 1 (NPC1L1) is the intestinal phytosterol and cholesterol transporter and a key modulator of whole-body cholesterol homeostasis. *J. Biol. Chem.* **2004**, *279*, 33586–33592. [CrossRef] [PubMed]

42. Takada, T.; Yamanashi, Y.; Konishi, K.; Yamamoto, T.; Toyoda, Y.; Masuo, Y.; Yamamoto, H.; Suzuki, H. NPC1L1 is a key regulator of intestinal vitamin K absorption and a modulator of warfarin therapy. *Sci. Transl. Med.* **2015**, *7*. [CrossRef] [PubMed]

43. Sato, Y.; Suzuki, R.; Kobayashi, M.; Itagaki, S.; Hirano, T.; Noda, T.; Mizuno, S.; Sugawara, M.; Iseki, K. Involvement of cholesterol membrane transporter Niemann-Pick C1-like 1 in the intestinal absorption of lutein. *J. Pharm. Pharm. Sci.* **2012**, *15*, 256–264. [CrossRef] [PubMed]

44. Drover, V.A.; Nguyen, D.V.; Bastie, C.C.; Darlington, Y.F.; Abumrad, N.A.; Pessin, J.E.; London, E.; Sahoo, D.; Phillips, M.C. CD36 mediates both cellular uptake of very long chain fatty acids and their intestinal absorption in mice. *J. Biol. Chem.* **2008**, *283*, 13108–13115. [CrossRef] [PubMed]

45. Hansen, G.H.; Niels-Christiansen, L.L.; Immerdal, L.; Danielsen, E.M. Scavenger receptor class B type I (SR-BI) in pig enterocytes: Trafficking from the brush border to lipid droplets during fat absorption. *Gut* **2003**, *52*, 1424–1431. [CrossRef] [PubMed]

46. Ge, L.; Wang, J.; Qi, W.; Miao, H.H.; Cao, J.; Qu, Y.X.; Li, B.L.; Song, B.L. The cholesterol absorption inhibitor ezetimibe acts by blocking the sterol-induced internalization of NPC1L1. *Cell Metab.* **2008**, *7*, 508–519. [CrossRef] [PubMed]

47. Kamishikiryo, J.; Haraguchi, M.; Nakashima, S.; Tasaka, Y.; Narahara, H.; Sugihara, N.; Nakamura, T.; Morita, T. N-terminal domain of the cholesterol transporter Niemann-Pick C1-like 1 (NPC1L1) is essential for alpha-tocopherol transport. *Biochem. Biophys. Res. Commun.* **2017**, *486*, 476–480. [CrossRef] [PubMed]

48. Tran, T.T.; Poirier, H.; Clement, L.; Nassir, F.; Pelsers, M.M.; Petit, V.; Degrace, P.; Monnot, M.C.; Glatz, J.F.; Abumrad, N.A.; et al. Luminal lipid regulates CD36 levels and downstream signaling to stimulate chylomicron synthesis. *J. Biol. Chem.* **2011**, *286*, 25201–25210. [CrossRef] [PubMed]

49. Beaslas, O.; Cueille, C.; Delers, F.; Chateau, D.; Chambaz, J.; Rousset, M.; Carriere, V. Sensing of dietary lipids by enterocytes: A new role for SR-BI/CLA-1. *PLoS ONE* **2009**, *4*, e4278. [CrossRef] [PubMed]

50. Buttet, M.; Traynard, V.; Tran, T.T.; Besnard, P.; Poirier, H.; Niot, I. From fatty-acid sensing to chylomicron synthesis: Role of intestinal lipid-binding proteins. *Biochimie* **2014**, *96*, 37–47. [CrossRef] [PubMed]

51. Briand, O.; Touche, V.; Colin, S.; Brufau, G.; Davalos, A.; Schonewille, M.; Bovenga, F.; Carriere, V.; de Boer, J.F.; Dugardin, C.; et al. Liver X Receptor Regulates Triglyceride Absorption Through Intestinal Down-regulation of Scavenger Receptor Class B, Type 1. *Gastroenterology* **2016**, *150*, 650–658. [CrossRef] [PubMed]

52. Goncalves, A.; Gontero, B.; Nowicki, M.; Margier, M.; Masset, G.; Amiot, M.J.; Reboul, E. Micellar lipid composition affects micelle interaction with class B scavenger receptor extracellular loops. *J. Lipid Res.* **2015**, *56*, 1123–1133. [CrossRef] [PubMed]

53. Anwar, K.; Kayden, H.J.; Hussain, M.M. Transport of vitamin E by differentiated Caco-2 cells. *J. Lipid Res.* **2006**, *47*, 1261–1273. [CrossRef] [PubMed]

54. Zimmer, S.; Stocker, A.; Sarbolouki, M.N.; Spycher, S.E.; Sassoon, J.; Azzi, A. A novel human tocopherol-associated protein—Cloning, in vitro expression, and characterization. *J. Biol. Chem.* **2000**, *275*, 25672–25680. [CrossRef] [PubMed]

55. Manor, D.; Atkinson, J. Is tocopherol associated protein a misnomer? *J. Nutr. Biochem* **2003**, *14*, 421–422. [CrossRef]

56. Zingg, J.M.; Kempna, P.; Paris, M.; Reiter, E.; Villacorta, L.; Cipollone, R.; Munteanu, A.; De Pascale, C.; Menini, S.; Cueff, A.; et al. Characterization of three human sec14p-like proteins: Alpha-tocopherol transport activity and expression pattern in tissues. *Biochimie* **2008**, *90*, 1703–1715. [CrossRef] [PubMed]

57. Anwar, K.; Iqbal, J.; Hussain, M.M. Mechanisms involved in vitamin E transport by primary enterocytes and in vivo absorption. *J. Lipid Res.* **2007**, *48*, 2028–2038. [CrossRef] [PubMed]

58. Reboul, E.; Trompier, D.; Moussa, M.; Klein, A.; Landrier, J.F.; Chimini, G.; Borel, P. ATP-binding cassette transporter A1 is significantly involved in the intestinal absorption of alpha- and gamma-tocopherol but not in that of retinyl palmitate in mice. *Am. J. Clin. Nutr.* **2009**, *89*, 177–184. [CrossRef] [PubMed]

59. Olivier, M.; Bott, G.R.; Frisdal, E.; Nowick, M.; Plengpanich, W.; Desmarchelier, C.; Roi, S.; Quinn, C.M.; Gelissen, I.; Jessup, W.; et al. ABCG1 is involved in vitamin E efflux. *Biochim. Biophys. Acta* **2014**, *1841*, 1741–1751. [CrossRef] [PubMed]

60. Nicod, N.; Parker, R.S. Vitamin E secretion by Caco-2 monolayers to APOA1, but not to HDL, is vitamer selective. *J. Nutr.* **2013**, *143*, 1565–1572. [CrossRef] [PubMed]

61. Ramasamy, I. Update on the molecular biology of dyslipidemias. *Clin. Chim. Acta* **2016**, *454*, 143–185. [CrossRef] [PubMed]

62. Cuerq, C.; Restier, L.; Drai, J.; Blond, E.; Roux, A.; Charriere, S.; Michalski, M.C.; Di Filippo, M.; Levy, E.; Lachaux, A.; et al. Establishment of reference values of alpha-tocopherol in plasma, red blood cells and adipose tissue in healthy children to improve the management of chylomicron retention disease, a rare genetic hypocholesterolemia. *Orphanet J. Rare Dis.* **2016**, *11*, 114. [CrossRef] [PubMed]

63. Burnett, J.R.; Hooper, A.J. Vitamin E and oxidative stress in abetalipoproteinemia and familial hypobetalipoproteinemia. *Free Radic. Biol. Med.* **2015**, *88*, 59–62. [CrossRef] [PubMed]

64. Reboul, E.; Borel, P. Proteins involved in uptake, intracellular transport and basolateral secretion of fat-soluble vitamins and carotenoids by mammalian enterocytes. *Prog. Lipid Res.* **2011**, *50*, 388–402. [CrossRef] [PubMed]

65. Goncalves, A.; Roi, S.; Nowicki, M.; Dhaussy, A.; Huertas, A.; Amiot, M.J.; Reboul, E. Fat-soluble vitamin intestinal absorption: Absorption sites in the intestine and interactions for absorption. *Food Chem.* **2015**, *172*, 155–160. [CrossRef] [PubMed]

66. Nassir, F.; Wilson, B.; Han, X.; Gross, R.W.; Abumrad, N.A. CD36 is important for fatty acid and cholesterol uptake by the proximal but not distal intestine. *J. Biol. Chem.* **2007**, *282*, 19493–19501. [CrossRef] [PubMed]

67. Bietrix, F.; Yan, D.; Nauze, M.; Rolland, C.; Bertrand-Michel, J.; Comera, C.; Schaak, S.; Barbaras, R.; Groen, A.K.; Perret, B.; et al. Accelerated lipid absorption in mice overexpressing intestinal SR-BI. *J. Biol. Chem.* **2006**, *281*, 7214–7219. [CrossRef] [PubMed]

68. Cai, S.F.; Kirby, R.J.; Howles, P.N.; Hui, D.Y. Differentiation-dependent expression and localization of the class B type I scavenger receptor in intestine. *J. Lipid Res.* **2001**, *42*, 902–909. [PubMed]
69. Masson, C.J.; Plat, J.; Mensink, R.P.; Namiot, A.; Kisielewski, W.; Namiot, Z.; Fullekrug, J.; Ehehalt, R.; Glatz, J.F.; Pelsers, M.M. Fatty acid- and cholesterol transporter protein expression along the human intestinal tract. *PLoS ONE* **2010**, *5*, e10380. [CrossRef] [PubMed]
70. Drevon, C.A. Absorption, transport and metabolism of vitamin E. *Free Radic. Res. Commun.* **1991**, *14*, 229–246. [CrossRef] [PubMed]
71. Traber, M.G.; Sies, H. Vitamin E in humans: Demand and delivery. *Annu. Rev. Nutr.* **1996**, *16*, 321–347. [CrossRef] [PubMed]
72. Cohn, W. Bioavailability of vitamin E. *Eur. J. Clin. Nutr.* **1997**, *51*, S80–S85. [PubMed]
73. Bruno, R.S.; Leonard, S.W.; Park, S.I.; Zhao, Y.; Traber, M.G. Human vitamin E requirements assessed with the use of apples fortified with deuterium-labeled alpha-tocopheryl acetate. *Am. J. Clin. Nutr.* **2006**, *83*, 299–304. [PubMed]
74. Traber, M.G.; Burton, G.W.; Ingold, K.U.; Kayden, H.J. RRR- and SRR-alpha-tocopherols are secreted without discrimination in human chylomicrons, but RRR-alpha-tocopherol is preferentially secreted in very low density lipoproteins. *J. Lipid Res.* **1990**, *31*, 675–685. [PubMed]
75. Traber, M.G.; Kayden, H.J. Preferential incorporation of alpha-tocopherol vs. gamma-tocopherol in human lipoproteins. *Am. J. Clin. Nutr.* **1989**, *49*, 517–526. [PubMed]
76. Porsgaard, T.; Hoy, C.E. Absorption by rats of tocopherols present in edible vegetable oils. *Lipids* **2000**, *35*, 1073–1078. [CrossRef] [PubMed]
77. Bardowell, S.A.; Ding, X.; Parker, R.S. Disruption of P450-mediated vitamin E hydroxylase activities alters vitamin E status in tocopherol supplemented mice and reveals extra-hepatic vitamin E metabolism. *J. Lipid Res.* **2012**, *53*, 2667–2676. [CrossRef] [PubMed]
78. Riedl, J.; Linseisen, J.; Hoffmann, J.; Wolfram, G. Some dietary fibers reduce the absorption of carotenoids in women. *J. Nutr.* **1999**, *129*, 2170–2176. [PubMed]
79. Greenwood, D.C.; Cade, J.E.; White, K.; Burley, V.J.; Schorah, C.J. The impact of high non-starch polysaccharide intake on serum micronutrient concentrations in a cohort of women. *Public Health Nutr.* **2004**, *7*, 543–548. [CrossRef] [PubMed]
80. Roodenburg, A.J.; Leenen, R.; van het Hof, K.H.; Weststrate, J.A.; Tijburg, L.B. Amount of fat in the diet affects bioavailability of lutein esters but not of alpha-carotene, beta-carotene, and vitamin E in humans. *Am. J. Clin. Nutr.* **2000**, *71*, 1187–1193. [PubMed]
81. Jeanes, Y.M.; Hall, W.L.; Ellard, S.; Lee, E.; Lodge, J.K. The absorption of vitamin E is influenced by the amount of fat in a meal and the food matrix. *Br. J. Nutr.* **2004**, *92*, 575–579. [CrossRef] [PubMed]
82. Kim, J.E.; Ferruzzi, M.G.; Campbell, W.W. Egg Consumption Increases Vitamin E Absorption from Co-Consumed Raw Mixed Vegetables in Healthy Young Men. *J. Nutr.* **2016**, *146*, 2199–2205. [CrossRef] [PubMed]
83. Mah, E.; Sapper, T.N.; Chitchumroonchokchai, C.; Failla, M.L.; Schill, K.E.; Clinton, S.K.; Bobe, G.; Traber, M.G.; Bruno, R.S. alpha-Tocopherol bioavailability is lower in adults with metabolic syndrome regardless of dairy fat co-ingestion: A randomized, double-blind, crossover trial. *Am. J. Clin. Nutr.* **2015**, *102*, 1070–1080. [CrossRef] [PubMed]
84. Preveraud, D.P.; Devillard, E.; Borel, P. Dietary fat modulates dl-alpha-tocopheryl acetate (vitamin E) bioavailability in adult cockerels. *Br. Poult. Sci.* **2015**, *56*, 94–102. [CrossRef] [PubMed]
85. Failla, M.L.; Chitchumronchokchai, C.; Ferruzzi, M.G.; Goltz, S.R.; Campbell, W.W. Unsaturated fatty acids promote bioaccessibility and basolateral secretion of carotenoids and alpha-tocopherol by Caco-2 cells. *Food Funct.* **2014**, *5*, 1101–1112. [CrossRef] [PubMed]
86. Koo, S.I.; Noh, S.K. Phosphatidylcholine inhibits and lysophosphatidylcholine enhances the lymphatic absorption of alpha-tocopherol in adult rats. *J. Nutr.* **2001**, *131*, 717–722. [PubMed]
87. Nishimukai, M.; Hara, H. Enteral administration of soybean phosphatidylcholine enhances the lymphatic absorption of lycopene, but reduces that of alpha-tocopherol in rats. *J. Nutr.* **2004**, *134*, 1862–1866. [PubMed]
88. Reboul, E.; Thap, S.; Perrot, E.; Amiot, M.J.; Lairon, D.; Borel, P. Effect of the main dietary antioxidants (carotenoids, gamma-tocopherol, polyphenols, and vitamin C) on alpha-tocopherol absorption. *Eur. J. Clin. Nutr.* **2007**, *61*, 1167–1173. [CrossRef] [PubMed]

89. Sklan, D.; Donoghue, S. Vitamin E response to high dietary vitamin A in the chick. *J. Nutr.* **1982**, *112*, 759–765. [PubMed]

90. Griffiths, D.W. The inhibition of digestive enzymes by polyphenolic compounds. *Adv. Exp. Med. Biol.* **1986**, *199*, 509–516. [PubMed]

91. Fardet, A.; Morise, A.; Kalonji, E.; Margaritis, I.; Mariotti, F. Influence of phytosterol and phytostanol food supplementation on plasma liposoluble vitamins and provitamin A carotenoid levels in humans: An updated review of the evidence. *Crit. Rev. Food Sci. Nutr.* **2017**, *57*, 1906–1921. [CrossRef] [PubMed]

92. Richelle, M.; Enslen, M.; Hager, C.; Groux, M.; Tavazzi, I.; Godin, J.P.; Berger, A.; Metairon, S.; Quaile, S.; Piguet-Welsch, C.; et al. Both free and esterified plant sterols reduce cholesterol absorption and the bioavailability of beta-carotene and alpha-tocopherol in normocholesterolemic humans. *Am. J. Clin. Nutr.* **2004**, *80*, 171–177. [PubMed]

93. Borel, P.; Desmarchelier, C. Genetic Variations Involved in Vitamin E Status. *Int. J. Mol. Sci.* **2016**, *17*. [CrossRef] [PubMed]

antioxidants

MDPI

Review

Long-Chain Metabolites of Vitamin E: Metabolic Activation as a General Concept for Lipid-Soluble Vitamins?

Martin Schubert [1,2,†], Stefan Kluge [1,2,†], Lisa Schmölz [1,2], Maria Wallert [1,3], Francesco Galli [4] ⓘ, Marc Birringer [5] ⓘ and Stefan Lorkowski [1,2,*] ⓘ

1   Department of Biochemistry and Physiology of Nutrition, Friedrich-Schiller-University Jena, 07743 Jena, Germany; m.schubert@uni-jena.de (M.S.); s.kluge@uni-jena.de (S.K.); lisa.schmoelz@uni-jena.de (L.S.); maria.wallert@uni-jena.de (M.W.)
2   Competence Center for Nutrition and Cardiovascular Health (nutriCARD), Halle-Jena-Leipzig, 07743 Jena, Germany
3   Baker IDI Heart and Diabetes Institute, Melbourne VIC 3004, Australia
4   Department of Pharmaceutical Sciences, Laboratory of Nutrition and Clinical Biochemistry, University of Perugia, 06123 Perugia, Italy; francesco.galli@unipg.it
5   Department of Nutrition, Food and Consumer Sciences, University of Applied Sciences Fulda, 36037 Fulda, Germany; marc.birringer@oe.hs-fulda.de
*   Correspondence: stefan.lorkowski@uni-jena.de; Tel.: +49-3641-949710
†   These authors contributed equally.

Received: 15 December 2017; Accepted: 11 January 2018; Published: 12 January 2018

Abstract: Vitamins E, A, D and K comprise the class of lipid-soluble vitamins. For vitamins A and D, a metabolic conversion of precursors to active metabolites has already been described. During the metabolism of vitamin E, the long-chain metabolites (LCMs) 13′-hydroxychromanol (13′-OH) and 13′-carboxychromanol (13′-COOH) are formed by oxidative modification of the side-chain. The occurrence of these metabolites in human serum indicates a physiological relevance. Indeed, effects of the LCMs on lipid metabolism, apoptosis, proliferation and inflammatory actions as well as tocopherol and xenobiotic metabolism have been shown. Interestingly, there are several parallels between the actions of the LCMs of vitamin E and the active metabolites of vitamin A and D. The recent findings that the LCMs exert effects different from that of their precursors support their putative role as regulatory metabolites. Hence, it could be proposed that the mode of action of the LCMs might be mediated by a mechanism similar to vitamin A and D metabolites. If the physiological relevance and this concept of action of the LCMs can be confirmed, a general concept of activation of lipid-soluble vitamins via their metabolites might be deduced.

Keywords: vitamin E; long-chain metabolites of vitamin E; 13′-hydroxychromanol (13′-OH); 13′-carboxychromanol (13′-COOH); vitamin E metabolism; biological activity

1. The Biological Significance of Vitamin E

The term vitamin E comprises eight lipophilic molecules, which can be classified as tocopherols (TOHs) and tocotrienols (T3). Both classes share two common features: (i) the phytyl-like side chain, which is bound to (ii) the chroman ring system. A saturated side chain characterizes the TOHs, while the T3s carry three double bonds in this substructure. Further, the methylation pattern of the chroman ring determines the classification as α-, β-, γ- or δ-TOH or T3, respectively. Vitamin E is found in oils, nuts, germs, seeds and a variety of other plant products. The naturally found vitamin E forms exist either in *RRR*-configuration (TOHs) or in *R*-configuration (T3s), whereas only synthetically produced forms contain a mixture of the different possible stereoisomers [1].

Vitamin E was discovered in 1922 as vital factor for the fertility of rats, indicating its essentiality for animal and human health, and was therefore classified as a vitamin [2]. Nevertheless, the benefits of vitamin E for human health are still a contentious issue. However, several disease conditions, such as anemia, erythrocyte rupture and neuronal degeneration, as well as muscle degeneration, are linked to vitamin E deficiency or malabsorption (extensively reviewed in [3]). Further, vitamin E was shown in human intervention trials to slow down the progression of age-related neurodegenerative pathologies such as Alzheimer's disease, maybe due to its antioxidative properties [4,5]. Vitamin E is also an essential factor for the development of the central nervous system and cognitive functions of the embryo [6,7]. Next, vitamin E may play a supportive role in the prevention of neural tube defects in humans along with folic acid [8,9]. Initially, the effects of vitamin E were only attributed to its antioxidant properties, however more recent work unveiled non-antioxidant regulatory effects. There is growing evidence that vitamin E modulates gene expression and enzyme activities and interferes with signaling cascades independent of its capacity as an antioxidant [10]. Over time, several functions of vitamin E, such as suppression of inflammatory mediators, reactive oxygen species, and adhesion molecules, the induction of scavenger receptors, and the activation of nuclear factor kappa-light-chain-enhancer of activated B cells (NF_kB) (reviewed in [11]) were revealed. Based on these observations, it was concluded that vitamin E likely plays a role in several inflammatory but also other diseases. However, further research is required, as the results obtained from clinical trials with TOHs are inconsistent with respect to beneficial effects on the development of chronic diseases such as cancer and cardiovascular diseases [12].

2. Absorption and Distribution of Vitamin E

Like for all macro- and micronutrients, intestinal absorption is the limiting factor for the bioavailability of vitamin E in humans. As a fat-soluble vitamin, intestinal absorption, hepatic metabolism and cellular uptake of vitamin E follows that of other lipophilic molecules [13]. The absorption rate of vitamin E varies between 20% and 80% [13,14], and is thus generally lower than for vitamins A and D [15,16]. Differences in the rates of absorption of vitamin E and the other fat-soluble vitamins may result also from the parallel intake of additional food ingredients. For example, retinoic acid [17], plant sterols [18], eicosapentaenoic acid [14], alcohol (chronic consumption) [14], and dietary fiber [19] are natural food components that may compete with the absorption of vitamin E. In addition, it has been shown that the supplied form of vitamin E, either as a free molecule or coupled to other compounds like acetate, is also crucial for its bioavailability [20].

For optimal absorption, fat must be consumed along with the ingested vitamin E. This is a general requirement for all types of fat-soluble vitamins and is therefore also applicable for vitamins A, D and K [16,21]. The absorption of triacylglycerides and esterified fat-soluble molecules starts with enzymatic processing in the stomach by the action of gastric lipases [15]. The following digestion of dietary lipids appears in the intestinal lumen by the action of various enzymes, including pancreatic lipase, carboxyl esterase and phospholipase A_2 [22]. Since most of the vitamin E in the human diet is not esterified, lipolytic degradation is scarce [14]. In contrast, the human diet contains significantly more esterified vitamin A and D, mostly in the form of retinyl-esters and vitamin D_3 oleate, which can be hydrolyzed by the above mentioned enzymes [16,21]. A key step of the intestinal absorption of fat-soluble vitamins is the emulsification, i.e., the incorporation into micelles formed with phospholipids and bile acids. Under normal conditions, bile salts facilitate the absorption of all three vitamins, but especially the vitamin D forms differ in their dependency for bile salt availability, i.e., vitamin D_3 absorption is more dependent on the presence of bile salts than 25-hydroxyvitamin D (OHD) [23]. After emulsification, vitamin E is taken up into the intestinal enterocytes by passive diffusion or receptor-mediated transport via scavenger receptor class B type 1 (SRB1) [24], or Niemann–Pick C1-like protein 1 [25], which is also involved in the uptake of the vitamins A, D and K as well as cholesterol [16,26,27]. Since no specific plasma transport protein for α-TOH is known, the subsequent transport of vitamin E in blood follows largely that of cholesterol [25], meaning that under normal physiological conditions, α-TOH is transported via chylomicrons. This transport is independent of the type of stereoisomer [28,29]. In addition, retinol, unconverted pro-retinoid carotenoids (β-carotene), non-pro-retinoid carotenoids (lycopene), vitamin D_3 and phylloquinone (representing the main dietary form of vitamin K) are also incorporated into chylomicrons [16,21,30]. After entering the circulation, chylomicrons undergo a process of remodeling that involves primarily the hydrolysis of triglycerides by lipoprotein lipase, resulting in the formation of chylomicron remnants [25]. Vitamins E, A, D and K are not affected by hydrolysis and remain in the lipoprotein particle for further transport to the liver [31]. The different forms of vitamin E are discriminated in the liver by the α-tocopherol transfer protein (α-TTP), which promotes the incorporation of 2R- or *RRR*-α-TOH into very low-density lipoproteins (VLDL) [32,33], whereas other forms and stereoisomers are either metabolized or secreted into bile [34]. Besides α-TTP, the TOH-associated protein and the TOH-binding protein are known mediators of the intracellular transport of vitamin E. Interestingly, α-TOH secretion from the liver is apparently not necessarily dependent on VLDL assembly and secretion, thus oxysterol-binding proteins [35] and ATP-binding cassette transporter A1 (ABCA1) [36] have been suggested to contribute to the release from the liver. Furthermore, ABCA1 mediates the efflux of vitamin E in the intestine, macrophages, and fibroblasts [36], and multidrug resistance P-glycoprotein has been identified as a transporter for the excretion of α-TOH via bile [37]. After the release of vitamin E-carrying VLDL into blood circulation and action of lipoprotein lipase as well as hepatic lipase, receptors such as SRB1, low-density lipoprotein (LDL) receptor as well as LDL receptor-related protein mediate the uptake of vitamin E into peripheral tissues and the liver [31,38].

3. Metabolism of Vitamin E

The metabolism of vitamin E is primarily localized in the liver (Figure 1) (reviewed in [39]), whereas extrahepatic pathways have been also suggested [40,41]. The degradation processes of hepatic metabolism remain poorly understood, but the initial mechanisms are generally accepted, i.e., all vitamers are degraded to vitamer-specific physiological metabolites with an intact chromanol ring and a shortened side-chain. Interestingly, accumulation of vitamin E to toxic levels is prevented by increased metabolism in response to higher vitamin E levels. Due to the preferential binding to α-TTP, α-TOH is the prevalent form of vitamin E in humans. It is speculated that α-TTP protects the α-form from degradation, thus leading to the accumulation of α-TOH. With the lower affinities of the other vitamin E forms to α-TTP taken into consideration, γ- and δ-forms are likely catabolized faster [42]. Despite of the different catabolic rates, all forms of vitamin E follow the same metabolic route, as confirmed by the detection of the respective end products of hepatic metabolism, α-, β-, γ-, and δ-carboxyethylhydroxychromanol (CEHC) [43,44]. However, the rate of catabolism is different for the vitamin E forms, possibly due to distinct affinities to key enzymes [42,45]. The chroman ring is not modified during catabolism (the catabolic end products are still classified as α-, β-, γ- and δ-forms); it is rather the aliphatic side chain where modifications are introduced. Metabolism of T3 follows the same principle, albeit further enzymes such as 2,4 dienoyl-coenzyme A (CoA) reductase and 3,2-enoyl-CoA isomerase (necessary for the metabolism of unsaturated fatty acids) are likely required for the degradation of the unsaturated side chain [46].

The catabolism of the vitamin E molecule takes place in different cell compartments: endoplasmic reticulum, peroxisomes, and mitochondria. However, the mechanism of metabolite transfer between the compartments is not well understood and requires further investigation. The initial step at the endoplasmic reticulum leads to the formation of 13′-hydroxychromanol (13′-OH) metabolites via ω-hydroxylation by cytochrome P450 (CYP) 4F2 or CYP3A4, respectively [45,47]. The following ω-oxidation, which is probably mediated by alcohol and aldehyde dehydrogenases (an aldehyde intermediate is formed), results in 13′-carboxychromanol (COOH) metabolites. In general, the resulting metabolites with carboxy function are degraded like branched-chain fatty acids. Hence, the side chain is shortened by β-oxidation, and the formed propionyl-CoA or acetyl-CoA is eliminated. The intermediate-chain metabolites 11′-COOH and 9′-COOH are formed in peroxisomes during the first two cycles of β-oxidation. Three additional cycles of β-oxidation are carried out in the mitochondria, resulting in the short-chain metabolites (SCMs) 7′-COOH and 5′-COOH as well as the end-product CEHC or 3′-COOH. Moreover, conjugation of the metabolites takes place during metabolism, resulting predominantly in sulfated and glucuronidated metabolites. However, glycine-, glycine–glucuronide-, and taurine-modified metabolites of vitamin E have also been identified [48].

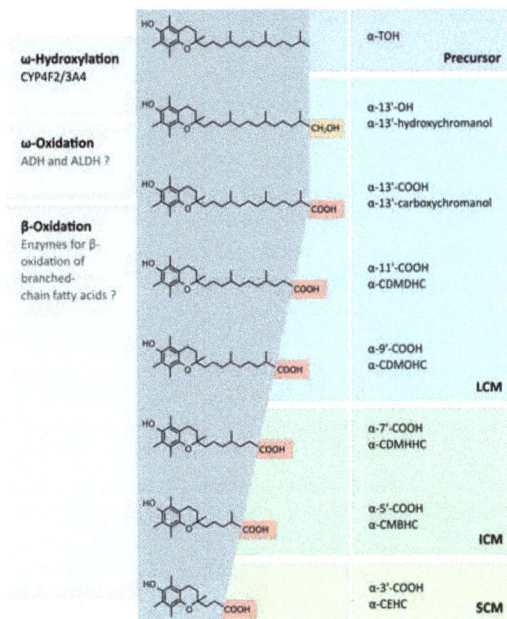

Figure 1. Metabolism of vitamin E. The metabolism of vitamin E is initiated by a terminal ω-hydroxylation of the side-chain via CYP4F2 and CYP3A4. The resulting hydroxychromanol is further modified by ω-oxidation, resulting in the formation of carboxychromanol, possibly by alcohol and aldehyde dehydrogenases. As a consequence, the metabolite can be subjected to β-oxidation. Five cycles of β-oxidation lead to the formation of the short-chain metabolite CEHC. However, this review focuses on the LCMs 13′-OH and 13′-COOH as these molecules have been synthesized in sufficient amounts for in vitro and in vivo investigations. The following abbreviations are used: ADH, alcohol dehydrogenase; ALDH, aldehyde dehydrogenase; CDMDHC, carboxydimethyldecylhydroxychromanol; CDMOHC, carboxymethyloctylhydroxychromanol; CDMHHC, carboxymethylhexylhydroxychromanol; CMBHC, carboxymethylbutylhydroxychromanol; CEHC, carboxyethylhydroxychromanol.

The conjugated SCMs are more hydrophilic and thus mainly found in glucuronidated form in human urine [44]. In contrast, the long-chain metabolites (LCMs) and their metabolic precursors are secreted via bile into the intestine and the metabolites in fecal samples are not conjugated. The fecal route is considered as the major pathway of vitamin E excretion [12,49].

Like vitamin E, other fat-soluble vitamins, such as the vitamins A (i), D (ii) and K (iii) are also metabolized in the human body:

(i).   Under physiological conditions, retinyl esters (in the intestinal lumen) and carotenoids (in enterocytes) are converted into retinol before or during their intestinal absorption, respectively. Inside the enterocytes, retinol is re-esterified by lecithin-retinol acyl transferase or acyl-CoA:retinol-acyltransferase and packed into chylomicrons for transport. The retinyl esters are transferred to the liver and stored in hepatic parenchymal and non-parenchymal cells.

Vitamin A is mobilized from liver stores by the retinol-binding protein, a specific transporter allowing the transport of retinol in blood circulation [50]. These results suggest that vitamin A has an active (retinol) and a storage form (retinyl ester). In addition, the oxidation of retinol leads to the formation of retinal, another active form of vitamin A, which is primarily bound to opsins in the photoreceptors of the retina [51]. More current research indicates that all-*trans* retinoic acid (ATRA), 9-*cis*-RA, and all-*trans*-4-oxo-RA are the vitamin A metabolites with the highest biological activity. These active vitamin A metabolites serve as ligands for nuclear receptors, called retinoic acid receptors (RARs) [52] and retinoid receptors (RXRs) [53], which act as ligand-activated transcription factors controlling the expression of their respective target genes. Therefore, hepatic retinol is transferred to extrahepatic tissues and metabolized to retinoic acid by different enzymatic systems. Lampen and co-workers found that ATRA is also formed in the small intestine via direct oxidation of vitamin A. Based on this result, they hypothesized that biologically active retinoids are formed in the gastrointestinal tract and act as retinoid-receptor ligands controlling various processes in the intestinal mucosa via RAR [53].

(ii). The human metabolism of vitamin D is primarily located in liver and kidney. Metabolism of vitamin D_2 and D_3 starts with the formation of 25-OHD, the major circulating vitamin D metabolite, by vitamin D-25 hydroxylase. Afterwards, 25-OHD is transferred to the kidney and further catabolized by 25-OHD-1α-hydroxylase to 1,25-dihydroxyvitamin $D_{2/3}$. These molecules serve as ligands for the vitamin D receptor (VDR), a transcription factor expressed in various tissues. Vitamin D receptor binds to specific regions in the promoter regions of genes, the so-called vitamin D responsive elements, thus controlling the expression of respective target genes. Therefore, 1,25-dihydroxyvitamin D is the active metabolic form of vitamin D [54,55].

(iii). Phylloquinone (vitamin K_1) and menaquinone (vitamin K_2) are summarized by the term vitamin K. Phylloquinone is synthesized in plants, while menaquinone is derived from animal and bacterial origins [30,56]. Both compounds share a 2-methyl-1,4-naphthoquinone structure, called menadione, and a side chain at the 3$'$-position. The side chain of phylloquinone is composed of three isopentyl units and one isopentenyl unit, while the side chain of menaquinone contains a variable number of only isopentenyl units (2–13) [30]. The metabolism of vitamin K is localized in the liver and has not been studied in detail so far [57]. Nevertheless, the metabolic pathway of phylloquinone and menaquinone degradation likely follows that of vitamin E. Hence, the degradation starts with an initial ω-oxidation, which is mediated by CYP. While the ω-oxidation of vitamin E is catalyzed primarily by CYP4F2, CYP3A4 has been described as the possible mediator for the ω-oxidation of vitamin K. Next, the following degradation of the side chain of vitamin K occurs via β-oxidation [30,56,58]. A 5-carbon carboxylic acid metabolite termed K acid 2 has been identified as the end-product of either phylloquinone or menaquinone metabolism and is excreted via urine and bile [30,58]. In addition to their metabolic degradation, it has been suggested that phylloquinones could also be converted to menaquinones [59,60]. For this, phylloquinone is likely transformed to the intermediate menadione by removing its side chain, which is subsequently replaced by a newly synthesized isopentenyl side chain to form menaquinone [30]. While menaquinone is considered as the physiologically active form of vitamin K in humans [56], almost nothing is known about a possible biological activity of the vitamin K metabolites. Further studies are needed to unravel whether vitamin K must be included into the general concept of a metabolic pre-activation of lipid-soluble vitamins.

Although the metabolisms of vitamin A and D differ in location and the involved enzymatic systems, the formation of active metabolites seems to be a key element of both metabolic pathways, i.e., both vitamins mediate their gene regulatory effects by metabolic pre-activation. Therefore, the discovery of vitamin E metabolism in animals and humans and the emerging evidence for important biological functions of vitamin E metabolites could indicate a general metabolic activation mechanism of fat-soluble vitamins in the human body.

In Vivo Verification of Systemic LCM Availability

Since the discovery of vitamin E by EVANS and BISHOP in 1922 [2], α-TOH has been accounted as an antioxidant capable to scavenge reactive oxygen species, and decreased α-TOH levels have been associated with several diseases including different types of cancer, cardiovascular diseases and diabetes [61]. It took 80 years since AZZI and co-workers set up the hypothesis for an additional gene regulatory role of α-TOH in the human body [62]. In addition, the discovery of vitamin E metabolism in animals and humans and the emerging evidence for important biological functions of the vitamin E metabolites [63,64], suggested that the TOHs may gain biological activity after metabolism (as confirmed for vitamin A and D). This prompted studies that investigated also the putative functions of the LCMs of TOH. In 2014, Wallert and co-workers showed the occurrence of α-13′-COOH in human serum, which has been confirmed later by others [65,66]. For these studies, serum obtained from a healthy, middle-aged (39 years), non-smoking male, who received a balanced diet with no additional vitamin E supplementation was used for the detection of α-13′-COOH via liquid chromatography coupled mass spectrometry [63]. The analyses revealed for the first time that α-TOH metabolites are transferred into blood circulation following metabolism of α-TOH in the liver. Furthermore, cell experiments showed that α-13′-OH and α-13′-COOH are more potent regulators of gene expression than their metabolic precursor α-TOH [63]. Taken together, the results of Wallert et al. provided the first evidence that the LCMs are an active form of their metabolic precursor [63], promoting regulatory effects in peripheral tissues of the human body. However, while the role of vitamin E as a lipophilic antioxidant in vitro is widely accepted, the relevance in vivo is still a matter of debate [67–69].

4. Biological Activity

Not much is known about the biological activity of the LCMs. However, the publications on this topic published during the last ten years can be categorized by the biological effects of the LCMs as follows: (i) anti-inflammatory actions [64,70–75]; (ii) anti-carcinogenic effects [72,76,77]; (iii) regulation of cellular lipid homeostasis [63,64]; (iv) interaction with pharmaceuticals [78]; and (v) regulation of their own metabolism [79] (Figure 2).

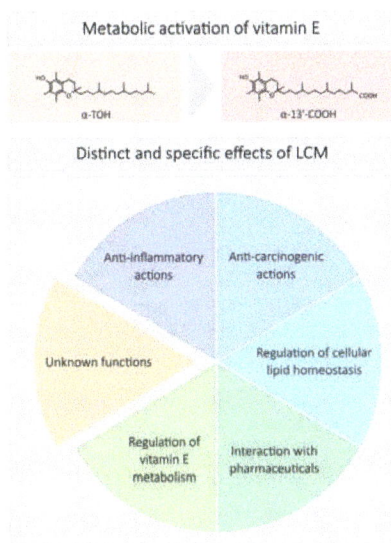

Figure 2. Reported biological functions of the LCMs of vitamin E.

4.1. Anti-Inflammatory Actions

Investigations on anti-inflammatory actions often focus on the regulation of pro-inflammatory enzymes, such as inducible cyclooxygenase 2 (COX2) [70–72,74], inducible nitric oxide synthase (iNOS or nitric oxide synthase, NOS2) [64,71,74,75], or 5-lipoxygenase (5-LO) [72,73], as well as mediators such as chemokines or cytokines. For this purpose, cells were treated with the LCMs and challenged with a pro-inflammatory stimulus or alternatively, isolated enzymes were used. Several LCMs (α-, γ-, δ-13′-COOH; δ-9′-COOH; α-13′-OH) have been tested and reduced the stimulus-induced expression (mRNA or protein) or enzyme activity. In general, 13′-COOH are more potent than the shorter LCMs and the conjugation of LCMs with sulfate abrogates their anti-inflammatory effects [64,70].

Jiang et al. gained first hints on the anti-inflammatory actions of LCMs [70]. A549 cells, which are capable of metabolizing vitamin E, were incubated with TOHs and an inhibition of the arachidonic acid-stimulated COX activity was reported. When the metabolism of vitamin E was suppressed by sesamin, the effects were less pronounced, indicating the involvement of the LCMs as regulatory molecules. For further experiments, the LCMs were extracted from the cell culture medium and their inhibitory capacity on COX activity was tested (half maximal inhibitory concentration (IC_{50}): δ-13′-COOH: 4 µM; δ-9′-COOH: 6 µM). The impact of conjugation was tested, and the sulfate LCM conjugates were unable to exert anti-inflammatory effects. In 2016, a comparison of the different types of LCMs was performed, and the LCMs showed similar effects regardless of their origin (isolated from cell culture medium or semisynthetic isolation from *Garcinia kola*) [72]. In RAW264.7 macrophages, the anti-inflammatory action on lipopolysaccharide (LPS)-stimulated COX2 mRNA and protein expression, as well as prostaglandin (PG) release was reported for α-13′-OH [71] and α-13′-COOH [74].

The regulation of iNos by the LCMs was studied in RAW264.7 macrophages [64,71,74,75]. The LPS-stimulated iNos mRNA and protein expression as well as release of nitric oxide were reduced by the LCMs tested (α- and δ-13′-OH, α- and δ-13′-COOH) [64]. The inhibitory effect of the LCMs was highly dependent on the structure of the LCMs. The 13′-COOH were more effective than the 13′-OH, while the substitution of the chromanol ring system (α- vs. δ-LCMs) had no influence.

The inhibition of ionophore-induced leukotriene release (leukotriene B_4) in HL-60 cells and neutrophils was reported with IC_{50} values of 4–7 µM [73]. Furthermore, the activity of isolated 5-LO was inhibited by δ-13′-COOH with IC_{50} values of 0.5–1 µM, which is more effective than the synthetic 5-LO inhibitor zileuton (IC_{50}: 3–5 µM) [73]. The inhibition of 5-LO activity by δ-13′-COOH was also confirmed by Jang et al. [72]. An overview of the known anti-inflammatory actions of the different LCMs of vitamin E studied so far is provided in Table 1.

Table 1. Overview of anti-inflammatory actions of the LCMs of vitamin E.

Targets	Cells	Effects	Substances	Refs.
COX2	A549 cells	Reduced activity in arachidonic acid-pre-induced cells	γ-13′-COOH δ-13′-COOH δ-9′-COOH	[70] [70,72] [70]
	Isolated enzyme	Inhibition of activity	δ-13′-COOH δ-9′-COOH	[70]
	RAW264.7	Inhibition of LPS-stimulated mRNA and protein expression, as well as reduced PG release	α-13′-OH α-13′-COOH	[71] [74]
iNos	RAW264.7	Inhibition of LPS-stimulated mRNA and protein expression, as well as reduced release of nitric oxide	α-13′-OH α-13′-COOH δ-13′-OH δ-13′-COOH	[64,71,74,75]
5-LO	Isolated enzyme	Inhibition of activity	δ-13′-COOH	[72,73]
	HL-60 neutrophils	Reduced activity and LT release in pre-induced cells	δ-13′-COOH	[73]

PG, prostaglandin; LT, leukotriene.

The metabolites of vitamin K have also been shown to exert anti-inflammatory functions. First experiments were carried out with a synthetic 7-carbon carboxylic acid vitamin K metabolite (2-methyl, 3-(2′methyl)-hexanoic acid-1,4-naphthoquinone; K acid 1), which was a more effective inhibitor of LPS-induced IL-6 release from fibroblast than the precursors phylloquinone and menaquinon-4 [80]. In LPS-challenged MG63 osteoblasts the 7-carbon carboxylic acid metabolite as well as the 5-carbon carboxylic acid metabolite (K acid 2) attenuated the expression of IL-6 [81]. Later, the long-chain metabolites of vitamin K (10 to 20-carbon carboxylic acid metabolites) were also synthesized and examined for their anti-inflammatory activity. In LPS-challenged mouse macrophages, these compounds reduced the induction of gene-expression of the inflammatory markers IL-1β, IL-6 and TNFα [82]. However, K acid 1 and K acid 2 were also effective in this study; and it is not possible to estimate, which vitamin K metabolite (either long-chain or short-chain) is the most effective [82]. Interestingly, the minor 7-carbon carboxylic acid metabolite was more effective in MG63 osteoblasts than the 5-carbon carboxylic acid metabolite, and a replacement of the carboxy function by a methyl group made the two metabolites less effective [81]. This is in line with findings for the LCMs of vitamin E. Here, the carboxy metabolite is more effective than the respective TOH precursor with respect to the anti-inflammatory actions (vide supra). However, the in vivo relevance of the regulatory activities of the vitamin K metabolites is a matter of debate, as they increase with vitamin K intake in urine [83], but have not yet been found in human blood or other tissues to the best of our knowledge.

4.2. Cancerogenesis and Chemoprevention

The metabolites of vitamin E were investigated with respect to putative anti-cancerogenic, i.e., anti-proliferative and pro-apoptotic, properties in several studies. First experiments revealed that the SCMs inhibit cell proliferation in different cell lines [84,85]. Interestingly, the metabolites as well as the precursor molecules showed different efficiencies, depending on the methylation pattern of the chroman ring and also on the cell type tested [84,85]. Based on the anti-proliferative effects of the SCMs, the interest in the effects of the LCMs aroused. Hence, Birringer et al. investigated the effects of the LCMs α-13′-COOH and δ-13′-COOH as well as α-13′-OH and δ-13′-OH on the proliferation of the human hepatocyte carcinoma cell line HepG2 [77]. Interestingly, both 13′-COOH metabolites effectively caused cell growth arrest, but the hydroxy metabolites did not exhibit anti-proliferative effects. Thus, the introduction of the carboxy group during TOH metabolism renders the molecule active with respect to cell growth arrest. This is supported by the finding that the metabolic precursors, i.e., TOHs, did not affect proliferation of HepG2 cells [77]. As mentioned above, the methylation of the chroman ring alters the efficiency of the molecules. With an effective concentration of 6.5 µM in HepG2 cells regarding the effects on cell growth, the δ-metabolite is more effective than its α-counterpart with 13.5 µM [77]. At first glance, contradictory results were reported for human prostate cancer cells. Here, not only δ-13′-COOH inhibited cell proliferation, but also the hydroxy metabolite α-13′-OH. The LCMs as well as the tested SCMs α-CEHC and γ-CEHC inhibited the proliferation by about 60% in a concentration of 10 µM [76]. Hence, the efficiency of the hydroxy metabolite is likely dependent on the cell type. It is possible that the differences in TOH metabolism in different cell types lead to divergent effects. Interestingly, even differences between different cancer and non-cancer cell lines have been described. The proliferation of the colon cancer cell lines HCT-116 and HT-29 was inhibited by δ-13′-COOH, with IC_{50} values of 8.9 µM and 8.6 µM, respectively [72]. While 10 µM of the LCMs reduce the viability of the cancer cells by around 60%, normal colon epithelial cells showed a reduction of 10–20% at this concentration. Comparable effects were found for the δ-T3 LCM δ-T3-13′-COOH (δ-garcinoic acid), which reduced the viability of the colon cancer cells by about 75%, but the viability of normal colon cells merely by 10–20% [72].

The actions of the vitamin E metabolites are comparable to that of the metabolites of vitamin D and vitamin A. The active vitamin D metabolite $1,25(OH)_2D_3$ has been shown to modulate differentiation and proliferation of colon cancer cells and prostate cancer cells [86]. However, $1,25(OH)_2D_3$ led to an arrest of most cells that express a functional vitamin D receptor in G0/G1 phase [87]. The actions are mediated by interference with several regulatory proteins, such as epidermal growth factor receptor (EGFR), insulin-like growth factors (IGFs), p21, p27 as well as cyclins and cyclin-dependent kinases (CDKs) [87]. The retinoids are also known for their modulation of the cell cycle. In several cancer cell lines, retinoic acid (RA) led to a cell cycle arrest in the G0/G1 phase via direct or indirect modulation of cyclins, CDKs and cell-cycle inhibitors [88]. Interestingly, TOHs and TOH SCMs have also been linked to cyclins and CDKs. In the human prostate cancer cell line PC3, γ-TOH as well as γ-CEHC led to a strong decrease in cyclin D1 protein expression. In line with this observation, CDK4 and p27 expression are reduced, albeit less pronounced [85]. Moreover, α-TOH and α-CEHC are ineffective with respect to anti-proliferative actions as well as suppression of cyclin D1 and CDK4 [85]. However, to date, no data is available on the action of the vitamin E LCMs on cell cycle regulators, although strong anti-proliferative effects have been shown for this class of metabolites.

More detailed investigations were carried out on the pro-apoptotic effects of the vitamin E LCMs. Birringer et al. found a significant induction of apoptosis in HepG2 cells treated with 20 μM of α-13'-COOH, δ-13'-COOH or δ-13'-OH [77]. The LCMs induced the cleavage of caspases 3, 7 and 9, and in line with this, the cleavage of the downstream mediator poly-ADP ribose polymerase-1 (PARP-1). Again, the 13'-COOH were more effective in caspase-cleavage and apoptosis induction than the hydroxy metabolite [77]. Moreover, induction of mitochondrial apoptosis by the LCMs was identified as the process leading to apoptosis. This process is accompanied by the formation of reactive oxygen species (ROS). Birringer et al. observed a significant increase in ROS production in cells treated with α- and δ-13'-COOH but not with the hydroxy metabolites and the TOHs [77]. The augmented ROS production was not only measured intracellularly but also intramitochondrial, hence providing evidence for mitochondrial-derived apoptosis. Alterations in the mitochondrial membrane potential supported this finding. Treatment with 20 μM of the LCMs led to a significant reduction of the mitochondrial membrane potential. Interestingly, in this particular case, the α-metabolite was more potent than the δ-metabolites with 60% reduction vs. 20% reduction [77]. The pro-apoptotic actions of the δ-LCMs of vitamin E were confirmed in colon cancer cells [72]. Early and late apoptosis were induced by δ-13'-COOH and δ-T3-13'-COOH. The activation of caspase-9 and cleavage of PARP found by Birringer et al. [77] were confirmed in colon cancer cells [72]. Moreover, an induction of the autophagy marker microtubule-associated protein 1A/1B-light chain 3 (LC3)-II was found. Jang et al. assumed that alterations in sphingolipid metabolism caused by the carboxy-LCMs are the reason for the induction of apoptosis. Indeed, both δ-13'-COOH and δ-T3-13'-COOH increased total ceramides, dihydroceramides and dihydrosphingosines, while all measured sphingomyelins were decreased. Inhibition of sphingosine biosynthesis revealed that LC3-II expression but not PARP-cleavage is modulated by the LCMs via alterations in sphingolipid metabolism [72].

Taken together, there are several similarities between the metabolites of vitamins A, D and E with respect to anti-cancerogenic properties. Data on anti-cancerogenic effects of vitamin K metabolites, however, are sparse. Merely synthetic carboxylic derivatives of menaquinone with different side-chain lengths have been studied [89]. The biologically most abundant 5-carbon carboxylic acid metabolite (K acid 2) was not included in this study and the 7-carbon carboxylic acid metabolite (K acid 1) was the structure with the shortest side-chain. Interestingly, the growth-suppressing effect on hepatocellular carcinoma cells increased with the length of the side chain of the carboxy derivatives, except for the full-length metabolite, which was as effective as the 7-carbon carboxylic acid metabolite. Conversely, menaquinone itself was completely ineffective, showing nicely that the introduction of a carboxy function activates the compound. Blocking of the effects with chemical antagonists suggested that the derivatives act through caspase/transglutaminase-related signaling [89]. The above mentioned disruption of mitochondrial function by the LCMs of vitamin E has also been described for the metabolites of vitamin A [90], and induction of apoptosis by $1,25(OH)_2D_3$ via mitochondrial pathways (e.g., via B-cell lymphoma (BCL)-2 and BCL-xL) in breast, colon and prostate cancer cells are also known [87]. Based on their anti-proliferative and pro-differentiation actions but also due to the induction of cell death, retinoids are used for treating certain types of cancer [91]. Vitamin A metabolites were successfully used in the treatment of acute promyelocytic leukemia (ATRA and 13-*cis*-RA, 13*c*RA), squamous cell skin cancer and neuroblastoma (13*c*RA), lung cancer (ATRA) and Kaposi's sarcoma (9-*cis*-RA, 9*c*RA). Beneficial effects of retinoids in cancer prevention have also been observed.

These properties can be explained by the targeting of regulators of cell cycle progression by retinoids. The expression of the CDK inhibitors p21 and p27 is regulated by ATRA via RARβ2 upregulation, and retinoic acid has been shown to stimulate the degradation of cyclin D1, leading to a suppression of CDK activity [91]. Interestingly, TOHs as well as SCMs of vitamin E modulate cyclins, CDKs and CDK inhibitors [85]. Albeit the LCMs of vitamin E efficiently suppress proliferation, the identification of effects on regulators of cell cycle progression is pending. However, given that 'decreased proliferation is one of the best biomarkers of a cancer preventive effect' [91], vitamin E and its metabolites are promising compounds for cancer prevention.

4.3. Cellular Lipid Homeostasis

To date, the effects of the LCMs of vitamin E on cellular lipid homeostasis have not been investigated extensively. However, the regulation of key metabolic pathways in foam cell development of macrophages by the LCMs were of particular interest in a study by Wallert et al. [63]. Here, the regulation of the expression of the cluster of differentiation 36 (CD36), the uptake of oxidized low density lipoprotein (oxLDL), phagocytosis and the intracellular storage of lipids were investigated [63]. For this, the monocytic THP-1 cell line, which can be differentiated to macrophage-like cells, was used. In differentiated macrophages, the LCMs α-13′-OH and α-13′-COOH induced the expression of CD36 mRNA and consequently CD36 protein levels. In contrast, the precursor α-TOH exerted opposite effects on CD36 mRNA and protein. Whereas α-TOH reduced the expression of CD36 at a concentration of 100 μM, the α-LCMs induced the expression of CD36 in concentrations of 5 and 10 μM, respectively [63]. Thus, the α-LCMs not only act in a different way than their precursors, but appeared to be also significantly more potent. Interestingly, similar effects were described for the lipid soluble vitamin A. Langmann et al. found that the precursor β-carotene is less effective in inducing expression of CD36 than its metabolites ATRA and 9cRA in human monocytes and macrophages [92]. The authors stated that the metabolites 9cRA and ATRA displayed high biological activity [92], while the precursors retinol and β-carotene were only marginally metabolized, an observation that parallels the characteristics of the LCMs of vitamin E with respect to their reported serum concentrations [63,93]. The effects of vitamin A metabolites are better characterized than that of the LCMs of vitamin E. It was repeatedly shown that the metabolites of vitamin A regulate CD36 expression in macrophage cell models. The metabolite 9cRA induced CD36 mRNA [94,95] and protein expression [95] in human THP-1 macrophages. ATRA increases expression of CD36 mRNA in THP-1 cells [96] and CD36 protein in THP-1 and HL60 macrophages [96,97]. The induction of CD36 expression by ATRA and 9cRA has been confirmed in primary human monocytes and macrophages [92,96] to show the physiological relevance in non-cancer cells. With the same intention, it was also shown that the LCMs of vitamin E induced CD36 expression in peripheral blood mononuclear cell (PBMC)-derived primary human macrophages [63].

The scavenger receptor CD36 mediates the uptake of the modified lipoprotein oxLDL [98], a process that in turn stimulates CD36 expression [99]. Given the induction of the expression of CD36 by the LCMs of vitamin E under basal conditions (vide supra), a further stimulation by oxLDL treatment could be expected. As the uptake of oxLDL is a hallmark of macrophage foam cell formation, Wallert et al. examined whether preincubation of THP-1 macrophages with the LCMs of vitamin E affects the oxLDL-induced expression of CD36 [63]. As expected, CD36 expression was induced by oxLDL treatment. Pre-treatment with α-TOH suppressed the induction by oxLDL. In contrast, the pre-incubation with the LCMs augmented the induction of CD36 expression by oxLDL. These findings resemble the reaction of the cells in the absence of oxLDL to α-TOH and its LCMs. Given the higher CD36 expression in the presence of the LCMs, the uptake of oxLDL should in turn be induced in LCM-treated macrophages. However, pre-incubation of the macrophages with the LCMs for 24 h led to decreased oxLDL uptake. Incubation with both, α-13′-OH or α-13′-COOH, decreased the uptake by about 20%. This effect was again confirmed in PBMC-derived macrophages. Here, oxLDL uptake was decreased by α-13′-OH pre-treatment by 24% and by α-13′-COOH pre-treatment by 20%, respectively [63]. The LCMs of vitamin E thus exerted unexpected effects on oxLDL uptake. As mentioned before, vitamin A metabolites also caused increased CD36 expression, but the metabolite 9cRA induced the binding and uptake of oxLDL in THP-1 macrophages as expected [94]. Generally, an activation of RXR leads to an augmented association of oxLDL to THP-1 macrophages [100]. However, 9cRA also promoted the degradation of oxLDL and the cholesterol efflux via ATP binding cassette transporters, thus leading to a net depletion of cholesterol esters. Triglyceride levels were apparently not affected, neither by oxLDL treatment nor combination with 9cRA [94]. In contrast, in the study of Wallert et al. on the LCMs of vitamin E, oxLDL treatment of the macrophages led to an increase of neutral lipids in the cells. Preincubation with the LCMs diminished the oxLDL-induced neutral lipid accumulation [63]. However, the contradictory results on the effects of the LCMs on CD36 expression and oxLDL uptake required an alternative explanation how the LCMs decrease oxLDL uptake. Thus, Wallert et al. focused on phagocytosis as an alternative uptake mechanism for oxLDL [101]. Indeed, treatment of the macrophages with α-13′-OH led to an inhibition of phagocytotic activity of 16% and with α-13′-COOH of 41%, respectively [63]. Hence, the inhibition of phagocytosis by the LCMs might explain the discrepancy between their effects on CD36 expression and oxLDL uptake in this study.

Taken together, the metabolites of vitamin E and vitamin A induce the expression of CD36 in macrophages. However, their effects on oxLDL uptake are different. While the vitamin A metabolite 9cRA induces oxLDL uptake, the LCMs of vitamin E reduce it. In contrast to vitamin A and vitamin E metabolites, the metabolite of vitamin D, $1,25(OH)_2D_3$ has been shown to reduce the expression of CD36 mRNA and protein in oxLDL-treated macrophages obtained from diabetic subjects. Concomitantly, oxLDL and cholesterol uptake are decreased [102,103]. Hence, the vitamin D metabolite as well as the vitamin E LCMs suppress macrophage foam cell formation and may thus exert positive effects in the context of atherosclerosis prevention.

4.4. Interaction with Pharmaceuticals

The interaction of the vitamin E LCMs with pharmaceuticals was tested by analyzing the regulation of P-glycoprotein (P-gp). P-gp regulates, inter alia, the intracellular concentration of pharmaceuticals and its expression is regulated by various transcription factors, including heat shock transcription factor 1, nuclear factor Y and the pregnane X receptor (PXR) [104,105].

Several vitamin E forms and their metabolites (α-TOH, α-T3, α-13′-COOH, α-CEHC, γ-TOH, γ-T3, γ-CEHC and plastochromanol-8) were used and the regulation of P-gp expression was analyzed in human epithelial-like colon LS180 cells [78]. Only α-13′-COOH and γ-T3 induced P-gp expression and α-T3, α-13′-COOH as well as γ-T3 induced the activity of PXR in a reporter gene assay. In case of vitamin E supplementation, an interaction with the metabolic handling of pharmaceuticals might be possible.

4.5. Regulation of LCM Formation

The regulatory processes, which modulate the metabolism of vitamin E, are largely unknown. In this context, two key issues are important: (i) Apart from CYP4F2 and CYP3A4, the full set of enzymes involved in the first steps of the catabolism of vitamin E remains to be identified, and (ii) the mechanisms by which vitamin E metabolism is regulated have not yet been sufficiently unraveled. However, the upregulation of CYP4F2 protein expression by α-13′-OH in human HepG2 liver cells was reported recently [79], pointing to a positive regulatory feedback loop. If this concept holds true, the enhancement of metabolism by products would be a new facet for the fat-soluble vitamins, as the metabolism of vitamin A and D is mainly regulated negatively by their metabolic products [54,106].

The aldehyde- and alcohol-dehydrogenases have been suggested to be responsible for the ω-oxidation steps and the enzymes for branched-chain fatty acids might catalyze the subsequent β-oxidation [107]. Following the identification of the specific set of enzymes required for vitamin E metabolism, a major aim will be the characterization of the regulatory factors, which modulate the metabolism of vitamin E.

5. Structure-Specific Effects

To get deeper insights into the specificity of the regulatory effects of the LCMs of vitamin E, a structure-activity study was conducted [64]. For this purpose, substances were used that represent specific substructures of the LCMs or their precursors. The chromanol ring system was mirrored by the SCM α-CEHC and the modified side-chain was represented by the branched-chain fatty acid pristanic acid. Furthermore, the α- and δ-forms of 13′-OH and 13′-COOH were used to study the influence of the side-chain modification. Overall, the application of α- and δ-forms of LCMs and their precursors (α-TOH, α-13′-OH, α-13′-COOH, δ-TOH, δ-13′-OH, δ-13′-COOH) should clarify the importance of the substitution of the ring-system. The regulation of CD36 and iNos by the test compounds was similar for all of the LCMs, but neither the precursors nor their substructures were able to cause the same effects on the expression of the target genes as the LCMs. The substitution of the chromanol ring system had no influence (α- and δ-forms), while the modification of the side-chain (oxidation of TOH to 13′-OH and 13′-COOH) was highly relevant for the effects. Overall, the 13′-COOH was most potent in this study. Based on these specific regulations the existence of specific regulatory molecular pathways for the LCMs has been suggested.

6. Receptors of Vitamin Metabolites

As indicated above, the lipid-soluble vitamins A and D need a conversion to their active metabolites to exert their effects. These metabolites are either bound intracellularly and transferred to the receptor or directly bind the receptor. The receptors for the vitamin A metabolites, RARs and RXRs, were identified in the late 1980s [108–111]. Evidence for binding proteins for the active vitamin D metabolite 1,25(OH)$_2$D$_3$ was already provided in the 1970's [112,113]; however, cloning of the human vitamin D receptor also succeeded in the late 1980's [114]. In contrast, no specific receptor for vitamin E and/or its metabolites has been identified yet. Interestingly, the metabolites of vitamin A and D act through nuclear receptors. This class of transcription factors can roughly be divided into more specific and rather unspecific members. The vitamin D receptor can be categorized as a more specific receptor, as it is activated by its endogenous ligand 1,25(OH)$_2$D$_3$ already at sub-nanomolar concentrations [115,116]. This feature is also shared by steroid hormone receptors (estrogen receptor, androgen receptor, ergosterone receptor, cortisol receptor), the thyroid hormone receptor and RARs. The RARs specifically bind ATRA, and also 9cRA with lower affinity [117]. The specificity of the nuclear receptors is mainly determined by the structure of the ligand binding pocket. Specific receptors have a relatively small ligand binding pocket, which allows only a limited number of molecules to interact. In contrast, the so-called adopted orphan receptors have a larger ligand binding pocket, allowing the activation of the receptor by a larger number of ligands [115]. Members of this group are the liver X receptors (LXRs), farnesoid X receptor (FXR), peroxisome proliferator-activated receptors (PPARs) and RXRs. The latter have been shown to bind the vitamin A metabolite 9cRA [118]. However, it is not entirely accepted that 9cRA represents the endogenous ligand for RXR [119]. Nonetheless, the example of 9cRA opens the possibility that vitamin metabolites act through highly specific receptors but also through rather unspecific ones.

Following the concept that the LCMs of vitamin E represent biologically active metabolites similar to 1,25(OH)$_2$D$_3$, ATRA and 9cRA, these molecules might also exert their effects through nuclear receptors. Indeed, Podszun et al. reported an activation of PXR by α-13′-COOH in the human colon adenocarcinoma cell line LS180 [78] (for detailed information, the reader is referred to the section 'Interaction with pharmaceuticals'). Interestingly, α-T3 and γ-T3 were also able to activate PXR, while α-TOH and γ-TOH as well as the SCMs α-CEHC and γ-CEHC failed to activate PXR [78]. These findings confirm earlier findings in HepG2 cells only in part. In HepG2 cells transfected with PXR and a CAT (chloramphenicol acetyltransferase) reporter gene, α-T3 and γ-T3 efficiently activated PXR-mediated gene transcription, but α-TOH, γ-TOH and δ-TOH were also able to induce the expression of the reporter gene via PXR [120]. In contrast, the SCMs α-CEHC and α-CMBHC were not able to activate PXR in this study and the LCMs were not tested [120]. Taken together, the T3s reliably activate PXR but the effects of the TOHs need further investigation. Possibly, LS180 and HepG2 metabolize TOH with different efficiency, in turn determining the amounts of LCMs formed as PXR-activating metabolites. Hence, the observed effects of TOHs in HepG2 might be explained by the intracellular formation of the LCMs. However, further investigations on the cell-type specific metabolism of TOH are needed to confirm this hypothesis. Further, with PXR a rather unspecific nuclear receptor is identified for TOHs and their LCMs. As a general sensor for toxic compounds and xenobiotics, PXR has a large ligand binding cavity, which allows the binding of a wide range of ligands [121]. Thus, it is not surprising that PXR has been described as a receptor of vitamin K [122,123], and it has been reported that several menaquinone derivatives activate PXR [124]. Unfortunately, the biologically occurring carboxy derivatives were not included in this study. Hence, merely speculations about the activity based on structure-function-relationships are possible. A reporter gene assay revealed that a terminal phenyl group enhances the activity of the derivatives, while a terminal hydroxy group diminished it compared to the unmodified menaquinone [124]. In conclusion, a more hydrophobic side chain leads to an increased activity on PXR. Hence, the natural metabolic products in humans bearing a terminal carboxy group are likely less potent with respect to the activation of PXR. However, this concept is in contrast to the findings for vitamin E. The TOH precursors are unable to activate PXR, while the LCM α-13'-COOH activates it [78]. Hence, further studies are needed to clarify whether vitamin K metabolites are physiological ligands for the rather unspecific nuclear receptor PXR, like their metabolic precursor menaquinone and the LCMs of vitamin E.

Given that RXR as a receptor for the vitamin A metabolite 9cRA is also rather unspecific, it might be possible that the LCMs of vitamin E also act through PXR. However, it is questionable whether all of the reported biological effects of the LCMs, i.e., anti-inflammatory actions, anti-cancerogenic features, and effects on cellular lipid homeostasis (please refer to the respective sections here) can be ascribed to PXR activation. Hence, further investigations aiming at the identification and characterization of receptors for the LCMs of vitamin E LCMs are highly required. Strategies for the identification of further receptors or a receptor specific for the LCMs of vitamin E might be the use of target fishing approaches, gene expression arrays, knockdown/knockout studies, as well as reporter gene assays and ligand binding studies.

7. Conclusions

With the detection of the LCMs of vitamin E in human serum, an important hint for the possible action of these metabolites as signaling molecules was provided. Several studies reinforced this hypothesis by the characterization of the biological effects of the LCMs, as summarized in Figure 2. Interestingly, the LCMs act more potent and in part even contrary to their metabolic precursors. Some of the controversial effects reported for vitamin E might be therefore explained by the action of the LCMs. The evidence of circulating α-LCM in human blood (nanomolar concentrations) provides a new perspective in vitamin E research [63]. Therefore, the LCMs must be seriously considered to correctly interpret the effects of vitamin E in humans, beside the better studied TOHs and T3s. So far, only a few studies have focused on this class of compounds. However, based on our current knowledge and our studies in progress, we speculate that the LCMs comprise a new class of regulatory molecules. These molecules can exert effects that are different from their metabolic precursors, complicating the interpretation of studies on the effects of vitamin E in vivo. Nevertheless, the LCMs share properties with their precursors but also exert unique or even adverse effects. It is evident that the LCMs and their precursors act in the same manner with respect to the modulation of COX2 and 5-LOX activity, but it is of note that the LCMs are significantly more potent than their precursors. Furthermore, the LCM can act in areas where the TOHs are virtually not effective. A prime example is the regulation of COX2 expression. Hence, the LCMs may indeed play a role in mediating some of the effects of vitamin E in the human body although blood concentrations are significantly lower than those of TOH. So far, blood concentrations are the only valid value for the systemic distribution of the LCMs of vitamin E in the human body. However, based on preliminary data of unpublished in vitro and in vivo studies of our group, we can hypothesize that the LCMs of vitamin E may also accumulate in different parts of the human body, where they reach concentrations higher than in blood. Further studies are required to study this issue in more detail and to differentiate between physiologic (at low concentrations) and pharmacologic (at high concentrations) actions of the LCMs.

To sum up, the LCMs could be regarded as the metabolically activated forms of vitamin E. This is in line with the metabolic activation of the other lipid-soluble vitamins A and D. Consequently, the concept of metabolic activation established for vitamin A and D could now be extended to vitamin E. Thus, a general concept for the biological activity and modes of action of the lipid-soluble vitamins could be defined.

Acknowledgments: The work of Stefan Lorkowski is supported by grants from the Federal Ministry of Education and Research (01EA1411A), the Deutsche Forschungsgemeinschaft (DFG; RTG 1715) and the German Ministry of Economics and Technology (AiF 16642 BR) via AiF (the German Federation of Industrial Research Associations) and FEI (the Research Association of the German Food Industry), and by the Free State of Thuringia and the European Social Fund (2016 FGR 0045). The work of Lisa Schmölz is supported by the Free State of Thuringia and the European Social Fund (2016 FGR 0045). The work of Maria Wallert is funded by the DFG (Wa 3836/1-1).

Author Contributions: Martin Schubert, Stefan Kluge, Lisa Schmölz, Maria Wallert, Francesco Galli, Marc Birringer and Stefan Lorkowski wrote the paper.

Conflicts of Interest: The authors declare no conflicts of interest.

References

1. Horn, M.; Gunn, P.; van Emon, M.; Lemenager, R.; Burgess, J.; Pyatt, N.A.; Lake, S.L. Effects of natural (*RRR* alpha-tocopherol acetate) or synthetic (all-*rac*-alpha-tocopherol acetate) vitamin E supplementation on reproductive efficiency in beef cows. *J. Anim. Sci.* **2010**, *88*, 3121–3127. [CrossRef] [PubMed]

2. Evans, H.M.; Bishop, K.S. On the existence of a hitherto unrecognized dietary factor essential for reproduction. *Science* **1922**, *56*, 650–651. [CrossRef] [PubMed]

3. Kluge, S.; Schubert, M.; Schmölz, L.; Birringer, M.; Wallert, M.; Lorkowski, S. Garcinoic Acid: A Promising Bioactive Natural Product for Better Understanding the Physiological Functions of Tocopherol Metabolites. *Stud. Nat. Prod. Chem.* **2016**, *51*, 435–481. [CrossRef]

4. Sano, M.; Ernesto, C.; Thomas, R.G.; Klauber, M.R.; Schafer, K.; Grundman, M.; Woodbury, P.; Growdon, J.; Cotman, C.W.; Pfeiffer, E.; et al. A controlled trial of selegiline, alpha-tocopherol, or both as treatment for Alzheimer's disease. The Alzheimer's Disease Cooperative Study. *N. Engl. J. Med.* **1997**, *336*, 1216–1222. [CrossRef] [PubMed]

5. Dysken, M.W.; Guarino, P.D.; Vertrees, J.E.; Asthana, S.; Sano, M.; Llorente, M.; Pallaki, M.; Love, S.; Schellenberg, G.D.; McCarten, J.R.; et al. Vitamin E and memantine in Alzheimer's disease: Clinical trial methods and baseline data. *Alzheimer Dement. J. Alzheimer Assoc.* **2014**, *10*, 36–44. [CrossRef] [PubMed]

6. Jishage, K.; Arita, M.; Igarashi, K.; Iwata, T.; Watanabe, M.; Ogawa, M.; Ueda, O.; Kamada, N.; Inoue, K.; Arai, H.; et al. Alpha-tocopherol transfer protein is important for the normal development of placental labyrinthine trophoblasts in mice. *J. Biol. Chem.* **2001**, *276*, 1669–1672. [CrossRef] [PubMed]

7. Shichiri, M.; Yoshida, Y.; Ishida, N.; Hagihara, Y.; Iwahashi, H.; Tamai, H.; Niki, E. α-Tocopherol suppresses lipid peroxidation and behavioral and cognitive impairments in the Ts65Dn mouse model of Down syndrome. *Free Radic. Biol. Med.* **2011**, *50*, 1801–1811. [CrossRef] [PubMed]

8. Czeizel, A.E.; Dudás, I. Prevention of the first occurrence of neural-tube defects by periconceptional vitamin supplementation. *N. Engl. J. Med.* **1992**, *327*, 1832–1835. [CrossRef] [PubMed]

9. Chandler, A.L.; Hobbs, C.A.; Mosley, B.S.; Berry, R.J.; Canfield, M.A.; Qi, Y.P.; Siega-Riz, A.M.; Shaw, G.M. Neural tube defects and maternal intake of micronutrients related to one-carbon metabolism or antioxidant activity. *Birth Defects Res. Part A Clin. Mol. Teratol.* **2012**, *94*, 864–874. [CrossRef] [PubMed]

10. Brigelius-Flohé, R. Vitamin E: The shrew waiting to be tamed. *Free Radic. Biol. Med.* **2009**, *46*, 543–554. [CrossRef] [PubMed]

11. Wallert, M.; Schmölz, L.; Galli, F.; Birringer, M.; Lorkowski, S. Regulatory metabolites of vitamin E and their putative relevance for atherogenesis. *Redox Biol.* **2014**, *2*, 495–503. [CrossRef] [PubMed]
12. Jiang, Q. Natural forms of vitamin E: Metabolism, antioxidant, and anti-inflammatory activities and their role in disease prevention and therapy. *Free Radic. Biol. Med.* **2014**, *72*, 76–90. [CrossRef] [PubMed]
13. Rigotti, A. Absorption, transport, and tissue delivery of vitamin E. *Mol. Asp. Med.* **2007**, *28*, 423–436. [CrossRef] [PubMed]
14. Bjørneboe, A.; Bjørneboe, G.E.; Drevon, C.A. Absorption, transport and distribution of vitamin E. *J. Nutr.* **1990**, *120*, 233–242. [CrossRef] [PubMed]
15. Borel, P.; Pasquier, B.; Armand, M.; Tyssandier, V.; Grolier, P.; Alexandre-Gouabau, M.C.; Andre, M.; Senft, M.; Peyrot, J.; Jaussan, V.; et al. Processing of vitamin A and E in the human gastrointestinal tract. *Am. J. Physiol.-Gastrointest. Liver Physiol.* **2001**, *280*, G95–G103. [CrossRef] [PubMed]
16. Reboul, E. Intestinal absorption of vitamin D: From the meal to the enterocyte. *Food Funct.* **2015**, *6*, 356–362. [CrossRef] [PubMed]
17. Bieri, J.G.; Wu, A.L.; Tolliver, T.J. Reduced intestinal absorption of vitamin E by low dietary levels of retinoic acid in rats. *J. Nutr.* **1981**, *111*, 458–467. [PubMed]
18. Richelle, M.; Enslen, M.; Hager, C.; Groux, M.; Tavazzi, I.; Godin, J.-P.; Berger, A.; Métairon, S.; Quaile, S.; Piguet-Welsch, C.; et al. Both free and esterified plant sterols reduce cholesterol absorption and the bioavailability of beta-carotene and alpha-tocopherol in normocholesterolemic humans. *Am. J. Clin. Nutr.* **2004**, *80*, 171–177. [PubMed]
19. Doi, K.; Matsuura, M.; Kawara, A.; Tanaka, T.; Baba, S. Influence of dietary fiber (konjac mannan) on absorption of vitamin B12 and vitamin E. *Tohoku J. Exp. Med.* **1983**, *141*, 677–681. [CrossRef] [PubMed]
20. Burton, G.W.; Ingold, K.U.; Foster, D.O.; Cheng, S.C.; Webb, A.; Hughes, L.; Lusztyk, E. Comparison of free alpha-tocopherol and alpha-tocopheryl acetate as sources of vitamin E in rats and humans. *Lipids* **1988**, *23*, 834–840. [CrossRef] [PubMed]
21. D'Ambrosio, D.N.; Clugston, R.D.; Blaner, W.S. Vitamin A metabolism: An update. *Nutrients* **2011**, *3*, 63–103. [CrossRef] [PubMed]
22. Weng, W.; Li, L.; van Bennekum, A.M.; Potter, S.H.; Harrison, E.H.; Blaner, W.S.; Breslow, J.L.; Fisher, E.A. Intestinal absorption of dietary cholesteryl ester is decreased but retinyl ester absorption is normal in carboxyl ester lipase knockout mice. *Biochemistry* **1999**, *38*, 4143–4149. [CrossRef] [PubMed]
23. Maislos, M.; Shany, S. Bile salt deficiency and the absorption of vitamin D metabolites. In vivo study in the rat. *Isr. J. Med. Sci.* **1987**, *23*, 1114–1117. [PubMed]
24. Reboul, E.; Klein, A.; Bietrix, F.; Gleize, B.; Malezet-Desmoulins, C.; Schneider, M.; Margotat, A.; Lagrost, L.; Collet, X.; Borel, P. Scavenger receptor class B type I (SR-BI) is involved in vitamin E transport across the enterocyte. *J. Biol. Chem.* **2006**, *281*, 4739–4745. [CrossRef] [PubMed]
25. Hacquebard, M.; Carpentier, Y.A. Vitamin E: Absorption, plasma transport and cell uptake. *Curr. Opin. Clin. Nutr. Metab. Care* **2005**, *8*, 133–138. [CrossRef] [PubMed]
26. Yamanashi, Y.; Takada, T.; Kurauchi, R.; Tanaka, Y.; Komine, T.; Suzuki, H. Transporters for the Intestinal Absorption of Cholesterol, Vitamin E, and Vitamin K. *J. Atheroscler. Thromb.* **2017**, *24*, 347–359. [CrossRef] [PubMed]
27. Reboul, E. Absorption of vitamin A and carotenoids by the enterocyte: Focus on transport proteins. *Nutrients* **2013**, *5*, 3563–3581. [CrossRef] [PubMed]
28. Traber, M.G.; Burton, G.W.; Ingold, K.U.; Kayden, H.J. RRR- and SRR-alpha-tocopherols are secreted without discrimination in human chylomicrons, but RRR-alpha-tocopherol is preferentially secreted in very low density lipoproteins. *J. Lipid Res.* **1990**, *31*, 675–685. [PubMed]
29. Traber, M.G.; Burton, G.W.; Hughes, L.; Ingold, K.U.; Hidaka, H.; Malloy, M.; Kane, J.; Hyams, J.; Kayden, H.J. Discrimination between forms of vitamin E by humans with and without genetic abnormalities of lipoprotein metabolism. *J. Lipid Res.* **1992**, *33*, 1171–1182. [PubMed]
30. Shearer, M.J.; Newman, P. Metabolism and cell biology of vitamin K. *Thromb. Haemost.* **2008**. [CrossRef]
31. Cooper, A.D. Hepatic uptake of chylomicron remnants. *J. Lipid Res.* **1997**, *38*, 2173–2192. [PubMed]
32. Kiyose, C.; Muramatsu, R.; Kameyama, Y.; Ueda, T.; Igarashi, O. Biodiscrimination of alpha-tocopherol stereoisomers in humans after oral administration. *Am. J. Clin. Nutr.* **1997**, *65*, 785–789. [CrossRef] [PubMed]

33. Weiser, H.; Riss, G.; Kormann, A.W. Selective biodiscrimination of alpha-tocopherol stereoisomers. Similar enrichment of all 2R forms in rat tissues after oral all-rac-alpha-tocopheryl acetate. *Ann. N. Y. Acad. Sci.* **1992**, *669*, 393–395. [CrossRef] [PubMed]

34. Traber, M.G.; Kayden, H.J. Alpha-tocopherol as compared with gamma-tocopherol is preferentially secreted in human lipoproteins. *Ann. N. Y. Acad. Sci.* **1989**, *570*, 95–108. [CrossRef] [PubMed]

35. Arita, M.; Nomura, K.; Arai, H.; Inoue, K. alpha-tocopherol transfer protein stimulates the secretion of alpha-tocopherol from a cultured liver cell line through a brefeldin A-insensitive pathway. *Proc. Natl. Acad. Sci. USA* **1997**, *94*, 12437–12441. [CrossRef] [PubMed]

36. Oram, J.F.; Vaughan, A.M.; Stocker, R. ATP-binding cassette transporter A1 mediates cellular secretion of alpha-tocopherol. *J. Biol. Chem.* **2001**, *276*, 39898–39902. [CrossRef] [PubMed]

37. Mustacich, D.J.; Shields, J.; Horton, R.A.; Brown, M.K.; Reed, D.J. Biliary secretion of alpha-tocopherol and the role of the mdr2 P-glycoprotein in rats and mice. *Arch. Biochem. Biophys.* **1998**, *350*, 183–192. [CrossRef] [PubMed]

38. Lemaire-Ewing, S.; Desrumaux, C.; Néel, D.; Lagrost, L. Vitamin E transport, membrane incorporation and cell metabolism: Is alpha-tocopherol in lipid rafts an oar in the lifeboat? *Mol. Nutr. Food Res.* **2010**, *54*, 631–640. [CrossRef] [PubMed]

39. Schmölz, L.; Birringer, M.; Lorkowski, S.; Wallert, M. Complexity of vitamin E metabolism. *World J. Biol. Chem.* **2016**, *7*, 14–43. [CrossRef] [PubMed]

40. Abe, C.; Uchida, T.; Ohta, M.; Ichikawa, T.; Yamashita, K.; Ikeda, S. Cytochrome P450-dependent metabolism of vitamin E isoforms is a critical determinant of their tissue concentrations in rats. *Lipids* **2007**, *42*, 637–645. [CrossRef] [PubMed]

41. Bardowell, S.A.; Ding, X.; Parker, R.S. Disruption of P450-mediated vitamin E hydroxylase activities alters vitamin E status in tocopherol supplemented mice and reveals extra-hepatic vitamin E metabolism. *J. Lipid Res.* **2012**, *53*, 2667–2676. [CrossRef] [PubMed]

42. Grebenstein, N.; Schumacher, M.; Graeve, L.; Frank, J. α-Tocopherol transfer protein is not required for the discrimination against γ-tocopherol in vivo but protects it from side-chain degradation in vitro. *Mol. Nutr. Food Res.* **2014**, *58*, 1052–1060. [CrossRef] [PubMed]

43. Chiku, S.; Hamamura, K.; Nakamura, T. Novel urinary metabolite of d-delta-tocopherol in rats. *J. Lipid Res.* **1984**, *25*, 40–48. [PubMed]

44. Swanson, J.E.; Ben, R.N.; Burton, G.W.; Parker, R.S. Urinary excretion of 2,7,8-trimethyl-2-(beta-carboxyethyl)-6-hydroxychroman is a major route of elimination of gamma-tocopherol in humans. *J. Lipid Res.* **1999**, *40*, 665–671. [PubMed]

45. Sontag, T.J.; Parker, R.S. Cytochrome P450 omega-hydroxylase pathway of tocopherol catabolism. Novel mechanism of regulation of vitamin E status. *J. Biol. Chem.* **2002**, *277*, 25290–25296. [CrossRef] [PubMed]

46. Birringer, M.; Pfluger, P.; Kluth, D.; Landes, N.; Brigelius-Flohé, R. Identities and differences in the metabolism of tocotrienols and tocopherols in HepG2 cells. *J. Nutr.* **2002**, *132*, 3113–3118. [PubMed]

47. Parker, R.S.; Sontag, T.J.; Swanson, J.E. Cytochrome P4503A-dependent metabolism of tocopherols and inhibition by sesamin. *Biochem. Biophys. Res. Commun.* **2000**, *277*, 531–534. [CrossRef] [PubMed]

48. Johnson, C.H.; Slanař, O.; Krausz, K.W.; Kang, D.W.; Patterson, A.D.; Kim, J.-H.; Luecke, H.; Gonzalez, F.J.; Idle, J.R. Novel metabolites and roles for α-tocopherol in humans and mice discovered by mass spectrometry-based metabolomics. *Am. J. Clin. Nutr.* **2012**, *96*, 818–830. [CrossRef] [PubMed]

49. Zhao, Y.; Lee, M.-J.; Cheung, C.; Ju, J.-H.; Chen, Y.-K.; Liu, B.; Hu, L.-Q.; Yang, C.S. Analysis of multiple metabolites of tocopherols and tocotrienols in mice and humans. *J. Agric. Food Chem.* **2010**, *58*, 4844–4852. [CrossRef] [PubMed]

50. Goodman, D.S. Overview of current knowledge of metabolism of vitamin A and carotenoids. *J. Natl. Cancer Inst.* **1984**, *73*, 1375–1379. [PubMed]

51. Zhong, M.; Kawaguchi, R.; Kassai, M.; Sun, H. Retina, retinol, retinal and the natural history of vitamin A as a light sensor. *Nutrients* **2012**, *4*, 2069–2096. [CrossRef] [PubMed]

52. Petkovich, M.; Brand, N.J.; Krust, A.; Chambon, P. A human retinoic acid receptor which belongs to the family of nuclear receptors. *Nature* **1987**, *330*, 444–450. [CrossRef] [PubMed]

53. Lampen, A.; Meyer, S.; Arnhold, T.; Nau, H. Metabolism of vitamin A and its active metabolite all-trans-retinoic acid in small intestinal enterocytes. *J. Pharmacol. Exp. Ther.* **2000**, *295*, 979–985. [PubMed]

54.	Bikle, D.D. Vitamin D metabolism, mechanism of action, and clinical applications. *Chem. Biol.* **2014**, *21*, 319–329. [CrossRef] [PubMed]

55.	Urrutia-Pereira, M.; Solé, D. Vitamin D deficiency in pregnancy and its impact on the fetus, the newborn and in childhood. *Rev. Paul. Pediatr.* **2015**, *33*, 104–113. [CrossRef]

56.	Landes, N.; Birringer, M.; Brigelius-Flohé, R. Homologous metabolic and gene activating routes for vitamins E and K. *Mol. Asp. Med.* **2003**, *24*, 337–344. [CrossRef]

57.	Hodges, S.J.; Pitsillides, A.A.; Ytrebø, L.M.; Soper, R. Anti-inflammatory actions of vitamin K. In *Vitamin K2—Vital for Health and Wellbeing*; Gordeladze, J.O., Ed.; InTech: Rijeka, Croatia, 2017.

58.	Traber, M.G. Vitamin E and K interactions—A 50-year-old problem. *Nutr. Rev.* **2008**, *66*, 624–629. [CrossRef] [PubMed]

59.	Yamamoto, R.; Komai, M.; Kojima, K.; Furukawa, Y.; Kimura, S. Menaquinone-4 accumulation in various tissues after an oral administration of phylloquinone in Wistar rats. *J. Nutr. Sci. Vitaminol.* **1997**, *43*, 133–143. [CrossRef] [PubMed]

60.	Okano, T.; Shimomura, Y.; Yamane, M.; Suhara, Y.; Kamao, M.; Sugiura, M.; Nakagawa, K. Conversion of phylloquinone (Vitamin K1) into menaquinone-4 (Vitamin K2) in mice: Two possible routes for menaquinone-4 accumulation in cerebra of mice. *J. Biol. Chem.* **2008**, *283*, 11270–11279. [CrossRef] [PubMed]

61.	Brigelius-Flohé, R.; Kelly, F.J.; Salonen, J.T.; Neuzil, J.; Zingg, J.-M.; Azzi, A. The European perspective on vitamin E: Current knowledge and future research. *Am. J. Clin. Nutr.* **2002**, *76*, 703–716. [PubMed]

62.	Azzi, A.; Ricciarelli, R.; Zingg, J.-M. Non-antioxidant molecular functions of α-tocopherol (vitamin E). *FEBS Lett.* **2002**, *519*, 8–10. [CrossRef]

63.	Wallert, M.; Mosig, S.; Rennert, K.; Funke, H.; Ristow, M.; Pellegrino, R.M.; Cruciani, G.; Galli, F.; Lorkowski, S.; Birringer, M. Long-chain metabolites of α-tocopherol occur in human serum and inhibit macrophage foam cell formation in vitro. *Free Radic. Biol. Med.* **2014**, *68*, 43–51. [CrossRef] [PubMed]

64.	Schmölz, L.; Wallert, M.; Rozzino, N.; Cignarella, A.; Galli, F.; Glei, M.; Werz, O.; Koeberle, A.; Birringer, M.; Lorkowski, S. Structure-Function Relationship Studies in vitro Reveal Distinct and Specific Effects of Long-Chain Metabolites of Vitamin E. *Mol. Nutr. Food Res.* **2017**. [CrossRef] [PubMed]

65.	Torquato, P.; Ripa, O.; Giusepponi, D.; Galarini, R.; Bartolini, D.; Wallert, M.; Pellegrino, R.; Cruciani, G.; Lorkowski, S.; Birringer, M.; et al. Analytical strategies to assess the functional metabolome of vitamin E. *J. Pharm. Biomed. Anal.* **2016**, *124*, 399–412. [CrossRef] [PubMed]

66.	Giusepponi, D.; Torquato, P.; Bartolini, D.; Piroddi, M.; Birringer, M.; Lorkowski, S.; Libetta, C.; Cruciani, G.; Moretti, S.; Saluti, G.; et al. Determination of tocopherols and their metabolites by liquid-chromatography coupled with tandem mass spectrometry in human plasma and serum. *Talanta* **2017**, *170*, 552–561. [CrossRef] [PubMed]

67.	Munteanu, A.; Zingg, J.-M.; Azzi, A. Anti-atherosclerotic effects of vitamin E—Myth or reality? *J. Cell. Mol. Med.* **2004**, *8*, 59–76. [CrossRef] [PubMed]

68.	Traber, M.G.; Atkinson, J. Vitamin E, antioxidant and nothing more. *Free Radic. Biol. Med.* **2007**, *43*, 4–15. [CrossRef] [PubMed]

69.	Brigelius-Flohé, R.; Davies, K.J.A. Is vitamin E an antioxidant, a regulator of signal transduction and gene expression, or a 'junk' food? Comments on the two accompanying papers: "Molecular mechanism of alpha-tocopherol action" by A. Azzi and "Vitamin E, antioxidant and nothing more" by M. Traber and J. Atkinson. *Free Radic. Biol. Med.* **2007**, *43*, 2–3. [CrossRef] [PubMed]

70.	Jiang, Q.; Yin, X.; Lill, M.A.; Danielson, M.L.; Freiser, H.; Huang, J. Long-chain carboxychromanols, metabolites of vitamin E, are potent inhibitors of cyclooxygenases. *Proc. Natl. Acad. Sci. USA* **2008**, *105*, 20464–20469. [CrossRef] [PubMed]

71.	Ciffolilli, S.; Wallert, M.; Bartolini, D.; Krauth, V.; Werz, O.; Piroddi, M.; Sebastiani, B.; Torquato, P.; Lorkowski, S.; Birringer, M.; et al. Human serum determination and in vitro anti-inflammatory activity of the vitamin E metabolite α-(13'-hydroxy)-6-hydroxychroman. *Free Radic. Biol. Med.* **2015**, *89*, 952–962. [CrossRef] [PubMed]

72.	Jang, Y.; Park, N.-Y.; Rostgaard-Hansen, A.L.; Huang, J.; Jiang, Q. Vitamin E metabolite 13'-carboxychromanols inhibit pro-inflammatory enzymes, induce apoptosis and autophagy in human cancer cells by modulating sphingolipids and suppress colon tumor development in mice. *Free Radic. Biol. Med.* **2016**, *95*, 190–199. [CrossRef] [PubMed]

73. Jiang, Z.; Yin, X.; Jiang, Q. Natural forms of vitamin E and 13′-carboxychromanol, a long-chain vitamin E metabolite, inhibit leukotriene generation from stimulated neutrophils by blocking calcium influx and suppressing 5-lipoxygenase activity, respectively. *J. Immunol.* **2011**, *186*, 1173–1179. [CrossRef] [PubMed]

74. Wallert, M.; Schmölz, L.; Koeberle, A.; Krauth, V.; Glei, M.; Galli, F.; Werz, O.; Birringer, M.; Lorkowski, S. Alpha-Tocopherol long-chain metabolite alpha-13′-COOH affects the inflammatory response of lipopolysaccharide-activated murine RAW264.7 macrophages. *Mol. Nutr. Food Res.* **2015**, *59*, 1524–1534. [CrossRef] [PubMed]

75. Schmölz, L.; Wallert, M.; Lorkowski, S. Optimized incubation regime for nitric oxide measurements in murine macrophages using the Griess assay. *J. Immunol. Methods* **2017**. [CrossRef] [PubMed]

76. Mazzini, F.; Betti, M.; Netscher, T.; Galli, F.; Salvadori, P. Configuration of the vitamin E analogue garcinoic acid extracted from Garcinia Kola seeds. *Chirality* **2009**, *21*, 519–524. [CrossRef] [PubMed]

77. Birringer, M.; Lington, D.; Vertuani, S.; Manfredini, S.; Scharlau, D.; Glei, M.; Ristow, M. Proapoptotic effects of long-chain vitamin E metabolites in HepG2 cells are mediated by oxidative stress. *Free Radic. Biol. Med.* **2010**, *49*, 1315–1322. [CrossRef] [PubMed]

78. Podszun, M.C.; Jakobi, M.; Birringer, M.; Weiss, J.; Frank, J. The long chain α-tocopherol metabolite α-13′-COOH and γ-tocotrienol induce P-glycoprotein expression and activity by activation of the pregnane X receptor in the intestinal cell line LS 180. *Mol. Nutr. Food Res.* **2017**, *61*. [CrossRef] [PubMed]

79. Torquato, P.; Bartolini, D.; Giusepponi, D.; Saluti, G.; Russo, A.; Barola, C.; Birringer, M.; Galarini, R.; Galli, F. a-13′-OH is the main product of a-tocopherol metabolism and influences CYP4F2 and PPARg: Gene expression in HepG2 human hepatocarcinoma cells. *Free Radic. Biol. Med.* **2016**, *96*, S19–S20. [CrossRef]

80. Reddi, K.; Henderson, B.; Meghji, S.; Wilson, M.; Poole, S.; Hopper, C.; Harris, M.; Hodges, S.J. Interleukin 6 production by lipopolysaccharide-stimulated human fibroblasts is potently inhibited by naphthoquinone (vitamin K) compounds. *Cytokine* **1995**, *7*, 287–290. [CrossRef] [PubMed]

81. Soper, R.J.; Oguz, C.; Emery, R.; Pitsillides, A.A.; Hodges, S.J. Vitamin K catabolite inhibition of ovariectomy-induced bone loss: Structure-activity relationship considerations. *Mol. Nutr. Food Res.* **2014**, *58*, 1658–1666. [CrossRef] [PubMed]

82. Fujii, S.; Shimizu, A.; Takeda, N.; Oguchi, K.; Katsurai, T.; Shirakawa, H.; Komai, M.; Kagechika, H. Systematic synthesis and anti-inflammatory activity of ω-carboxylated menaquinone derivatives—Investigations on identified and putative vitamin K₂ metabolites. *Bioorg. Med. Chem.* **2015**, *23*, 2344–2352. [CrossRef] [PubMed]

83. Harrington, D.J.; Soper, R.; Edwards, C.; Savidge, G.F.; Hodges, S.J.; Shearer, M.J. Determination of the urinary aglycone metabolites of vitamin K by HPLC with redox-mode electrochemical detection. *J. Lipid Res.* **2005**, *46*, 1053–1060. [CrossRef] [PubMed]

84. Conte, C.; Floridi, A.; Aisa, C.; Piroddi, M.; Floridi, A.; Galli, F. Gamma-tocotrienol metabolism and antiproliferative effect in prostate cancer cells. *Ann. N. Y. Acad. Sci.* **2004**, *1031*, 391–394. [CrossRef] [PubMed]

85. Galli, F.; Stabile, A.M.; Betti, M.; Conte, C.; Pistilli, A.; Rende, M.; Floridi, A.; Azzi, A. The effect of alpha- and gamma-tocopherol and their carboxyethyl hydroxychroman metabolites on prostate cancer cell proliferation. *Arch. Biochem. Biophys.* **2004**, *423*, 97–102. [CrossRef] [PubMed]

86. Kennel, K.A.; Drake, M.T. Vitamin D in the cancer patient. *Curr. Opin. Support. Palliat. Care* **2013**, *7*, 272–277. [CrossRef] [PubMed]

87. Vanoirbeek, E.; Krishnan, A.; Eelen, G.; Verlinden, L.; Bouillon, R.; Feldman, D.; Verstuyf, A. The anti-cancer and anti-inflammatory actions of 1,25(OH)₂D₃. *Best Pract. Res. Clin. Endocrinol. Metab.* **2011**, *25*, 593–604. [CrossRef] [PubMed]

88. Tang, X.-H.; Gudas, L.J. Retinoids, retinoic acid receptors, and cancer. *Ann. Rev. Pathol.* **2011**, *6*, 345–364. [CrossRef] [PubMed]

89. Qin, X.-Y.; Fujii, S.; Shimizu, A.; Kagechika, H.; Kojima, S. Carboxylic Derivatives of Vitamin K2 Inhibit Hepatocellular Carcinoma Cell Growth through Caspase/Transglutaminase-Related Signaling Pathways. *J. Nutr. Sci. Vitaminol.* **2015**, *61*, 285–290. [CrossRef] [PubMed]

90. De Oliveira, M.R. Vitamin A and Retinoids as Mitochondrial Toxicants. *Oxid. Med. Cell. Longev.* **2015**, *2015*, 140267. [CrossRef] [PubMed]

91. Uray, I.P.; Dmitrovsky, E.; Brown, P.H. Retinoids and rexinoids in cancer prevention: From laboratory to clinic. *Semin. Oncol.* **2016**, *43*, 49–64. [CrossRef] [PubMed]

92. Langmann, T.; Liebisch, G.; Moehle, C.; Schifferer, R.; Dayoub, R.; Heiduczek, S.; Grandl, M.; Dada, A.; Schmitz, G. Gene expression profiling identifies retinoids as potent inducers of macrophage lipid efflux. *Biochim. Biophys. Acta* **2005**, *1740*, 155–161. [CrossRef] [PubMed]

93. Galli, F.; Azzi, A.; Birringer, M.; Cook-Mills, J.M.; Eggersdorfer, M.; Frank, J.; Cruciani, G.; Lorkowski, S.; Özer, N.K. Vitamin E: Emerging aspects and new directions. *Free Radic. Biol. Med.* **2017**, *102*, 16–36. [CrossRef] [PubMed]

94. Argmann, C.A.; Sawyez, C.G.; McNeil, C.J.; Hegele, R.A.; Huff, M.W. Activation of peroxisome proliferator-activated receptor gamma and retinoid X receptor results in net depletion of cellular cholesteryl esters in macrophages exposed to oxidized lipoproteins. *Arterioscler. Thromb. Vasc. Biol.* **2003**, *23*, 475–482. [CrossRef] [PubMed]

95. Han, S.; Sidell, N. Peroxisome-proliferator-activated-receptor gamma (PPARgamma) independent induction of CD36 in THP-1 monocytes by retinoic acid. *Immunology* **2002**, *106*, 53–59. [CrossRef] [PubMed]

96. Wuttge, D.M. Induction of CD36 by all-trans retinoic acid: Retinoic acid receptor signaling in the pathogenesis of atherosclerosis. *FASEB J.* **2001**. [CrossRef]

97. Barber, N.; Belov, L.; Christopherson, R.I. All-trans retinoic acid induces different immunophenotypic changes on human HL60 and NB4 myeloid leukaemias. *Leuk. Res.* **2008**, *32*, 315–322. [CrossRef] [PubMed]

98. Endemann, G.; Stanton, L.W.; Madden, K.S.; Bryant, C.M.; White, R.T.; Protter, A.A. CD36 is a receptor for oxidized low density lipoprotein. *J. Biol. Chem.* **1993**, *268*, 11811–11816. [PubMed]

99. Silverstein, R.L.; Li, W.; Park, Y.M.; Rahaman, S.O. Mechanisms of cell signaling by the scavenger receptor CD36: Implications in atherosclerosis and thrombosis. *Trans. Am. Clin. Climatol. Assoc.* **2010**, *121*, 206–220. [PubMed]

100. Tontonoz, P.; Nagy, L.; Alvarez, J.G.; Thomazy, V.A.; Evans, R.M. PPARgamma promotes monocyte/macrophage differentiation and uptake of oxidized LDL. *Cell* **1998**, *93*, 241–252. [CrossRef]

101. Schrijvers, D.M.; De Meyer, G.R.Y.; Herman, A.G.; Martinet, W. Phagocytosis in atherosclerosis: Molecular mechanisms and implications for plaque progression and stability. *Cardiovasc. Res.* **2007**, *73*, 470–480. [CrossRef] [PubMed]

102. Oh, J.; Weng, S.; Felton, S.K.; Bhandare, S.; Riek, A.; Butler, B.; Proctor, B.M.; Petty, M.; Chen, Z.; Schechtman, K.B.; et al. 1,25(OH)$_2$ vitamin D inhibits foam cell formation and suppresses macrophage cholesterol uptake in patients with type 2 diabetes mellitus. *Circulation* **2009**, *120*, 687–698. [CrossRef] [PubMed]

103. Riek, A.E.; Oh, J.; Bernal-Mizrachi, C. Vitamin D regulates macrophage cholesterol metabolism in diabetes. *J. Steroid Biochem. Mol. Biol.* **2010**, *121*, 430–433. [CrossRef] [PubMed]

104. Dewanjee, S.; Dua, T.K.; Bhattacharjee, N.; Das, A.; Gangopadhyay, M.; Khanra, R.; Joardar, S.; Riaz, M.; Feo, V.D.; Zia-Ul-Haq, M. Natural products as alternative choices for P-Glycoprotein (P-gp) inhibition. *Molecules* **2017**, *22*, 871. [CrossRef] [PubMed]

105. Silva, R.; Vilas-Boas, V.; Carmo, H.; Dinis-Oliveira, R.J.; Carvalho, F.; de Lourdes Bastos, M.; Remião, F. Modulation of P-glycoprotein efflux pump: Induction and activation as a therapeutic strategy. *Pharmacol. Ther.* **2015**, *149*, 1–123. [CrossRef] [PubMed]

106. Henry, H.L. Regulation of vitamin D metabolism. *Best Pract. Res. Clin. Endocrinol. Metab.* **2011**, *25*, 531–541. [CrossRef] [PubMed]

107. Mustacich, D.J.; Leonard, S.W.; Patel, N.K.; Traber, M.G. Alpha-tocopherol beta-oxidation localized to rat liver mitochondria. *Free Radic. Biol. Med.* **2010**, *48*, 73–81. [CrossRef] [PubMed]

108. Benbrook, D.; Lernhardt, E.; Pfahl, M. A new retinoic acid receptor identified from a hepatocellular carcinoma. *Nature* **1988**, *333*, 669–672. [CrossRef] [PubMed]

109. Brand, N.; Petkovich, M.; Krust, A.; Chambon, P.; de Thé, H.; Marchio, A.; Tiollais, P.; Dejean, A. Identification of a second human retinoic acid receptor. *Nature* **1988**, *332*, 850–853. [CrossRef] [PubMed]

110. Giguere, V.; Ong, E.S.; Segui, P.; Evans, R.M. Identification of a receptor for the morphogen retinoic acid. *Nature* **1987**, *330*, 624–629. [CrossRef] [PubMed]

111. Zelent, A.; Krust, A.; Petkovich, M.; Kastner, P.; Chambon, P. Cloning of murine alpha and beta retinoic acid receptors and a novel receptor gamma predominantly expressed in skin. *Nature* **1989**, *339*, 714–717. [CrossRef] [PubMed]

112. Brumbaugh, P.F.; Hughes, M.R.; Haussler, M.R. Cytoplasmic and nuclear binding components for 1alpha25-dihydroxyvitamin D3 in chick parathyroid glands. *Proc. Natl. Acad. Sci. USA* **1975**, *72*, 4871–4875. [CrossRef] [PubMed]

113. Tsai, H.C.; Norman, A.W. Studies on calciferol metabolism. 8. Evidence for a cytoplasmic receptor for 1,25-dihydroxy-vitamin D3 in the intestinal mucosa. *J. Biol. Chem.* **1973**, *248*, 5967–5975. [PubMed]

114. Baker, A.R.; McDonnell, D.P.; Hughes, M.; Crisp, T.M.; Mangelsdorf, D.J.; Haussler, M.R.; Pike, J.W.; Shine, J.; O'Malley, B.W. Cloning and expression of full-length cDNA encoding human vitamin D receptor. *Proc. Natl. Acad. Sci. USA* **1988**, *85*, 3294–3298. [CrossRef] [PubMed]

115. Carlberg, C.; Campbell, M.J. Vitamin D receptor signaling mechanisms: Integrated actions of a well-defined transcription factor. *Steroids* **2013**, *78*, 127–136. [CrossRef] [PubMed]

116. Gross, C.; Krishnan, A.V.; Malloy, P.J.; Eccleshall, T.R.; Zhao, X.Y.; Feldman, D. The vitamin D receptor gene start codon polymorphism: A functional analysis of FokI variants. *J. Bone Miner. Res.* **1998**, *13*, 1691–1699. [CrossRef] [PubMed]

117. Allenby, G.; Janocha, R.; Kazmer, S.; Speck, J.; Grippo, J.F.; Levin, A.A. Binding of 9-*cis*-retinoic acid and all-trans-retinoic acid to retinoic acid receptors alpha, beta, and gamma. Retinoic acid receptor gamma binds all-trans-retinoic acid preferentially over 9-cis-retinoic acid. *J. Biol. Chem.* **1994**, *269*, 16689–16695. [PubMed]

118. Heyman, R.A.; Mangelsdorf, D.J.; Dyck, J.A.; Stein, R.B.; Eichele, G.; Evans, R.M.; Thaller, C. 9-*cis*-retinoic acid is a high affinity ligand for the retinoid X receptor. *Cell* **1992**, *68*, 397–406. [CrossRef]

119. Wolf, G. Is 9-*cis*-retinoic acid the endogenous ligand for the retinoic acid-X receptor? *Nutr. Rev.* **2006**, *64*, 532–538. [CrossRef] [PubMed]

120. Landes, N.; Pfluger, P.; Kluth, D.; Birringer, M.; Rühl, R.; Böl, G.-F.; Glatt, H.; Brigelius-Flohé, R. Vitamin E activates gene expression via the pregnane X receptor. *Biochem. Pharmacol.* **2003**, *65*, 269–273. [CrossRef]

121. Kliewer, S.A.; Goodwin, B.; Willson, T.M. The nuclear pregnane X receptor: A key regulator of xenobiotic metabolism. *Endocr. Rev.* **2002**, *23*, 687–702. [CrossRef] [PubMed]

122. Tabb, M.M.; Sun, A.; Zhou, C.; Grün, F.; Errandi, J.; Romero, K.; Pham, H.; Inoue, S.; Mallick, S.; Lin, M.; et al. Vitamin K2 regulation of bone homeostasis is mediated by the steroid and xenobiotic receptor SXR. *J. Biol. Chem.* **2003**, *278*, 43919–43927. [CrossRef] [PubMed]

123. Ichikawa, T.; Horie-Inoue, K.; Ikeda, K.; Blumberg, B.; Inoue, S. Steroid and xenobiotic receptor SXR mediates vitamin K2-activated transcription of extracellular matrix-related genes and collagen accumulation in osteoblastic cells. *J. Biol. Chem.* **2006**, *281*, 16927–16934. [CrossRef] [PubMed]

124. Suhara, Y.; Watanabe, M.; Nakagawa, K.; Wada, A.; Ito, Y.; Takeda, K.; Takahashi, K.; Okano, T. Synthesis of novel vitamin K2 analogues with modification at the ω-terminal position and their biological evaluation as potent steroid and xenobiotic receptor (SXR) agonists. *J. Med. Chem.* **2011**, *54*, 4269–4273. [CrossRef] [PubMed]

antioxidants

MDPI

Review

Tocotrienols: A Family of Molecules with Specific Biological Activities

Raffaella Comitato, Roberto Ambra and Fabio Virgili *

Council for Agricultural Research and Economics, Research Centre for Food and Nutrition (CREA-AN) via Ardeatina 546, 00178 Rome, Italy; raffaella.comitato@crea.gov.it (R.C.); roberto.ambra@crea.gov.it (R.A.)
* Correspondence: fabio.virgili@crea.gov.it; Tel.: +39-06-51494517

Received: 24 October 2017; Accepted: 16 November 2017; Published: 18 November 2017

Abstract: Vitamin E is a generic term frequently used to group together eight different molecules, namely: α-, β-, γ- and δ-tocopherol and the corresponding tocotrienols. The term tocopherol and eventually Vitamin E and its related activity was originally based on the capacity of countering foetal re-absorption in deficient rodents or the development of encephalomalacia in chickens. In humans, Vitamin E activity is generally considered to be solely related to the antioxidant properties of the tocolic chemical structure. In recent years, several reports have shown that specific activities exist for each different tocotrienol form. In this short review, tocotrienol ability to inhibit cancer cell growth and induce apoptosis thanks to specific mechanisms, not shared by tocopherols, such as the binding to Estrogen Receptor-β (ERβ) and the triggering of endoplasmic reticulum (EndoR) stress will be described. The neuroprotective activity will also be presented and discussed. We propose that available studies strongly indicate that specific forms of tocotrienols have a distinct mechanism and biological activity, significantly different from tocopherol and more specifically from α-tocopherol. We therefore suggest not pooling them together within the broad term "Vitamin E" on solely the basis of their putative antioxidant properties. This option implies obvious consequences in the assessment of dietary Vitamin E adequacy and, probably more importantly, on the possibility of evaluating a separate biological variable, determinant in the relationship between diet and health.

Keywords: tocopherols; tocotrienols; estrogen receptors; endoplasmic reticulum stress; neuroprotection

1. Introduction

The term "natural Vitamin E" is commonly used to group together eight different molecules, namely α-, β-, γ- and δ-tocopherol and the corresponding tocotrienols. The original description of Vitamin E is: "a fat soluble vitamin that inhibits oxidative destruction of biological membranes and is necessary for fertility and to prevent hemolysis in rats and muscle dystrophy in poultry" [1]. If we held that this definition is biologically true and descriptive of the vitamin functions, we must face the evidence that, with the exception of a "generic" putative antioxidant activity in protecting biological membranes from peroxidation, in comparison to α-tocopherol, β-, γ- and δ-tocopherol and all tocotrienols have very low (if any) biological activity in the fetal re-absorption test. Their relative efficiency, in fact, ranges from zero to about 40% for β-tocopherol, which is the only alternative form significantly active in this test [1]. Accordingly, the units definition provided by the United States Pharmacopoeia of the different Vitamin E analogues is standardized on the efficiency of α-tocopherol, whereas β-, γ- and δ-tocopherol and the tocotrienols are much less active (either inactive) in this assay [2].

If we agree in considering that the only biological function of Vitamin E is acting as an "antioxidant", the inclusion of tocotrienols in the Vitamin E family should not be questioned but, accordingly, we should include in the "Vitamin E family" several other molecules displaying a "lipid peroxidation chain breaking activity" within biological membranes. Conversely, if we accept that the biological role of α-tocopherol is more complex and goes beyond its activity as a membrane antioxidant, we must accept that this role is not shared by other tocols, and solely assign the term Vitamin E to α-tocopherol.

In fact, it is possible to describe similarities and differences between the members of the Vitamin E family. Dietary α-tocopherol, non-α-tocopherols and tocotrienols are all absorbed by intestinal cells by passive diffusion; and receptor-mediated transport and delivered to the lymph and then to the liver via chylomicron [3]. However, the efficiency of absorption is not the same, being apparently higher for tocotrienols than for α-tocopherol. This latter has been reported to significantly inhibit the cellular uptake of δ-tocotrienol (and probably of other tocotrienols), at least in endothelial cells by a still unknown mechanism [4].

Once in the liver, α-tocopherol is immediately transferred to the α-tocopherol transfer protein (α-TTP) (see below) and further disposed to the peripheral tissues after the incorporation into VLDL/LDL/HDL [3]. If not delivered to peripheral tissues, α-tocopherol, non-α-tocopherols and tocotrienols are metabolized by phase I and II enzymes and excreted as glucuronide or sulfate [3], but tocotrienols have been reported to be degraded to a larger extent than their counterparts with saturated side chains. The significant quantitative differences in the metabolism between each tocopherol and between tocotrienols and tocopherols that have been reported in vitro, suggest that similar differences may exist also in vivo [5].

Very importantly, the biological activity of Vitamin E is highly dependent upon regulatory mechanisms exerted by the intracellular α-TTP that enrich the plasma with α-tocopherol while non-α-tocopherols molecules are directed to metabolism. α-TTP specifically recognizes α-tocopherol by the three methyl groups on the chromanol ring, by the hydroxyl group on the chromanol ring and the structure and orientation of the phytyl side chain. Thanks to these multiple "recognition mechanisms", α-TPP preferably and efficiently only binds α-tocopherol while all the other Vitamin E forms display a very low or null binding activity [6,7]. As the result of this specificity, only α-tocopherol, once taken up from the liver from dietary derived chilomicrons, can be released to peripheral extrahepatic tissues via lipoprotein VLDL trafficking, while the other tocopherols and the tocotrienols are rapidly metabolized by phase I and phase II enzymes and finally excreted in the bile, feces or urine [3]. Interestingly, a linear relationship between the relative affinity and the known biological activity obtained from the rat resorption-gestation assay exists, clearly indicating that the ability to bind to α-TTP is critical to determine the biological activity. However, it has been demonstrated that long oral supplementation of tocotrienol to mice and rats results in the delivery to vital organs including the brain, liver, heart, skin, lungs, adipose tissue, and whole blood independently of α-TTP expression [8,9]. These findings strongly demonstrate the existence of TTP-independent, still partially unknown, mechanisms of transport for oral tocotrienol. Moreover, the presence of α-tocopherol leads to decreased binding and to an acceleration of metabolism of non-α-tocopherols and tocotrienols [9].

Significant differences also exist in plasma concentration of α-tocopherol and tocotrienols. The administration of high doses (750 mg and 1000 mg) of a tocotrienol mixture from Annatto (*Bixa orellana*), results in a maximum plasma concentration levels at 3–4 h for all isomers, while α-tocopherol is reported to peak at 6 h, suggesting a different distribution mechanism [10]. Even though γ- and δ-tocotrienols have been probably studied more in the detail, α-tocotrienol has been found as the most abundant form circulating in plasma, chilomicrons, LDL, and HDL after the dietary supplementation with a mixture from palm (*Elaeis guineensis*) oil, also containing α-tocopherol. According to the study mentioned above [10], concentrations are in the order of μM and peak at less than 5 h for γ- and δ-tocotrienol and at 6 h for α-tocotrienol [11].

Overall, in conclusion, it is quite surprising that α-, β-, γ- and δ-tocotrienol are still pooled together with tocopherols in spite of a significantly different steric hindrance due to the unsaturated phytyl tail, and of several evidences indicating that they have a distinct metabolism and different routes of tissue delivery and storage [12]. All these evidences suggest that tocotrienol should be a candidate to play a different biological role than α-tocopherol [13]. A "side-to side" visualization of D-α-tocopherol and D-α-tocotrienols (Figure 1) highlights the evident differences between these two molecules at the tridimensional level that are frequently ignored in a two-dimensional, planar visualization. A dynamic view of these two molecules would further underline their different spatial seizures, due to the presence of two *trans* double bonds in the phytyl tail of tocotrienols limiting the rotational freedom of carbon-carbon bonds.

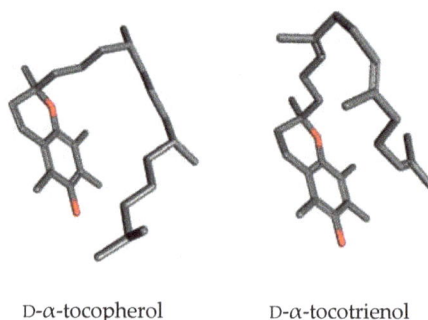

D-α-tocopherol   D-α-tocotrienol

Figure 1. Structural differences between α-tocotrienol and α-tocopherol (from https://pubchem.ncbi. nlm.nih.gov/).

A plethora of papers, also published by high impact factor journals (including those by the authors) encourage a kind of confusion. In fact, the term Vitamin E is frequently introduced as (e.g.,): " … *Tocotrienols and tocopherols are natural forms of the Vitamin E family* … ", even though it is always admitted that (e.g.,) " … *Vitamin E deficiency syndromes cannot be prevented by supplying non-α-tocopherols or tocotrienols* … " and that (e.g.,) "*Although all natural forms of Vitamin E display potent antioxidant activity, the activity of tocotrienols is mediated independently of their antioxidant activity..*" and also: " … *current studies of the biological functions of Vitamin E indicate that members in the Vitamin E family possess unique biological functions often not shared by other family members*".

Starting from the "contradictions" described above, this review focuses on some of the most evident specific activities reported for tocotrienols, not shared by α-tocopherol or by other tocopherols, namely the estrogen receptor-β (ERβ) binding activity and endoplasmic reticulum (EndoR) response activation both leading to a pro-apoptotic cellular response by γ- and δ-tocotrienol. The specificity of α-tocotrienol in protecting neuronal cells after ischemia will also be presented and discussed together with other tocotrienol-specific mechanisms on cell functions and survival.

2. Tocotrienols as Ligands of ERβ

Interestingly, one of the first studies reporting the antiproliferative/pro-apoptotic effects of tocotrienols, excluded the possibility of a mechanism related to the binding to estrogen receptors [14]. This early paper reported a study that was conducted somehow before the discovery and characterization of the β form of ERs [15]. The experimental design was in fact based on the assumption that the human breast cancer cell line estrogen-independent (MDA-MB-231) was void of any ER form and utilized to compare the effects of tocotrienols on the estrogen responsive breast cancer cell line, MCF7. Conversely, MDA cells have been eventually demonstrated to express a functional β form, while MCF7 express both α- and β-ER. In this study, authors observed that a tocotrienol-rich fraction (TRF) of palm (*Elaeis guineensis*) oil, containing α-tocopherol and α-, γ- and δ-tocotrienol, inhibited MCF7 cells growth in both the presence and absence of estradiol with a nonlinear dose-response. MDA-MB-231 cells were also inhibited by TRF but with a linear dose-response. In the same study, the authors reported, after fractionation of TRF, γ- and δ-tocotrienols containing fractions were the most inhibitory ones. On the other hand, and according to the matter of the present paper, α-tocopherol had no effect on the growth rate of both cell lines. The authors ruled out the contribution of ER activity on the basis that TRF (and therefore tocotrienols) did not affect the expression of a gene, the breast cancer estrogen inducible sequence-trefoil factor 1 (pS2-TFF1), known to be driven by estradiol due to the presence of an ERα responsive elements in its promoter. The original conclusion has been reconsidered under the light of an "*omic*" approach utilized in subsequent studies performed by our laboratory. In a first study [16], utilizing cultured cells as an experimental model, we performed a cDNA array analysis of cancer-related gene expression in estrogen-dependent (MCF-7) and estrogen-independent (MDA-MB-231) human breast cancer cells. In this study, we utilized the best transcriptomic platform available at that time (1200 gene) that allowed us to conclude that the supplementation tocotrienol rich fraction form palm oil (TRF) was associated to the modulation of a number of genes encoding for proteins involved in cell cycle and therefore to inhibitory effects on cell growth and differentiation of the tumor cell lines. Previous studies had already demonstrated that α-tocopherol, the only tocopherol form present in TRF, had no effect on the induction of apoptosis in both cell lines, the only one having some pro-apoptotic activity being δ-tocopherol [17].

In a second study, based on a more sophisticated in vivo model [18], MCF-7 breast cancer cells were injected into athymic nude mice also fed with TRF. At the end of 20 wk dietary treatment there was a significant delay in the onset, incidence, and size of the tumors in nude mice supplemented with TRF compared with the controls. In addition, a cDNA array technique providing the differential expression of 1200 genes was performed on excised tumor tissues and, in agreement with the study performed on cultured cells, a significant number of genes was affected by TRF treatment. According to a "gene-ontology" analysis we identified a set of genes involved in the regulation of immune response and in the functional class of intracellular transducers/effectors/modulators. Data obtained in the course of these studies, further interrogated in silico provided a consistent hypothesis for a direct interaction of tocotrienols with estrogen pathway. In silico docking analysis, and in vitro binding-displacement test to purified ER protein demonstrated a high affinity of tocotrienols for ERβ but not for ERα. We also demonstrated that in ERβ-containing MDA-MB-231 breast cancer cells, tocotrienols, but not α-tocopherol, increase ERβ translocation into the nucleus and the expression of a spectrum of pro-apoptotic estrogen-responsive genes such as the Macrophage-inhibiting Cytokine-1 (MIC-1), the Early Growth Response-1 (EGR-1) and Cathepsin-D and accompanied by typically apoptotic-like alterations of cell morphology, DNA fragmentation, and caspase-3 activation [19]. These results have been corroborated and completed by a second study conducted on MCF-7 breast cancer cell, expressing both ERα and ERβ [20]. Furthermore, in this cell line, treatment with TRF and in particular with γ-tocotrienol, but not α-tocopherol, was associated with ERβ nuclear translocation and ER-dependent genes expression (MIC-1, EGR-1 and Cathepsin-D). At same time, the treatment induced a very evident inhibition of ERα activity finally leading to DNA fragmentation and apoptosis [20]. cDNA-array data obtained within this study also suggested the presence of an alternative pathway activated by γ- and δ-tocotrienols that will be presented below.

It is well known that ERβ activation can produce different cellular outcomes according to the balance between the "non genomic" signaling triggered by ERα and ERβ and other specific characteristics of cellular and tissue environment [21,22]. Accordingly, Nakaso and coworkers have more recently reported a cytoprotective, rather than pro-apoptotic effect of γ- and δ-tocotrienol in SH-SY5Y neuroblastoma cells, a cellular model addressing the pathogenesis of human Parkinson's disease [23]. In agreement with our original studies, this study confirmed that purified γ- and δ-tocotrienol bind to ERβ in vitro and that γ- and δ-tocotrienol were cytoprotective against Parkinson's disease-related toxicities such as 1-methyl-4-phenylpyridinium ion (MPP$^+$) thanks to a marked activation of the phosphoinositide-3-kinase/serine/threonine kinase-1 (PI3K/Akt) signaling pathway downstream to ERβ binding. The pivotal and functional role of γ- and δ-tocotrienol binding to ERβ was confirmed by ERβ silencing that was associated to the abrogation of cytoprotection and Akt phosphorylation [23].

The same authors also reported that the silencing of caveolin-1 and/or caveolin-2, candidates for the early events of signal transduction, prevented the cytoprotective effects of γ- and δ-tocotrienol, but not Akt phosphorylation. In agreement with us, the authors conclude that tocotrienols, in particular γ- and δ-tocotrienol, have distinct biological activities, unrelated to their putative antioxidant capacity, and that they can exert a specific neuro-protective activity mediated by ERβ binding and PI3K/Akt signaling activation [23].

Remarkably, one early study reported that tocotrienols induce apoptosis and a significant delay in cell growth in normal mammary cells obtained from midpregnant BALB/c mice [24]. However, neither the same authors, nor other research groups, have eventually confirmed these observations. Conversely, the same authors more recently, reported that γ-tocotrienol has a potent antiproliferative and cytotoxic effects by autophagy in MCF-7 and MDA-MD-231 cancer cell [25]. In contrast to the reports of Nakaso and coworkers mentioned above [23], γ-tocotrienol is reported to induce a reduction in PI3K/Akt/ mechanistic target of rapamycin kinase (mTOR) signaling and a corresponding increase in the Bax/Bcl-2 ratio, cleaved caspase-3, and cleaved poly (ADP-ribose) polymerase (PARP) levels in these cancer cell lines. These events suggest that γ-tocotrienol-induced autophagy may be involved in the initiation of apoptosis. In contrast, the same treatment was not found to increase autophagy marker expression in immortalized mouse (CL-S1) and human (MCF-10 A) normal mammary epithelial cell lines [25].

It seems therefore possible to speculate that γ-tocotrienol induces opposite effects in cancer and normal (or pseudo-normal) cells in agreement with the established perturbation of PI3K/Akt/mTOR signaling pathway in cancer cells [26]. Similarly, a specific effect of tocotrienols on Bax/Bcl2 mediated apoptosis has been also reported in prostate tumorigenesis in the transgenic adenocarcinoma mouse prostate (TRAMP) mouse model. In this study, a dietary supplementation with of a tocotrienol mixture induced a decrease in the levels of high-grade neoplastic lesions associated with an increased expression of proapoptotic proteins BAD a Bcl2 antagonist of cell death and cleaved caspase-3 and cell cycle regulatory proteins cyclin dependent kinase inhibitors p21 and p27 [27].

3. Tocotrienols as Inducers of Apoptosis via Endoplasmic Reticulum (EndoR) Stress

The first report of an EndoR stress mediated apoptosis by tocotrienols dates back about 10 years. Wali and coworkers [28] tested the effect of γ-tocotrienol in highly metastatic +SA rodent mammary epithelial cells. This study identified that 15–40 μM γ-tocotrienol induces a dose dependent apoptotic response paralleled by an increase of poly (ADP-ribose) polymerase (PARP)-cleavage and activation of a pathway distinctive of EndoR stress response, the protein kinase-like endoplasmic reticulum kinase/eukaryotic translational initiation factor/activating transcription factor 4 (PERK/eIF2α/ATF-4). γ-tocotrienol treatment also caused a large increase of key components of EndoR stress mediated apoptosis, tribbles 3 (TRB3) and C/emopamil binding protein (EBP) homologous protein (CHOP). The silencing of this latter gene significantly quenched γ-tocotrienol-induced PARP-cleavage and TRB3 expression. The same study [28] also reports that γ-tocotrienol treatment was associated with a decrease of full-length caspase-12 levels, indicating an activation of caspase-12 cleavage.

Park and collaborators have later confirmed these observations in a syngeneic mouse mammary tumor model and in cultured human breast cancer cells [29]. In the animal study, after the subcutaneous implantation of 66cl-4-GFP murine mammary tumor cells, female BALB/c mice were fed a diet enriched with purified α- and γ-tocotrienol to approximately provide 0.625 mg of each tocotrienol/mouse/day. The authors excluded a possible interference from Vitamin E supplementation, by adjusting Vitamin E in the diet with 30 IU/kg diet of DL-α-tocopheryl acetate to fully meet the animal's requirement. Dietary γ-tocotrienol suppressed tumor growth by inhibiting cell proliferation and inducing apoptosis. In fact, tumors excised from the γ-tocotrienol fed animals had significant more TUNEL (terminal deoxynucleotidyl transferase mediated nick end labeling assay) and less Ki-67 (a biomarker for cell division) positive cells (386% and 55%, respectively) compared to the animal fed the basal diet group. The same study reports no effects on tumor vascularization, and that the feeding with α-tocotrienol supplementation was not associated to comparable activities potential, at the tested dose. When utilizing cultured cells (MDA-MB-231 and MCF-7 human breast cancer cells and 66cl-4-GFP murine mammary tumor cells) the authors could confirm that α-, γ- and δ-tocotrienols induce apoptosis, independently of cell estrogen receptors' expression profile [29]. γ-tocotrienol was the most active form and was studied in deeper detail. The authors observed a significant activation of c-Jun NH(2)-terminal kinase (JNK) and p38-mitogen activated kinase-like protein (MAPK) accompanied by the upregulation of the expression death receptor 5 (DR5) and C/EBP homologous protein (CHOP), an endoplasmic reticulum (EndoR) stress marker, these events finally leading to the PARP, caspase-8, -9, and -3 cleavage. The silencing of JNK or p38 MAPK was associated to a reduced increase of DR5 and CHOP and partially inhibiting apoptosis. α-tocotrienol did not reduce tumor burden in vivo, but had inhibitory effects on colony formation in all three cell lines at relatively high concentrations in comparison to γ- and δ-tocotrienols, corroborating the evidence that the observed effects are strongly specific.

A further indication for the involvement of EndoR stress in tocotrienol-induced apoptosis was provided by Patacsil and coworkers [30] in MDA-MB 231 and MCF-7 breast cancer cells. This study demonstrates that 40 μM γ-tocotrienol induces PARP cleavage and caspase-7 activation in MCF-7 cells, accompanied by alterations in the expression of genes involved in cell growth and proliferation, cell death, cell cycle, cellular development, cellular movement and gene regulation. Among the categories studied by means of in silico Pathway Analysis the paper reports the modulation of signal transduction mediated by NRF-2-mediated "hormetic" oxidative stress response, transforming Growth Factor-β (TGF-β) signaling and EndoR stress response. Moreover, in agreement with other studies conducted on the same cellular models [31], the same study reports that MCF-7 and MDA-MB 231 cells respond to γ-tocotrienol by inducing the activation of PERK and pIRE1α pathways. The strong up-regulation of the Activating transcription factor 3 (ATF3) (16.8-fold) associated to EndoR stress and the abrogation of the apoptotic response after ATF3 silencing suggests this protein as a potential molecular target for γ-tocotrienol in breast cancer cells.

The specific activity of γ-tocotrienol on CHOP expression and EndoR stress has been also reported in several human malignant mesothelioma H2052 (sarcomatoid), H28 (epithelioid), H2452 (bi-phasic), and MSTO-211H (biphasic). In these cells, statins (atorvastatin and simvastatin) and 20 μM γ-tocotrienol have been observed to have synergistic effect on cell growth inhibition and acting through the inhibition of mevalonate pathway [32]. The authors report that, in H2052 and MSTO-211H cells, the treatment with γ-tocotrienol alone was sufficient to induce a significant increase of the expression of the endoplasmic reticulum stress markers CHOP and glucose regulated protein-78 (GRP-78). Conversely, the intrinsic apoptotic marker, caspase 3 activation, was induced only in the presence of statins. It is well known that the Bcl-2 family plays a pivotal role in apoptosis, either as an activator (through Bax and Bak) or as an inhibitor (Bcl-2 and Bcl-xL), Bcl-2 to Bax ratio is recognized as a key factor in the regulation of the apoptotic process or cell death [33]. According to this notion, this study also considered the possibility that γ-tocotrienol could affect Bax to Bcl2 ratio. Consistently with a previous report indicating that γ-tocotrienol induced a mitochondrial disruption pathway without affecting Bax/Bcl-2 expression in human breast cancer MDA-MB-231 cells [34], the Bax/Bcl2 ratio was not affected by any of the treatments.

We have also studied the involvement of EndoR stress on the effects of δ-tocotrienol on the growth of two different lines of human melanoma cells, BLM and A375 [35]. In agreement to observations obtained utilizing different tumor lines, the treatment with 5–20 μM δ-tocotrienol had a significant proapoptotic effect on both cell lines, involving the intrinsic apoptosis pathway. Very importantly, we observed no effect on the viability of normal human melanocytes. δ-tocotrienol effects were associated to the activation of the PERK/p-eIF2α/ATF4/CHOP, Inositol-requiring enzyme-1α (IRE1α) and caspase-4 strongly suggesting the involvement of EndoR stress upstream to the apoptotic response. We confirmed this hypothesis observing the quenching of the apoptotic response after a treatment with Salubrinal, an inhibitor of the EndoR stress. In the same study, in disagreement with observations reported by others [32,34], but in agreement with the observations by Barve and coworkers [27], we observed cytochrome c release from mitochondria, indicating a disruption of the mitochondrial outer membrane potential, associated to a significant increase of Bax/Bcl2 ratio. In vivo experiments performed in nude mice bearing A375 cells xenografts confirmed that a supplementation with dietary δ-tocotrienol (100 mg/kg daily, 5 days/week) up to 35 days, results in a reduced tumor volume and tumor mass and in significant delay of tumor progression [35].

A final confirmation of the ability of specific tocotrienol forms to activate EndoR stress is reported in a very recent study from our laboratory [31]. We reported a series of experiments based on the analysis of transcriptomic data obtained within our previous studies [18,19]. These data, interrogated by different bio-informatics tools, suggested the existence of an alternative pathway, activated by specific tocotrienols forms and leading to apoptosis, also in tumor cells not expressing ERs. This hypothesis was, in fact, confirmed utilizing HeLa cells, a line of human cervical cancer cells void of any canonical ER form. Once synchronized and treated either with the tocotrienol-rich fraction (α, γ and δ form) from palm oil (10–20 μg/mL) or with purified α-, γ- and δ-tocotrienol (5–20 μg/mL), HeLa cells underwent apoptosis which was accompanied by a significant expression of caspase 8, caspase 10 and caspase 12. Following the interrogation of additional data obtained from transcriptomic platforms, we considered the hypothesis that the administration of γ- and δ-tocotrienol could induce a release of Ca^{2+} from the EndoR. Accordingly, in living cells, we observed a significant activation of Ca-dependent signals. This event was followed by the expression and activation of IRE-1α and by the splicing and activation of X-box binding protein 1 (XBP-1), another molecule involved in the unfolded protein response, the core pathway coping with EndoR stress in eukaryotic cells, finally leading to apoptosis. Very importantly, and not very commonly indeed, our study considered α-tocopherol as a negative control, confirming that only γ- and δ-tocotrienol, not the proper Vitamin E D-α-tocopherol, are responsible for the observed effects. In the same paper [31], addressing the molecular mechanism underlying tocotrienols' activity, we wanted to speculate about the possible presence of a putative (orphan) receptor, possibly located at the level of the cellular membrane and able to accept tocotrienols and other estrogen mimetics as selective ligands. We based this speculation on the observation that the treatment with the specific ER inhibitor ICI-182,780 weakens the effects of tocotrienols on the upregulation of pro-apoptotic genes and the apoptotic response in cells lacking ERs. According to the chemo-physical characteristics of tocotrienols, the candidate downstream target(s) of the activity of this receptor could reasonably be located at the level of EndoR. The activation of this hypothetical (orphan) receptor would sequentially trigger EndoR stress, IRE-1 activation and XBP-1 splicing, finally inducing apoptosis.

The ability of γ-tocotrienol to induce apoptosis through activation of both the intrinsic and extrinsic pathway has been also confirmed in Jurkat cells, a human T-cell lymphoma [36]. In this study, γ-tocotrienol but not α-tocotrienol inhibited proliferation and induced apoptosis in this cell line in a dose dependent manner. The administration of 10–50 μM γ-tocotrienol resulted in elevated mitochondrial ROS production, JNK activation and suppression of extracellular regulated MAP kinase (ERK) and p38-MAPK. In agreement with our observations [31], γ-tocotrienol induced intracellular calcium release and, in agreement with others [27] a loss of mitochondrial membrane potential and cytochrome c release. These changes were found to be associated with an increase Bax/Bcl-xL expression rate and to an increased expression of Fas and FasL. Similarly to other studies [24,25,35], γ-tocotrienol had no effects on normal human peripheral blood mononuclear cells suggesting a specific cytotoxicity towards transformed lymphoma cells [36].

γ-Tocotrienol has also been demonstrated to induce paraptosis, a type of caspase independent programmed cell death, morphologically distinct from apoptosis, in that it displays cytoplasmic vacuolation and lacks of the typical apoptotic morphology. Zhang and collaborators [37] focused on the effects of δ-tocotrienol on human colon cancer SW620 cells and observed that δ-tocotrienol inhibits proliferation in a dose-dependent manner, correlated with cytoplasmic vacuolation possibly resulting from welling and fusion of mitochondria and/or EndoR. No changes in caspase 3 activation and other morphological (blebbing) or molecular markers of apoptosis were observed. The authors also report that δ-tocotrienol treatment (10 to 40 μM) was associated to a reduced β-catenin and wnt-1 expression and to lower cyclin D1, c-jun and matrix metallopeptidase 7 (MMP-7) protein levels, indicating the triggering of paraptosis-like cell death, with the suppression of the Wnt signaling pathway. The same authors have more recently reported similar results after the treatment with γ-tocotrienol both in SW620 and HCT-8 cells, another type of human ileocecal colorectal adenocarcinoma also associated to the suppression of Wnt signaling pathway [38].

4. Neuroprotection and Lipoxygenase Inhibition by α-Tocotrienol

While γ- and δ-tocotrienol have been demonstrated to induce specific responses in cancer cells with little or no effects on normal cells, α-tocotrienol has been frequently reported to play a protective activity on normal neuronal cells and, in general, on central nervous system. A study by Fukui and collaborators [39] utilized neuro2a cells, a line obtained from a spontaneous neuroblastoma in an albino strain A mouse frequently utilized as a model to study neurite outgrowth, neurotoxicity, Alzheimer disease and in general for neuronal tumourigenicity studies. This study, even appearing somehow an oversimplification in comparison with other studies based on a more complex experimental design, demonstrated that 5 μM α-tocotrienol counters the effects of the water-soluble free radical generator 2,2′-azobis(2-methylpropionamide) dihydrochloride (AAPH) on neurite degeneration and dynamics.

More specific research, directly targeting the protective effects of α-tocotrienol against neurodegeneration, has been conducted by the group of Sen and collaborators [40]. These authors originally reported the ability of nanomolar concentrations of α-tocotrienol, but not α-tocopherol to block glutamate-induced cell death. Glutamate is one of the most important neurotransmitters but, at high concentrations, it induces a rise of intracellular Ca^{2+} and mitochondrial dysfunction followed by cell death [41]. α-tocotrienol inhibits cell death by suppressing the activation of c-Src kinase and ERK phosphorylation in HT4, an embryonal carcinoma from metastatic lung with neural characteristics and expressing Simian vacuolating polyoma virus 40 (SV40) [42]. Very interestingly, the concentrations utilized in this study were 4–10-fold lower than levels detected in plasma of supplemented humans, indicating that α-tocotrienol regulates a specific signal transduction pathway insensitive to comparable concentrations of tocopherol, clearly suggesting that this activity is completely independent of its antioxidant capacity.

In a following study [43], the same group demonstrated that, in the same tumor cell line (HT4) and in immature primary cortical neurons, glutamate toxicity is mediated by the activation of neuronal 12-lipoxygenase that precede the production of peroxides, increased influx of Ca^{2+} and cell death. At lower physiologically achievable concentrations (nanomolar), α-tocotrienol displayed potent neuroprotective properties in both cell lines challenged by glutamate apparently interacting with 12-lipoxygenase, suppressing arachidonic acid metabolism. This observation was confirmed in vitro, by testing the activity of a purified enzyme in the presence of α-tocotrienol and by an in silico docking study suggesting that α-tocotrienol hampers the access of lipoxygenase substrate, to the enzyme catalytic site [43]. Even though a 12-lipoxygenase inhibiting activity has been reported also for α-tocopherol [44], the IC50 was in the order of μM, which is an order of magnitude higher than that reported for α-tocotrienol and, therefore, possibly due to a different mechanism.

The same group has further investigated the neuroprotective activity of α-tocotrienol in a sophisticated model based on single neuron microinjection technique on HT4 cells. The same study also reports the effects of dietary supplementation of α-tocotrienol in 12-lipoxigenase deficient mice undergoing surgically induced stroke and in spontaneously hypertensive rats [45]. The authors observed that very low quantities (sub-attomole) of α-tocotrienol, but not of α-tocopherol, protected isolated neurons from glutamate challenge. In agreement with the observations of a rapid 12-lipoxygenase tyrosine phosphorylation catalyzed by c-Src, lipoxygenase-deficient mice were more resistant to stroke-induced brain injury than their wild-type controls. Oral supplementation of α-tocotrienol to spontaneously hypertensive rats was associated to increased levels in the brain and to a protection against stroke-induced injury, associated with lower c-Src activation and 12-lipoxygenase phosphorylation at the stroke site.

In a study that followed [46], the same group addressed the comparison between the antioxidant-independent and -dependent neuroprotective properties of α-tocotrienol in HT4 cells cytotoxicity elicited by homocysteic acid and linoleic acid, respectively. Homocysteic acid administration was associated to a significant neurodegeneration presenting features similar to those previously observed in glutamate-induced neurotoxicity, in particular, the activation of c-Src and 12-lipoxygenase phosphorylation as early events. The administration of both homocysteic and linoleic acid was associated to an increased ratio of oxidized to reduced glutathione and to the triggering of an EndoR stress, as indicated by raised intracellular Ca^{2+} concentration and altered mitochondrial membrane potential eventually followed by cell death [46]. As the oxidative stress is considered a late event in apoptosis induced by homocysteic acid, the antioxidant component of α-tocotrienol protection was verified by inducing oxidative stress and cell death by linoleic acid. In all cases, the presence of nanomolar concentrations of α-tocotrienol, but not of α-tocopherol, was significantly protective. The authors concluded that α-tocotrienol protection against neurotoxicity is attributable to a combination of an antioxidant-independent and to antioxidant-dependent mechanisms.

The neuroprotective activity tocotrienols and, in particular, of α-tocotrienol has been confirmed by Osakada and coworkers [47]. In their study, the authors observed that a tocotrienol mixture from palm oil, further purified in order to eliminate α-tocopherol, purified α-, γ- and δ-tocotrienol (0.1–10 μM) significantly protects primary neuronal cells from rat striatum challenged by different pro-oxidants. Namely, hydrogen peroxide, a superoxide generating molecule (paraquat), nitric oxide donors (*S*-nitrosocysteine and 3-morpholinosydnonimine) and an inhibitor of glutathione synthesis L-buthionine-[*S*,*R*]-sulfoximine were utilized. In the same paper, only α-tocotrienol, but not γ- and δ-tocotrienol prevented the apoptosis induced by a protein kinase inhibitor, staurosporine, suggesting a specific mechanism underlying α-tocotrienol activity, totally independent of its putative antioxidant capacity. Conversely, α-tocopherol was ineffective in all cases, corroborating the evidence of a specific tocotrienol activity unrelated to their nucleophilic, antioxidant, properties.

It is worth noting that Wang and coworkers have elegantly demonstrated that the arylating oxidative product of γ-tocopherol, γ-tocopherol quinone, but not the non-arylating α-tocopherol quinone, induces EndoR stress affecting activation of PERK/CHOP proteins in N2A neuroblastoma cells [48]. As mentioned above, γ- and δ-tocotrienol have been also reported to affect this pathway in melanoma [35], mesothelioma H2052 and MSTO-211H cells [32], while our recent study indicated that EndoR stress induced by tocotrienols was mediated by IRE1 activation and not by PERK, suggesting a separate mechanism of action [31]. However, the paper by Wang and collaborators [48] clearly identifies the arylating activity of γ-tocopherol quinone as the mechanism underlying the activation of EndoR stress, while other investigations and in particular our study [31], did not take this event into consideration, on the basis of the indications of a receptor-mediated mechanism.

Finally, it is possible that the relative intracellular concentration and the specific localization of α-tocopherol and γ- and δ-tocotrienol can affect the cell response, eventually determining the final outcome. At present, very few studies [45] addressed this important issue and more investigation is surely warranted.

5. Other Tocotrienol-Triggered Cellular Responses

Besides ERβ activation and EndoR stress mediated apoptosis, γ-tocotrienol has also recently been observed to significantly alter sphingolipids composition in various types of cancer cells such as Human colon HCT-116, pancreatic PANC-1 and breast MCF-7 [49]. In particular, the authors observed a rapid elevation of dihydrosphingosine and dihydroceramides associated with increased cellular stress, phosphorylation of the mitogen-activated protein kinase, JNK and apoptosis. The inhibition of the de novo synthesis of sphingolipids was associated to a parallel inhibition of γ-tocotrienol-induced apoptosis and autophagy. γ-tocotrienol was demonstrated to inhibit dihydroceramide desaturase

(DEGS) activity but not its protein expression or de novo synthesis of sphingolipids. The increase of dihydroceramides paralleled by a relative decrease of ceramides coincides with the induction of apoptosis and autophagy. Overall, the authors conclude that γ-tocotrienol inhibits ceramide desaturase activity inducing an early elevation of dihydro-sphingolipids and late increase of saturated ceramide suggesting a role of these sphingolipid in tocotrienol-induced cell stress and death [49].

According to the evidence that several chemopreventive agents negatively affect cell growth by targeting p53 pathway, Agarwal and coworkers [50], utilized the human colon carcinoma RKO cell line, to investigate the effects of a TRF on the components of p53 signaling network. The authors report that the treatment with TRF results in a dose- and time- dependent inhibition of growth and colony formation associated to the induction of cyclin dependent kinase inhibitor 1A (WAF1)/p21, independent of cell cycle regulation and is transcriptionally upregulated in p53 dependent fashion.

In agreement with observations from our group and by others [27,35], but in disagreement with other reports [32,34], Agarwal and collaborators [50] reported that TRF induces the release of cytochrome c and induction of apoptotic protease-activating factor-1 upon a significant alteration of Bax/Bcl2 ratio. The altered ratio between the members of the Bax family triggered in turn by the activation of initiator caspase-9 followed by activation of effector caspase-3 finally lead to apoptosis characterized by chromatin condensation, DNA fragmentation and shrinkage of cell membrane.

Other biological activities have been reported specifically for γ- and δ-tocotrienol. A recent paper by Wong and coworkers [51] reports that the supplementation with 85 mg/kg/day γ- and δ-tocotrienol, but not with α-tocotrienol and α-tocopherol, improves cardiovascular functions in rats fed for 16 weeks a diet high in simple carbohydrates (fructose) and fats (beef tallow), in comparison to a corn starch based diet. In the same study [51], only δ-tocotrienol was associated with improved glucose tolerance, insulin sensitivity, lipid profile and abdominal adiposity. In the liver, these interventions reduced lipid accumulation, inflammatory infiltrates and plasma liver enzyme activities. In agreement with previous investigations [8,9], despite low or no detection of tocotrienols in plasma, the authors found detectable levels in heart, liver and adipose tissue confirming that chronic oral administration is associated to tocotrienols delivery and accumulation to these organs [51].

6. Conclusions

We have provided some references that strongly indicate that specific tocotrienols forms, and in particular α-, γ- and δ-, have biological activities distinct from that of tocopherols and in particular of D-α-tocopherol, the only molecule that, in our opinion should be considered as "Vitamin E".

In this review, we focused our interest on the ability of γ- and δ-tocotrienol to induce cell growth arrest and apoptosis in different tumor cell types, by specific mechanisms apparently not shared by tocopherols. Even though other specific effects have been reported, the ability to act as "productive" ligands of ERβ and to trigger Ca^{2+} release from EndoR, seem the most evident features of δ- and γ-tocotrienol, eventually associated with the expression of pro-apoptotic genes, finally leading to apoptosis or paraptosis. Interestingly, the pathway associated to cell death depends, at least in part, on the tumor type and pro-apoptotic effects have not been reported in non-tumor cells.

These observations not only suggest that specific forms of tocotrienols (in this case, δ- and γ-tocotrienol) have a specific capacity in inducing programmed death in tumor cells, but also that their specific activity occurs through diverse specific mechanisms, according to the cell type. Under this perspective, specific forms of tocotrienols and, in particular, δ-tocotrienol could be proposed as an expedient, potentially effective option for novel chemopreventive/therapeutic strategies in different types of solid tumors. On the other hand, α-tocotrienol displays a specific neuroprotective capacity due to its ability to inhibit 12-lipoxygenase, a key enzyme in the execution of cellular death induced by glutamate and by homocysteic acid. Taken together, current findings strongly indicate that tocotrienols may have a significant role in different specific pathological conditions, which are not necessarily affected by Vitamin E (α-tocopherol). Figure 2 shows the different molecular targets of tocotrienols in the representative cell types we have considered in this review, namely cancer cells expressing

ER, cancer cells not expressing ER and normal neuronal cells submitted to specific stressors such as glutamate or homocysteic acid.

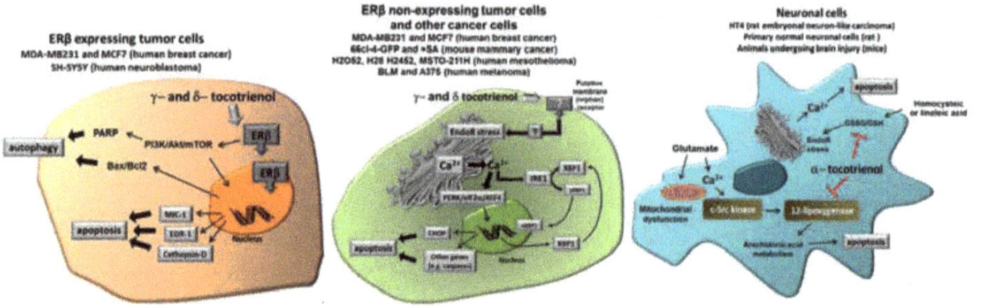

Figure 2. Different molecular targets of tocotrienols in the representative cell types we have considered in this review, namely cancer cells expressing Estrogen Receptor (ER), cancer cells not expressing ER and normal neuronal cells submitted to specific stressors such as glutamate or homocysteic acid. See text for more details. (Modified from [31]).

In spite of the relative abundance of studies dealing with the effects associated to the administration of tocotrienols, either in cultured cells or in experimental animals, very few studies considered the mechanism of action. In fact, very often, the terms "mechanism" and "effects" are utilized interchangeably, when they should refer to different concepts. In our opinion, the term mechanism should be solely utilized to describe the molecular events leading to the observed effects, such as (e.g.,) in the case of the tocotrienol binding to ERβ (mechanism) leading to the transcriptional activation of gene expression (effects). To our knowledge, our laboratory and few others [20,43] are the only ones that addressed this issue by the combination of in silico *docking* analysis, to build up the hypothesis of ER (either other nuclear receptors) binding or the inhibition of 12-lipoxygenase, respectively, eventually confirmed in vitro, in cultured cells and in animals. Further studies, dealing with the understanding of the molecular (mechanistic) basis underlying the observed effects of specific tocotrienol forms are surely warranted, in order to define the real limits of these molecules as a possible alternative for cancer treatment or co-adjuvant in cancer treatment.

Besides this, a major problem in the correct interpretation of some of studies reported herein is, in general, the absence of a treatment, either in cultured cells or in animals, with α-tocopherol. In fact, for instance, studies utilizing mixtures administered to cultured cells or to animals contained significant amounts of α-tocopherol. This experimental weakness make it unclear if the observed effects can be totally ascribed to tocotrienols or also to α-tocopherol, as several authors tend to claim. However, studies correctly performing this specific "negative" control published by our laboratory [19,20,31] and by others [29,40,42,43,45–47] provide indirect but solid indications about the differential ability of α-, γ- and δ-tocotrienol, but not α-tocopherols to trigger specific cellular response in different cell types.

7. Practical Consequences at Nutritional Level

If we accept that each tocotrienol form has a distinct, specific biological activity, different cellular targets and molecular mechanisms of action, significantly different from tocopherol and more specifically from α-tocopherol, we must necessarily stop pooling them together on solely the basis of their putative antioxidant properties. Consequently, tocotrienols should be excluded from the "Vitamin E family". This option would probably imply some consequences in the assessment of dietary Vitamin E adequacy even though more studies are surely needed to investigate their relative body distribution and plasma levels in normal dietary conditions. More importantly, the distinction between Vitamin E (α-tocopherol) and specific non-α-tocopherol molecules will allow for investigations aiming to evaluate a separate biological variable determinant in the relationship between diet and health.

Finally, a system biology approach, accompanied by high throughput methodologies, appears an indispensable tool for a better understanding of the role of molecules of nutritional interest in human health and disease. Classical "hypothesis driven studies", even when very well conducted and designed, may risk consolidating original weaknesses, possibly leading to a "*narrowing of perspective and from there false expectation*".

Conflicts of Interest: The authors declare no conflicts of interest. The views and opinions expressed in this article are those of the authors and do not necessarily reflect the official policy or position of our institution.

References

1. Machlin, L.J. *Vitamin E*, 2nd ed.; Marcel Dekker, Inc.: New York, NY, USA, 1991.
2. Board, F.A.N.; Medicine, I.O. *Dietary Reference Intakes for Vitamin C, Vitamin E, Selenium, and Carotenoids*; The National Academies Press: Washington, DC, USA, 2000.
3. Schmölz, L.; Birringer, M.; Lorkowski, S.; Wallert, M. Complexity of vitamin E metabolism. *World J. Biol. Chem.* **2016**, *7*, 14–43. [CrossRef] [PubMed]
4. Shibata, A.; Nakagawa, K.; Tsuduki, T.; Miyazawa, T. Alpha-Tocopherol suppresses antiangiogenic effect of delta-tocotrienol in human umbilical vein endothelial cells. *J. Nutr. Biochem.* **2015**, *26*, 345–350. [CrossRef] [PubMed]
5. Jiang, Q. Natural forms of vitamin E: Metabolism, antioxidant, and anti-inflammatory activities and their role in disease prevention and therapy. *Free Radic. Biol. Med.* **2014**, *72*, 76–90. [CrossRef] [PubMed]
6. Hosomi, A.; Arita, M.; Sato, Y.; Kiyose, C.; Ueda, T.; Igarashi, O.; Arai, H.; Inoue, K. Affinity for alpha-tocopherol transfer protein as a determinant of the biological activities of vitamin E analogs. *FEBS Lett.* **1997**, *409*, 105–108. [CrossRef]
7. Manor, D.; Morley, S. The alpha-tocopherol transfer protein. *Vitam. Horm.* **2007**, *76*, 45–65. [PubMed]
8. Patel, V.; Khanna, S.; Roy, S.; Ezziddin, O.; Sen, C.K. Natural vitamin E alpha-tocotrienol: Retention in vital organs in response to long-term oral supplementation and withdrawal. *Free Radic. Res.* **2006**, *40*, 763–771. [CrossRef] [PubMed]
9. Khanna, S.; Patel, V.; Rink, C.; Roy, S.; Sen, C.K. Delivery of orally supplemented alpha-tocotrienol to vital organs of rats and tocopherol-transport protein deficient mice. *Free Radic. Biol. Med.* **2005**, *39*, 1310–1319. [CrossRef] [PubMed]
10. Qureshi, A.A.; Khan, D.A.; Silswal, N.; Saleem, S.; Qureshi, N. Evaluation of Pharmacokinetics, and Bioavailability of Higher Doses of Tocotrienols in Healthy Fed Humans. *J. Clin. Exp. Cardiol.* **2016**, *7*, 434. [CrossRef] [PubMed]
11. Fairus, S.; Nor, R.M.; Cheng, H.M.; Sundram, K. Alpha-tocotrienol is the most abundant tocotrienol isomer circulated in plasma and lipoproteins after postprandial tocotrienol-rich vitamin E supplementation. *Nutr. J.* **2012**, *11*, 5. [CrossRef] [PubMed]
12. Patel, V.; Rink, C.; Gordillo, G.M.; Khanna, S.; Gnyawali, U.; Roy, S.; Shneker, B.; Ganesh, K.; Phillips, G.; More, J.L.; et al. Oral tocotrienols are transported to human tissues and delay the progression of the model for end-stage liver disease score in patients. *J. Nutr.* **2012**, *142*, 513–519. [CrossRef] [PubMed]
13. Peh, H.Y.; Tan, W.S.; Liao, W.; Wong, W.S. Vitamin E therapy beyond cancer: Tocopherol versus tocotrienol. *Pharmacol. Ther.* **2016**, *162*, 152–169. [CrossRef] [PubMed]
14. Nesaretnam, K.; Stephen, R.; Dils, R.; Darbre, P. Tocotrienols inhibit the growth of human breast cancer cells irrespective of estrogen receptor status. *Lipids* **1998**, *33*, 461–469. [CrossRef] [PubMed]
15. Mosselman, S.; Polman, J.; Dijkema, R. ER beta: Identification and characterization of a novel human estrogen receptor. *FEBS Lett.* **1996**, *392*, 49–53. [CrossRef]
16. Nesaretnam, K.; Ambra, R.; Selvaduray, K.R.; Radhakrishnan, A.; Reimann, K.; Razak, G.; Virgili, F. Tocotrienol-rich fraction from palm oil affects gene expression in tumors resulting from MCF-7 cell inoculation in athymic mice. *Lipids* **2004**, *39*, 459–467. [CrossRef] [PubMed]
17. Yu, W.; Simmons-Menchaca, M.; Gapor, A.; Sanders, B.G.; Kline, K. Induction of apoptosis in human breast cancer cells by tocopherols and tocotrienols. *Nutr. Cancer* **1999**, *33*, 26–32. [CrossRef] [PubMed]

18. Nesaretnam, K.; Ambra, R.; Selvaduray, K.R.; Radhakrishnan, A.; Canali, R.; Virgili, F. Tocotrienol-rich fraction from palm oil and gene expression in human breast cancer cells. *Ann. N. Y. Acad. Sci.* **2004**, *1031*, 143–157. [CrossRef] [PubMed]
19. Comitato, R.; Nesaretnam, K.; Leoni, G.; Ambra, R.; Canali, R.; Bolli, A.; Marino, M.; Virgili, F. A novel mechanism of natural vitamin E tocotrienol activity: Involvement of ERbeta signal transduction. *Am. J. Physiol. Endocrinol. Metab.* **2009**, *297*, E427–E437. [CrossRef] [PubMed]
20. Comitato, R.; Leoni, G.; Canali, R.; Ambra, R.; Nesaretnam, K.; Virgili, F. Tocotrienols activity in MCF-7 breast cancer cells: Involvement of ERbeta signal transduction. *Mol. Nutr. Food Res.* **2010**, *54*, 669–678. [CrossRef] [PubMed]
21. Acconcia, F.; Totta, P.; Ogawa, S.; Cardillo, I.; Inoue, S.; Leone, S.; Trentalance, A.; Muramatsu, M.; Marino, M. Survival versus apoptotic 17beta-estradiol effect: Role of ER alpha and ER beta activated non-genomic signaling. *J. Cell. Physiol.* **2005**, *203*, 193–201. [CrossRef] [PubMed]
22. Lewis-Wambi, J.S.; Jordan, V.C. Estrogen regulation of apoptosis: How can one hormone stimulate and inhibit? *Breast Cancer Res.* **2009**, *11*, 206. [CrossRef] [PubMed]
23. Nakaso, K.; Tajima, N.; Horikoshi, Y.; Nakasone, M.; Hanaki, T.; Kamizaki, K.; Matsura, T. The estrogen receptor beta-PI3K/Akt pathway mediates the cytoprotective effects of tocotrienol in a cellular Parkinson's disease model. *Biochim. Biophys. Acta* **2014**, *1842*, 1303–1312. [CrossRef] [PubMed]
24. McIntyre, B.S.; Briski, K.P.; Tirmenstein, M.A.; Fariss, M.W.; Gapor, A.; Sylvester, P.W. Antiproliferative and apoptotic effects of tocopherols and tocotrienols on normal mouse mammary epithelial cells. *Lipids* **2000**, *35*, 171–180. [CrossRef] [PubMed]
25. Tiwari, R.V.; Parajuli, P.; Sylvester, P.W. Gamma-Tocotrienol-induced autophagy in malignant mammary cancer cells. *Exp. Biol. Med.* **2014**, *239*, 33–44. [CrossRef] [PubMed]
26. Porta, C.; Paglino, C.; Mosca, A. Targeting PI3K/Akt/mTOR Signaling in Cancer. *Front. Oncol.* **2014**, *4*, 64. [CrossRef] [PubMed]
27. Barve, A.; Khor, T.O.; Reuhl, K.; Reddy, B.; Newmark, H.; Kong, A.N. Mixed tocotrienols inhibit prostate carcinogenesis in TRAMP mice. *Nutr. Cancer* **2010**, *62*, 789–794. [CrossRef] [PubMed]
28. Wali, V.B.; Bachawal, S.V.; Sylvester, P.W. Endoplasmic reticulum stress mediates gamma-tocotrienol-induced apoptosis in mammary tumor cells. *Apoptosis* **2009**, *14*, 1366–1377. [CrossRef] [PubMed]
29. Park, S.K.; Sanders, B.G.; Kline, K. Tocotrienols induce apoptosis in breast cancer cell lines via an endoplasmic reticulum stress-dependent increase in extrinsic death receptor signaling. *Breast Cancer Res. Treat.* **2010**, *124*, 361–375. [CrossRef] [PubMed]
30. Patacsil, D.; Tran, A.T.; Cho, Y.S.; Suy, S.; Saenz, F.; Malyukova, I.; Ressom, H.; Collins, S.P.; Clarke, R.; Kumar, D. Gamma-tocotrienol induced apoptosis is associated with unfolded protein response in human breast cancer cells. *J. Nutr. Biochem.* **2012**, *23*, 93–100. [CrossRef] [PubMed]
31. Comitato, R.; Guantario, B.; Leoni, G.; Nesaretnam, K.; Ronci, M.B.; Canali, R.; Virgili, F. Tocotrienols induce endoplasmic reticulum stress and apoptosis in cervical cancer cells. *Genes Nutr.* **2016**, *11*, 32. [CrossRef] [PubMed]
32. Tuerdi, G.; Ichinomiya, S.; Sato, H.; Siddig, S.; Suwa, E.; Iwata, H.; Yano, T.; Ueno, K. Synergistic effect of combined treatment with gamma-tocotrienol and statin on human malignant mesothelioma cells. *Cancer Lett.* **2013**, *339*, 116–127. [CrossRef] [PubMed]
33. Lindqvist, L.M.; Vaux, D.L. BCL2 and related prosurvival proteins require BAK1 and BAX to affect autophagy. *Autophagy* **2014**, *10*, 1474–1475. [CrossRef] [PubMed]
34. Takahashi, K.; Loo, G. Disruption of mitochondria during tocotrienol-induced apoptosis in MDA-MB-231 human breast cancer cells. *Biochem. Pharmacol.* **2004**, *67*, 315–324. [CrossRef] [PubMed]
35. Montagnani Marelli, M.; Marzagalli, M.; Moretti, R.M.; Beretta, G.; Casati, L.; Comitato, R.; Gravina, G.L.; Festuccia, C.; Limonta, P. Vitamin E delta-tocotrienol triggers endoplasmic reticulum stress-mediated apoptosis in human melanoma cells. *Sci. Rep.* **2016**, *6*, 30502. [CrossRef] [PubMed]
36. Wilankar, C.; Khan, N.M.; Checker, R.; Sharma, D.; Patwardhan, R.; Gota, V.; Sandur, S.K.; Devasagayam, T.P. γ-Tocotrienol induces apoptosis in human T cell lymphoma through activation of both intrinsic and extrinsic pathways. *Curr. Pharm. Des.* **2011**, *17*, 2176–2189. [CrossRef] [PubMed]
37. Zhang, J.S.; Li, D.M.; He, N.; Liu, Y.H.; Wang, C.H.; Jiang, S.Q.; Chen, B.Q.; Liu, J.R. A paraptosis-like cell death induced by delta-tocotrienol in human colon carcinoma SW620 cells is associated with the suppression of the Wnt signaling pathway. *Toxicology* **2011**, *285*, 8–17. [CrossRef] [PubMed]

38. Zhang, J.S.; Li, D.M.; Ma, Y.; He, N.; Gu, Q.; Wang, F.S.; Jiang, S.Q.; Chen, B.Q.; Liu, J.R. Gamma-Tocotrienol induces paraptosis-like cell death in human colon carcinoma SW620 cells. *PLoS ONE* **2013**, *8*, e57779.

39. Fukui, K.; Sekiguchi, H.; Takatsu, H.; Koike, T.; Urano, S. Tocotrienol prevents AAPH-induced neurite degeneration in neuro2a cells. *Redox Rep.* **2013**, *18*, 238–244. [CrossRef] [PubMed]

40. Sen, C.K.; Khanna, S.; Roy, S. Tocotrienol: The natural vitamin E to defend the nervous system? *Ann. N. Y. Acad. Sci.* **2004**, *1031*, 127–142. [CrossRef] [PubMed]

41. Kritis, A.A.; Stamoula, E.G.; Paniskaki, K.A.; Vavilis, T.D. Researching glutamate—Induced cytotoxicity in different cell lines: A comparative/collective analysis/study. *Front. Cell. Neurosci.* **2015**, *9*, 91. [CrossRef] [PubMed]

42. Sen, C.K.; Khanna, S.; Roy, S.; Packer, L. Molecular basis of vitamin E action. Tocotrienol potently inhibits glutamate-induced pp60(c-Src) kinase activation and death of HT4 neuronal cells. *J. Biol. Chem.* **2000**, *275*, 13049–13055. [CrossRef] [PubMed]

43. Khanna, S.; Roy, S.; Ryu, H.; Bahadduri, P.; Swaan, P.W.; Ratan, R.R.; Sen, C.K. Molecular basis of vitamin E action: Tocotrienol modulates 12-lipoxygenase, a key mediator of glutamate-induced neurodegeneration. *J. Biol. Chem.* **2003**, *278*, 43508–43515. [CrossRef] [PubMed]

44. Reddanna, P.; Rao, M.K.; Reddy, C.C. Inhibition of 5-lipoxygenase by vitamin E. *FEBS Lett.* **1985**, *193*, 39–43. [CrossRef]

45. Khanna, S.; Roy, S.; Slivka, A.; Craft, T.K.; Chaki, S.; Rink, C.; Notestine, M.A.; DeVries, A.C.; Parinandi, N.L.; Sen, C.K. Neuroprotective properties of the natural vitamin E alpha-tocotrienol. *Stroke* **2005**, *36*, 2258–2264. [CrossRef] [PubMed]

46. Khanna, S.; Roy, S.; Parinandi, N.L.; Maurer, M.; Sen, C.K. Characterization of the potent neuroprotective properties of the natural vitamin E alpha-tocotrienol. *J. Neurochem.* **2006**, *98*, 1474–1486. [CrossRef] [PubMed]

47. Osakada, F.; Hashino, A.; Kume, T.; Katsuki, H.; Kaneko, S.; Akaike, A. Alpha-tocotrienol provides the most potent neuroprotection among vitamin E analogs on cultured striatal neurons. *Neuropharmacology* **2004**, *47*, 904–915. [CrossRef] [PubMed]

48. Wang, X.; Thomas, B.; Sachdeva, R.; Arterburn, L.; Frye, L.; Hatcher, P.G.; Cornwell, D.G.; Ma, J. Mechanism of arylating quinone toxicity involving Michael adduct formation and induction of endoplasmic reticulum stress. *Proc. Natl. Acad. Sci. USA* **2006**, *103*, 3604–3609. [CrossRef] [PubMed]

49. Jang, Y.; Rao, X.; Jiang, Q. Gamma-tocotrienol profoundly alters sphingolipids in cancer cells by inhibition of dihydroceramide desaturase and possibly activation of sphingolipid hydrolysis during prolonged treatment. *J. Nutr. Biochem.* **2017**, *46*, 49–56. [CrossRef] [PubMed]

50. Agarwal, M.K.; Agarwal, M.L.; Athar, M.; Gupta, S. Tocotrienol-rich fraction of palm oil activates p53, modulates Bax/Bcl2 ratio and induces apoptosis independent of cell cycle association. *Cell Cycle* **2004**, *3*, 205–211. [CrossRef] [PubMed]

51. Wong, W.Y.; Ward, L.C.; Fong, C.W.; Yap, W.N.; Brown, L. Anti-inflammatory gamma- and delta-tocotrienols improve cardiovascular, liver and metabolic function in diet-induced obese rats. *Eur. J. Nutr.* **2017**, *56*, 133–150. [CrossRef] [PubMed]

antioxidants

MDPI

Review

Vitamin E as an Antioxidant in Female Reproductive Health

Siti Syairah Mohd Mutalip [1,*], Sharaniza Ab-Rahim [2] and Mohd Hamim Rajikin [2]

[1] Faculty of Pharmacy, Universiti Teknologi MARA (UiTM) Puncak Alam Campus, Selangor 42300, Malaysia
[2] Faculty of Medicine, Universiti Teknologi MARA (UiTM) Sg. Buloh Campus, Selangor 42300, Malaysia;
    sharaniza_abrahim@salam.uitm.edu.my (S.A.-R.); hamim400@salam.uitm.edu.my (M.H.R.)
* Correspondence: syairah@puncakalam.uitm.edu.my; Tel.: +60-03-3258-4840

Received: 11 December 2017; Accepted: 25 January 2018; Published: 26 January 2018

Abstract: Vitamin E was first discovered in 1922 as a substance necessary for reproduction. Following this discovery, vitamin E was extensively studied, and it has become widely known as a powerful lipid-soluble antioxidant. There has been increasing interest in the role of vitamin E as an antioxidant, as it has been discovered to lower body cholesterol levels and act as an anticancer agent. Numerous studies have reported that vitamin E exhibits anti-proliferative, anti-survival, pro-apoptotic, and anti-angiogenic effects in cancer, as well as anti-inflammatory activities. There are various reports on the benefits of vitamin E on health in general. However, despite it being initially discovered as a vitamin necessary for reproduction, to date, studies relating to its effects in this area are lacking. Hence, this paper was written with the intention of providing a review of the known roles of vitamin E as an antioxidant in female reproductive health.

Keywords: vitamin E; reproduction; antioxidant; tocopherol; tocotrienol

---

1. Vitamin E

Vitamin E was first discovered by Evans and Bishop in 1922, and it was initially denoted as an "anti-sterility factor X" that was necessary for reproduction [1]. Since then, vitamin E has been well characterized as a powerful lipid-soluble antioxidant through extensive research. The antioxidant activities of vitamin E were reported following findings on its ability to scavenge reactive oxygen species (ROS) in cellular membranes [2–4].

1.1. Sources of Vitamin E

Vitamin E, which consists of a mixture of tocopherols (TOCs) and tocotrienols (TCTs), is available in a number of foods and plants, ranging from edible oils to nuts. Some vitamin E-containing foods include wheat, rice bran, barley, oat, coconut, palm and annatto [5,6]. Other sources include rye, amaranth, walnut, hazelnut, poppy, safflower, maize and the seeds of grape and pumpkins. Vitamin E derivatives have also been detected in human milk [7] and palm dates (*Phoenix canariensis*) [8]. Among the many sources of vitamin E, rice bran, palm oil and annatto oil have been described as the richest sources of TCTs [9].

1.2. Structure of Vitamin E

Vitamin E consists of a mixture of tocopherols (TOCs) and tocotrienols (TCTs) that are synthesized by plants from homogenestic acid [10]. These substances are present in eight different homologues; namely, α-tocopherol, β-tocopherol, γ-tocopherol, δ-tocopherol, α-tocotrienol, β-tocotrienol, γ-tocotrienol and δ-tocotrienol [11]. The four TOC homologues (α-, β-, γ-, δ-TOC) have a fully saturated 16-carbon isoprenoid sidechain, while TCT homologues have a similar isoprenoid chain, containing three double bonds (an unsaturated side chain). The TOC homologues are named with respect to the position and number of the methyl groups on the phenol ring. The α-, β-, γ- and δ-homologues contain three, two, two and one methyl groups, respectively (Figure 1). These structural differences and the isomerism determine the biological activity, with α-homologues being the most biologically active [12]. However, it has been reported that light, temperature, and oxygen availability could promote rancidity in vegetable oils [13]. According to a study [13], soybean oil that was stored in the dark for 56 days had increased peroxide value. In addition, its exposure to light in a 12 h light/darkness cycle over for 56 days resulted in an increase in peroxide values of around 1473%.

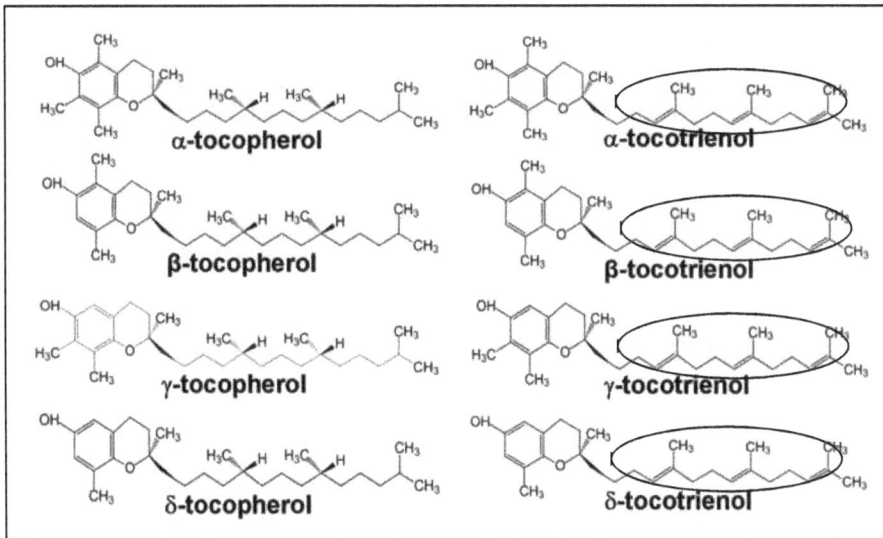

Figure 1. Structure differences between tocopherols (TOCs) and tocotrienols (TCTs). TOCs have saturated side chains, while TCTs have unsaturated side chains. The latter are shown by the presence of three double bonds in TCTs (circled) [14].

2. Reproductive Disorders: The Risk Factors

A number of risk factors contributing to reproductive- and pregnancy-related disorders have been previously reported [15–17]. These factors are generally categorized into two major groups: environmental and lifestyle factors. Examples of major environmental pollutants include hazardous man-made chemicals, industrial discharge, agricultural run-off, human and animal waste, municipal and domestic effluents, and spillage of vessels and oil spills [17]. Exposure to these pollutants during the time of periconceptional period (periconceptional period refers to the time of preconception, conception, implantation, placentation and embryogenesis (or organogenesis) stages of pregnancy) were reported to have adverse effects on the development of conceptus and the neonatal health [15]. These include the risks of embryonic mortality and fetal loss, intrauterine growth restriction (IUGR), birth defects, childhood diseases, premature sexual maturation and a few types of adult cancers [15]. Additionally, Rider et al. [16] also reported that exposure of conceptus to multiple environmental pollutants *in utero* during pregnancy could affect embryonic implantation and the developmental course in a cumulative dose-additive manner.

Exposure to multicomponent mixtures of endocrine-disturbing chemicals may act as hormone mimics or antagonists, leading to the disruption of estrogen, androgen and other hormonal pathways [18]. Furthermore, exposure to multiple environmental pollutants may also result in reactive oxygen species (ROS)-induced oxidative stress (OS) [19–21]. The presence of high levels of OS may be a risk factor for a number of pregnancy-related disorders, such as embryonic mortality, early spontaneous abortion, IUGR, fetal death, premature delivery and low birth weight [22,23].

Lifestyle factors represent another category of major risk factors for reproductive and pregnancy-related disorders. Unhealthy lifestyle behaviors, including cigarette smoking, alcohol consumption, and/or drug abuse, have negative impacts, particularly on female fertility [24,25]. The underlying mechanism of the developmental defects following these unhealthy lifestyle behaviors is mainly a result of an increase in ROS production and associated OS-induced cellular damage [26]. There are also extensive epidemiological studies which have reported on a number of factors such as exposure to tobacco and alcohol, diet, stress, and gestational diabetes as the factors influencing fetal development including miscarriages [27–29].

Much evidence-based epidemiological, clinical, and experimental data on the adverse effects of cigarette smoking on female reproductive health has been reported [30]. The effects of smoking on steroidogenesis, folliculogenesis embryo transport, endometrial receptivity, endometrial angiogenesis, uterine blood flow, and uterine myometrium, all of which are related to delayed or failed implantation and pregnancy loss, have been reported. This is in line with an animal study on the effects of alcohol on reproductive health and pregnancy that indicated that prenatal exposure to ethanol in rats induced hypothalamic OS and neuroendocrine alterations in offspring [31]. Furthermore, excess ethanol administration to pregnant mice [32] and rats [33] caused disturbances in embryogenesis and increased the rate of malformations and fetal death by inducing high levels of OS. Medication use or drug abuse during pregnancy has also been associated with OS [34,35]. Phenytoin [36], thalidomide [37], valproic acid [38], almokalant, dofetilide, cisapride and astemizole [39] are the examples of identified medical drugs known to induce OS and affect the embryonic development leading to birth defects.

To explain further, maternal smoking during pregnancy has been widely recognized as one of the most common factors of reproductive- and pregnancy-related disorders. Cigarette smoke contains a complex mixture of numerous toxic constituents including nicotine, polycyclic aromatic hydrocarbons, and cadmium [30,40]. The different constituents of the mixture cause an increased level of OS and adversely affect the cell proliferation and differentiation during embryonic development in pregnant female smokers [41]. This is supported by studies on the effects of maternal smoking during pregnancy, showing that cigarette smoking is associated with spontaneous abortion [42], placenta previa and placental abruption [43–45], low birth weight and preterm birth [46–48], stillbirth [49,50] and sudden infant death syndrome (SIDS) [51].

One of the most important cigarette smoke constituents, nicotine, has been reported to reduce fertility during adulthood in women [52]. In addition, cotinine (a metabolite of nicotine), cadmium, and benzo[a]pyrene have also been detected in the follicular fluid of smoking women [53–55], suggesting that the chemicals present in cigarette smoke can accumulate in the ovary. The results of these studies suggested that smoking women might develop impaired fertility, resulting from the combination of deteriorated oocyte function and viability [53,56,57].

In addition, laboratories studies have indicated that maternal exposure to cigarette smoke or cigarette smoke condensate (CSC) for 4 weeks results in the increased oocyte fragmentation or delayed fertilization, thus reducing the embryonic development to blastocysts in vitro [58]. Additionally, fragmented oocytes also showed increased production of ROS. Another study on the effects of nicotine on early embryogenesis in murine embryos reported that embryos treated with 3–6 µM of nicotine were smaller than control embryos [59]. Meanwhile, embryos treated with 6 µM of nicotine showed severe defects in the posterior trunk, resembling caudal dysplasia [59]. In addition, excessive apoptosis was also observed in the deformed structures and this was associated with the increased levels of ROS [59]. Nicotine exposure during fetal and neonatal development was also reported to cause reduction in fertility, dysregulation in ovarian steroidogenesis, and alterations in follicle dynamics in female offspring [60]. This has been further supported by another study which reported that treatment with 5 mg/mL nicotine beginning from day 1 of pregnancy throughout gestation decreased the pregnancy rates by 33.3% in Sprague-Dawley rats [61]. Another study by Rajikin et al. [62] reported that the ultrastructure of oocytes from nicotine-exposed mice showed a non-spherical shape with rough surface and torn *zona pellucida*. In addition, treatment with 5 mg/kg nicotine for 30 days increased the apoptosis rate in oocytes [63]. Meanwhile, productions of hatched blastocysts were decreased following injection with 1 mg/kg and 3 mg/kg of nicotine, and embryonic development ceased at the morula stage following exposure to 5 mg/kg of nicotine [64]. This was in line with the work of Phoebe et al. [65], which showed that after 12 weeks of cigarette smoking (directly to the lungs) in mice, the retrieved oocytes had a significantly thicker *zona pellucida*, and also shorter and wider meiotic spindles.

Oxidative Stress (OS) as One of the Risk Factors in Reproductive Disorders

Oxidative stress (OS) is widely recognized as the key element in the pathogenesis of most of the diseases [66], and occurs when there is an imbalance in the presence of antioxidants and pro-oxidants [20,22,67]. Excess pro-oxidants induce OS by either generating reactive oxygen species (ROS) or by inhibiting antioxidant systems [68]. ROS are highly reactive and unstable. They acquire electrons from nucleic acids, lipids, proteins, carbohydrates, or any other nearby molecule causing a string of chain reactions to become stable. These chain reactions result in cellular damage and diseases [69].

In the female reproductive system, ROS can impair cellular functions and subsequently interrupt intracellular homeostasis and furthermore lead to cell damages. The presence of excess ROS can influence early embryonic development through modification of the key transcription factors that modify gene expressions [70]. High concentrations of ROS in the female reproductive tract could also negatively affect the fertilization of oocytes and cause inhibition of embryonic implantation [71,72]. Additionally, earlier studies reported that OS is involved in defective and retarded embryonic development due to OS-induced cell-membrane damage, DNA damage, and apoptosis [73,74]. Apoptosis results in the formation of fragmented embryos which have limited chances of implantation and growth [75].

Previous studies on the effect of OS during the periconceptional period have shown that the placenta could be the key source of OS because of the high metabolic rate and increase in the mitochondrial activities [76,77]. During the first trimester, placental tissues contain low concentrations and activities of principle antioxidant enzymes including catalase, glutathione peroxidase, and superoxide dismutase. This condition may expose the embryonic trophoblast cells to oxygen-mediated damage [78]. An earlier study reported that due to the increase in the oxygen tension during the onset of maternal arterial flow during the beginning of second trimester, a burst of OS was observed in the placenta [79]. The study suggested that this oxidative injury could adversely impair placental remodeling and functions that would subsequently affect the course of gestation [79]. This was further supported by Jauniaux et al. [80], who found that high production of ROS and reduced antioxidant defense capability might cause the developing fetus to be exposed to increased OS.

According to other reports, macromolecule damage mediated by OS has been suggested as a mechanism of thalidomide-induced embryopathy and other embryopathies [81,82]. This suggestion was supported by an experimental finding on untreated pregnant mutant mice with a hereditary glucose-6-phosphate dehydrogenase (G6PD) deficiency that resulted in decreased litter size at birth and increased pre- and post-natal (pre-weaning) death. G6PD is a cytoprotective enzyme for OS. This result indicated that a physiological level of endogenous OS due to a dysfunctional G6PD enzyme during development can cause embryopathy that might lead to both infertility and death [83].

Oxidative stress in the female reproductive system is generally reported in most reproductive- and pregnancy-related disorders. For instance, OS has been associated with endometriosis. Although there is no established information on the involvement of OS in endometriosis, a number of studies have reported on the increased level of OS markers in patients with endometriosis [84–90]. Additionally, OS also has been reported to be involved in cases of spontaneous abortion and idiopathic recurrent pregnancy loss [66,78,80,91], unexplained infertility [92–94], preeclampsia [95–97], intrauterine growth restriction (IUGR) [98,99] and preterm labor [100–103].

3. Antioxidants and Their Roles in Reproductive Disorders

Antioxidants regulate the overproduction of ROS. They are present in two types, enzymatic and non-enzymatic forms. Enzymatic antioxidants, including superoxide dismutase (SOD), catalase, glutathione (GSH) peroxidase and glutathione (GSH) reductase are also known as natural antioxidants or endogenous antioxidants [66,104]. The non-enzymatic antioxidants, also known as exogenous antioxidants, are obtained from dietary fruits and vegetables. These include taurine, hypotaurine, β-carotene, selenium, zinc, vitamin C and vitamin E [66].

The roles of antioxidants during the periconceptional period have been previously reported [105,106]. Endogenous antioxidants play important roles within the placenta as well as in the protection of trophoblast cells from OS [106]. It has been reported that SOD has a primary role in cellular protection, metabolizing two molecules of superoxide (O_2^-) to produce hydrogen peroxide (H_2O_2) and molecular oxygen (O_2). Meanwhile, catalase (predominantly located in the peroxisomes) catalyzes the conversion of H_2O_2 to O_2 and water (H_2O). GSH peroxidase and GSH reductase are involved in oxidizing glutathione peroxides by removing H_2O_2 and lipid hydroperoxides [106].

Another antioxidant system that is highly available in the placental cells is the thioredoxin system [105]. This system consists of three antioxidant enzymes; namely, thioredoxin peroxidase, thioredoxin, and thioredoxin reductase. Thioredoxin peroxidase catalyzes the conversion of H_2O_2 and alkyl hydroperoxides to H_2O and corresponding alcohols. This reaction results in the oxidation of thioredoxin peroxidases to an inactive state requiring reduction by thioredoxin [105]. Thioredoxins have been reported to be involved in a number of cellular functions, including cell growth [107], reduction of thioredoxin peroxidase [105], inhibition of apoptosis through the binding of apoptosis signal-regulating kinase-1 (ASK-1) [108], and the supply of electrons for the synthesis of deoxyribonucleotides by ribonucleotide reductase [109].

Exogenous antioxidants, in line with their endogenous counterparts, also play a prime role in cellular defense against OS. The effects of maternal taurine deficiency, including growth retardation of the offspring, impaired perinatal development of the central nervous and pancreatic endocrine systems, impaired glucose tolerance, and vascular dysfunction, were reported by Aerts and Van [110]. Another exogenous antioxidant, zinc (Zn) is used in assisting the fetal brain development and also as an aid to the mothers in labor [111]. According to one study, an early and progressive decline in serum Zn occurs during pregnancy, and therefore the capacity for metabolic adaptation of pregnant mothers may be limited if the maternal Zn status is poor [112]. This is supported by a meta-analysis study on zinc-supplementation in women, which resulted in 14% of reduction in premature delivery [113].

Vitamin C acts as a reducing agent to protect cells against the adverse effects of OS [114]. Zhang et al. [115] reported that pregnant women who consumed vitamin C at levels lower than the recommended daily allowance (85 mg) had a 2-fold higher risk of developing preeclampsia, suggesting the importance of vitamin C supplementation in pregnant women. One randomized controlled clinical trial on patients with luteal phase defects reported that pregnancy rates were higher in the group supplemented with vitamin C (750 mg/day) than in the control group (no treatment) [116]. Another double-blinded, placebo-controlled pilot study on the effect of supplementation containing vitamin E, iron, zinc, selenium and L-arginine resulted in an increase in ovulation and pregnancy rates [117].

Maternal (preeclampsia, abortion, and hypertension) and neonatal outcomes following antioxidant supplementation for 8 to 12 weeks in pregnancy for women with low antioxidant status were reported by Rumiris et al. [118]. This study was a randomized, double-blind, placebo-controlled trial of daily antioxidant supplementation. The supplementation included vitamins A (1000 international unit (IU)), B6 (2.2 mg), B12 (2.2 μg), C (200 mg), and E (400 IU), folic acid (400 μg), N-acetylcysteine (200 mg), Cu (2 mg), Zn (15 mg), Mn (0.5 mg), Fe (30 mg), Ca (800 mg), and selenium

(100 μg). Meanwhile, the control subjects were given ferum (30 mg) and folic acid (400 μg). Results from this study indicated that antioxidant supplementation was associated with better maternal and perinatal outcomes in pregnant women with low antioxidant status as compared to control supplementation with iron and folate alone [118].

In addition, vitamin E functioning as a chain-breaking antioxidant was reported to protect cellular membranes against ROS, for example through defending polyunsaturated fatty acids (PUFAs) from auto-oxidation [119]. Antioxidants such as vitamin C and vitamin E have been reported to be efficient, and their uses in reproductive- and pregnancy-related disorders have been the subject of significant clinical trials [120]. For instance, a randomized clinical trial was conducted from January 2007 to February 2008 at the Women's Hospital of Tabriz University of Medical Sciences, Iran. This study was conducted in response to the inadequate available evidence about the role of supplementary vitamin E in normal pregnancy, and assessed the potential benefit of vitamin E supplementation on health in pregnancy [121]. This trial involved 104 pregnant women who were treated with vitamin E supplementation, and 168 women (control) who were not treated with the supplementation. Treated women were administered 400 IU vitamin E from week 14 to the end of the pregnancy. The study result indicated a non-significant relationship between supplementation and maternal and perinatal outcomes and birth weight, in which preeclampsia was reported to occur in 1% of treated women as compared to 1.78% of women in the control group. From these results, the authors concluded that the administration of supplementary vitamin E starting from the second trimester of pregnancy did not show any risks with respect to pregnancy outcomes and the occurrence of preeclampsia [121].

This is also supported by earlier studies on the possible beneficial effects of supplementary vitamin E during pregnancy, which investigated the changes in vitamin E levels in normal versus problematic pregnancies. Oxidative stability of vitamin E levels was shown to increase in maternal blood during normal pregnancies [122]. Moreover, it has also been shown that vitamin E requirements may increase in some circumstances, such as in smoking during pregnancy [123]. In a comparative study between abnormal and normal pregnancies, the mean levels of vitamin E were reported to increase from 12.9 μg/mL in early pregnancy to 22.5 μg/mL at term in normal pregnancies. However, vitamin E levels were lower than in normal pregnancies at the corresponding gestational age in abnormal pregnancies [124]. Another study by Tamura et al. [125] on 289 pregnant women in Birmingham, United Kingdom reported that there were no significant associations between vitamin E serum concentrations and pregnancy outcomes. All of these reports suggest that vitamin E is essential for normal and healthy pregnancy, and supplementation of vitamin E does not cause any detrimental effects on pregnancy outcomes.

Another recent study was conducted on the effects of vitamin E on the treatment outcomes of women with unexplained infertility who were undergoing controlled ovarian stimulation and intrauterine insemination (IUI) [126]. The study was conducted between June 2011 and December 2011 in Zekai Tahir Burak Women's Training and Research Hospital, Reproductive Endocrinology and Infertility Department, Ankara, Turkey. The study groups were divided into Group A ($n = 53$) and Group B ($n = 50$). Group A underwent controlled ovarian stimulation with clomiphene citrate with vitamin E administration at 400 IU/day, while Group B (control) underwent ovulation induction without the vitamin E administration. The results of the study showed that the difference in the endometrial thickness on the day of human chorionic gonadotropin (hCG) administration was significant between the two groups; however, there was no significant association observed between vitamin E administration and implantation and pregnancy rates. Based on these results, it was concluded that vitamin E administration could improve endometrial response in women with unexplained infertility through the antioxidant and anticoagulant effects. Vitamin E may also modulate the anti-estrogenic effect of clomiphene citrate. Moreover, the issue of thin endometrium in patients may also be improved by vitamin E [126].

As discussed above, vitamin E has been proven to be beneficial in pregnancy and neonatal health. This is in line with previous studies that reported, for instance, plasma α-tocopherol concentrations that is below than 12 mmol/L are associated with increased infection, anemia, growth retardation and poor pregnancy outcomes in both mothers and infants (reviewed in [127]). These problems occur mainly because when low dietary amounts of α-tocopherol are consumed, the requirements for tissue α-tocopherol will exceed the available amounts, resulting in increased damages of the tissues [127].

Vitamin E as an Antioxidant in Female Reproduction: The Reported Studies

Following the first publication by Evans and Bishop [1], a later report discussed the role of vitamin E in reproduction after observations in which a vitamin-E deficient diet resulted in uterine discolorations in rats [128]. Decades later, research on the role of vitamin E in reproductive physiology was re-initiated, and it was reported to have beneficial effects against stress-induced oxidative stress (OS) [129–133].

In another study, a population of women suffering habitual abortion was observed to have high levels of lipid peroxidation and decreased levels of plasma vitamin E [129]. Another study conducted in Egypt also reported that vitamin E to be a key missing micronutrient in children with stunted growth [130]. The study showed that 78.2% of children with stunted growth had vitamin E deficiency, where plasma α-TOC concentrations were recorded at 7.7 μmol/L as compared to 14.1 μmol/L in control (normal) children. In addition, a recent report by [131] also indicated that vitamin A, E, and D deficiencies were very common in very-low-birthweight Tunisian neonates and were associated with preeclampsia.

In more detailed experiments using in vivo laboratory animal models supplemented with palm-tocotrienol rich fractions (TRF), Mokhtar et al. [61] reported that co-administration with 5 mg/kg body weight (bw) of nicotine and 60 mg/kg of tocotrienol-rich fraction (TRF) increased the rates of pregnancy to 83.3% in rats, compared to those treated with nicotine alone, who had pregnancy rates of 33.3%. About 25.7% of the embryos developed into 2- and 4-cell stage in rats treated with both nicotine and TRF [61]. In addition, there was also a report stating that supplementation with γ-TCT in nicotine-induced mice reduced the detrimental effects of nicotine on the ultrastructure of the oocytes [62]. Another study conducted using concurrent treatment with corticosterone (CORT) and TCT reported that the numbers of abnormal embryos were reduced following supplementation with 90 mg/kg and 120 mg/kg of TCT [134]. Meanwhile, co-administration with γ-TCT improved the embryonic development in nicotine-induced mice [64]. Moreover, as reported in more recent findings, using co-incubation in media supplemented with γ-TCT and hydrogen peroxide (H_2O_2), γ-TCT improved the development of porcine embryos through modulation of the apoptotic BCL-XL and BAX genes [135]. The beneficial effects of TCTs were also supported by the reports on the concomitant supplementation of TRF with the anti-cancer prodrug, cyclophosphamide (CPA) on ovarian cells, which was reported to provide protection against OS-induced apoptosis in the ovaries [132,133].

An earlier study using supplementation with annatto-TCTs in pregnant Wistar rats reported that no adverse effects, no increase in embryo lethality and no reduction in fetal body weight were observed [136]. These findings were in line with our recent findings, together with our observations on the anti-survival effects of annatto-delta tocotrienol and soy alpha-tocopherol on the preimplantation embryos of nicotine-treated female mice [137–140]. Furthermore, a recent study also reported that annatto-TCTs suppressed cell growth in human prostate cancer cells through inhibition of the *Src* and *Stat3* genes [141].

In addition to the studies in human and laboratory animals, the benefits of vitamin E have also been studied in domestic animals. An earlier study using the culture of bovine embryos (embryos were derived from the in vitro matured-and-fertilized oocytes) with vitamin E, vitamin C, and ethylenediaminetetraacetic acid (EDTA) showed that more zygotes were developed to the expanded blastocyst stage in culture medium containing 100 μM of vitamin E compared to the control medium. The development to the early, expanded, and hatched blastocyst stages were also lower in the culture medium supplemented with both vitamin E and C, compared to the medium supplemented with vitamin E alone [142]. Moreover, the in vitro-produced embryos were cultured for 5.5 days in medium with or without 100 μM of vitamin E and were non-surgically transferred to recipient cows. After 7 days of transfer, the embryos were non-surgically collected, and the results indicated that embryos cultured with vitamin E were approximately 63% larger in surface area than in the control embryos [142]. Another study by [143] also reported on the effects of antioxidants such as beta-mercaptoethanol (beta-ME) and vitamin E where both suppressed oxidative damage and improved the developmental ability in the porcine embryos.

The beneficial effects of vitamin E were also studied in buffalos. A study was conducted to find whether the supplementation of vitamin E in the culture medium could ameliorate the developmental competence of preimplantation buffalo embryos. The study results indicated that under the culture condition of 20% of O_2 level, the frequency of blastocyst formation and the total cell count were enhanced, and the formation of comet tail (DNA fragmentation) was significantly reduced following supplementation with 100 μM of vitamin E [144]. Another similar study was also conducted in sheep with the aim of determining the effects of α-TOC supplementation of the oocyte maturation media and embryo culture media on the yield of the embryos. Findings from the study showed that supplementation with 200 μM of α-TOC in the embryo culture medium at 20% of O_2 level significantly increased the rates of cleavage, formation of morula and blastocysts, and the total cell number of blastocysts, as compared to the control groups [145].

4. Conclusions

Vitamin E has received much attention in recent years due to its ability to improve reproductive health. As discussed in the present paper, vitamin E has been reported to exert beneficial effects as an antioxidant against the reproductive disorders. Hence, it is highly recommended for women to consume vitamin E regularly, especially those who are in their reproductive age. However, available study reports on the effects of vitamin E on reproduction, pregnancy, and preimplantation embryonic development are still lacking. Many future studies are necessary in order to gain a greater understanding of the antioxidative role of vitamin E, especially with respect to female reproductive health.

Acknowledgments: We are deeply grateful to the **Ministry of Higher Education (MOHE)** Malaysia for providing the financial support through the Fundamental Research Grant Scheme (FRGS) (600-IRMI/FRGS 5/3 (037/2017). Thanks are given to all staff and members of the Faculty of Pharmacy, UiTM Selangor, Puncak Alam Campus, and the Faculty of Medicine, UiTM Selangor, Sg. Buloh Campus, Malaysia for all the assistance and support given in the completion of this paper.

Author Contributions: All authors contributed equally to the completion of this paper. Siti Syairah Mohd Mutalip conceived the idea, collected sources of information, and initiated and contributed to manuscript writing. Mohd Hamim Rajikin and Sharaniza Ab-Rahim contributed to manuscript analysis and review; as well as contributed to the manuscript writing.

Conflicts of Interest: The authors declare no conflict of interest.

References

1. Evans, H.M.; Bishop, K.S. On the existence of a hitherto unrecognized dietary factor essential for reproduction. *Science* **1922**, *56*, 650–651. [CrossRef] [PubMed]
2. Tappel, A.L. Vitamin E as the biological lipid antioxidant. *Vitam. Horm.* **1962**, *20*, 493–510.
3. Burton, G.W.; Ingold, K.U. Vitamin E application of the principles of physical organic chemistry to the exploration of its structure and function. *Acc. Chem. Res.* **1986**, *19*, 194–201. [CrossRef]
4. Esterbauer, H.; Dieber-Rotheneder, M.; Striegl, G.; Waeg, G. Role of vitamin E in preventing the oxidation of low density lipoprotein. *Am. J. Clin. Nutr.* **1991**, *53*, 314S–321S. [CrossRef] [PubMed]
5. Sheppard, A.J.; Pennington, J.A.T.; Weihrauch, J.L. Analysis and distribution of vitamin E in vegetable oils and foods. In *Vitamin E in Health and Disease*; Packer, L., Fuchs, J., Eds.; Marcel Dekker: New York, NY, USA, 1993; pp. 9–31, ISBN 0-8247-8692-0.
6. Ramaswamy, K.; Subash, C.G.; Ji, H.K.; Bharat, B.A. Tocotrienols fight cancer by targeting multiple cell signaling pathways. *Genes Nutr.* **2012**, *7*, 43–52. [CrossRef]

7. Kobayashi, H.; Kanno, C.; Yamauchi, K.; Tsugo, T. Identification of alpha-, beta-, gamma-, and delta-tocopherols and their contents in human milk. *Biochim. Biophys. Acta* **1975**, *380*, 282–290. [PubMed]

8. Nehdi, I.; Omri, S.; Khalil, M.I.; Al-Resayes, S.I. Characteristics and chemical composition of date palm (*Phoenix canariensis*) seeds and seed oil. *Ind. Crops Prod.* **2010**, *32*, 360–365. [CrossRef]

9. Tan, B. Vitamin E: Tocotrienols—The Science behind Tocotrienols. Available online: https://assets.kyani. net/documents/us/Tocotrienols_Science_White_Paper-1.12-EN-ALL.pdf (accessed on 14 August 2017).

10. Rimbach, G.; Jennifer, M.; Patricia, H.; John, K.L. Gene-Regulatory Activity of α-Tocopherol. *Molecules* **2010**, *15*, 1746–1761. [CrossRef] [PubMed]

11. IUPAC-IUB Joint Commission on Biochemical Nomenclature. Nomenclature of tocopherols and related compounds. (Recommendations 1981). *Eur. J. Biochem.* **1982**, *123*, 473–475.

12. Rimbach, G.; Minihane, A.M.; Majewicz, J.; Fischer, A.; Pallauf, J.; Virgli, F.; Weinberg, P.D. Regulation of cell signalling by vitamin E. *Proc. Nutr. Soc.* **2002**, *61*, 415–425. [CrossRef] [PubMed]

13. Pignitter, M.; Stolze, K.; Gartner, S.; Dumhart, B.; Stoll, C.; Steiger, G.; Kraemer, K.; Somoza, V. Cold fluorescent light as major inducer of lipid oxidation in soybean oil stored at household conditions for eight weeks. *J. Agric. Food Chem.* **2014**, *62*, 2297–2305. [CrossRef] [PubMed]

14. The Structure of Vitamin, E. Available online: https://www.omicsonline.org/articles-images/2155-9899-4-137-g001.html (accessed on 23 January 2018).

15. Wigle, D.T.; Arbuckle, T.E.; Turner, M.C.; Bérubé, A.; Yang, Q.; Liu, S.; Krewski, D. Epidemiologic evidence of relationships between reproductive and child health outcomes and environmental chemical contaminants. *J. Toxicol. Environ. Health B Crit. Rev.* **2008**, *11*, 373–517. [CrossRef] [PubMed]

16. Rider, C.V.; Furr, J.R.; Wilson, V.S.; Gray, L.E., Jr. Cumulative effects of in utero administration of mixtures of reproductive toxicants that disrupt common target tissues via diverse mechanisms of toxicity. *J. Androl.* **2010**, *33*, 443–462. [CrossRef] [PubMed]

17. Al-Gubory, K.H. Environmental pollutants and lifestyle factors induce oxidative stress and poor prenatal development. *Reprod. Biomed. Online* **2014**, *29*, 17–31. [CrossRef] [PubMed]

18. Kortenkamp, A. Ten years of mixing cocktails: A review of combination effects of endocrine-disrupting chemicals. *Environ. Health Perspect.* **2007**, *115* (Suppl. 1), 98–105. [CrossRef] [PubMed]

19. Luo, Z.C.; Liu, J.M.; Fraser, W.D. Large prospective birth cohort studies on environmental contaminants and child health—Goals, challenges, limitations and needs. *Med. Hypotheses* **2010**, *74*, 318–324. [CrossRef] [PubMed]

20. Ruder, E.H.; Hartman, T.J.; Blumberg, J.; Goldman, M.B. Oxidative stress and antioxidants: Exposure and impact on female fertility. *Hum. Reprod. Update* **2008**, *14*, 345–357. [CrossRef] [PubMed]

21. Wells, P.G.; McCallum, G.P.; Chen, C.S.; Henderson, J.T.; Lee, C.J.; Perstin, J.; Preston, T.J.; Wiley, M.J.; Wong, A.W. Oxidative stress in developmental origins of disease: Teratogenesis, neurodevelopmental deficits, and cancer. *Toxicol. Sci.* **2009**, *108*, 4–18. [CrossRef] [PubMed]

22. Al-Gubory, K.H.; Fowler, P.A.; Garrel, C. The roles of cellular reactive oxygen species, oxidative stress and antioxidants in pregnancy outcomes. *Int. J. Biochem. Cell Biol.* **2010**, *42*, 1634–1650. [CrossRef] [PubMed]

23. Agarwal, A.; Aponte-Mellado, A.; Premkumar, B.J.; Shaman, A.; Gupta, S. The effects of oxidative stress on female reproduction: A review. *Reprod. Biol. Endocrinol.* **2012**, *10*, 49. [CrossRef] [PubMed]

24. Anderson, K.; Nisenblat, V.; Norman, R. Lifestyle factors in people seeking infertility treatment—A review. *Aust. N. Z. J. Obstet. Gynaecol.* **2010**, *50*, 8–20. [CrossRef] [PubMed]

25. Sharma, R.; Biedenharn, K.R.; Fedor, J.M.; Agarwal, A. Lifestyle factors and reproductive health: Taking control of your fertility. *Reprod. Biol. Endocrinol.* **2013**, *11*, 66. [CrossRef] [PubMed]

26. Kovacic, P.; Somanathan, R. Mechanism of teratogenesis: Electron transfer, reactive oxygen species, and antioxidants. *Birth Defects Res. C Embryo Today* **2006**, *78*, 308–325. [CrossRef] [PubMed]

27. Rasch, V. Cigarette, alcohol, and caffeine consumption: Risk factors for spontaneous abortion. *Acta Obstet. Gynecol. Scand.* **2003**, *82*, 182–188. [CrossRef] [PubMed]

28. Weng, X.; Odouli, R.; Li, D.K. Maternal caffeine consumption during pregnancy and the risk of miscarriage: A prospective cohort study. *Am. J. Obstet. Gynecol.* **2008**, *198*, 279.e1–279.e8. [CrossRef] [PubMed]

29. Barua, S.; Junaid, M.A. Lifestyle, pregnancy and epigenetic effects. *Epigenomics* **2015**, *7*, 85–102. [CrossRef] [PubMed]

30. Dechanet, C.; Anahory, T.; Mathieu Daude, J.C.; Quantin, X.; Reyftmann, L.; Hamamah, S.; Hedon, B.; Dechaud, H. Effects of cigarette smoking on reproduction. *Hum. Reprod. Update* **2011**, *17*, 76–95. [CrossRef] [PubMed]

31. Dembele, K.; Yao, X.H.; Chen, L.; Nyomba, B.L. Intrauterine ethanol exposure results in hypothalamic oxidative stress and neuroendocrine alterations in adult rat offspring. *Am. J. Physiol. Regul. Integr. Comp. Physiol.* **2006**, *291*, R796–R802. [CrossRef] [PubMed]

32. Wentzel, P.; Eriksson, U.J. Ethanol-induced fetal dysmorphogenesis in the mouse is diminished by high antioxidative capacity of the mother. *Toxicol. Sci.* **2006**, *92*, 416–422. [CrossRef] [PubMed]

33. Wentzel, P.; Rydberg, U.; Eriksson, U.J. Antioxidative treatment diminishes ethanol-induced congenital malformations in the rat. *Alcohol. Clin. Exp. Res.* **2006**, *30*, 1752–1760. [CrossRef] [PubMed]

34. Van Gelder, M.M.; van Rooij, I.A.; Miller, R.K.; Zielhuis, G.A.; de Jong-van den Berg, L.T.; Roeleveld, N. Teratogenic mechanisms of medical drugs. *Hum. Reprod. Update* **2010**, *16*, 378–394. [CrossRef] [PubMed]

35. Deavall, D.G.; Martin, E.A.; Horner, J.M.; Roberts, R. Drug-induced oxidative stress and toxicity. *J. Toxicol.* **2012**. [CrossRef] [PubMed]

36. Liu, L.; Wells, P.G. In vivo phenytoin-initiated oxidative damage to proteins and lipids in murine maternal hepatic and embryonic tissue organelles: Potential molecular targets of chemical teratogenesis. *Toxicol. Appl. Pharmacol.* **1994**, *125*, 247–255. [CrossRef] [PubMed]

37. Hansen, J.M.; Harris, C. A novel hypothesis for thalidomide-induced limb teratogenesis: Redox misregulation of the NF-kappaB pathway. *Antioxid. Redox Signal.* **2004**, *6*, 1–14. [CrossRef] [PubMed]

38. Defoort, E.N.; Kim, P.M.; Winn, L.M. Valproic acid increases conservative homologous recombination frequency and reactive oxygen species formation: A potential mechanism for valproic acid-induced neural tube defects. *Mol. Pharmacol.* **2006**, *69*, 1304–1310. [CrossRef] [PubMed]

39. Danielsson, B.R.; Danielsson, C.; Nilsson, M.F. Embryonic cardiac arrhythmia and generation of reactive oxygen species: Common teratogenic mechanism for IKr blocking drugs. *Reprod. Toxicol.* **2007**, *24*, 42–56. [CrossRef] [PubMed]

40. Borgerding, M.; Klus, H. Analysis of complex mixtures—Cigarette smoke. *Exp. Toxicol. Pathol.* **2005**, *57*, 43–73. [CrossRef] [PubMed]

41. Feltes, B.C.; de Faria Poloni, J.; Notari, D.L.; Bonatto, D. Toxicological effects of the different substances in tobacco smoke on human embryonic development by a systems chemo-biology approach. *PLoS ONE* **2013**, *8*, e61743. [CrossRef] [PubMed]

42. George, L.; Granath, F.; Johansson, A.L.; Anneren, G.; Cnattingius, S. Environmental tobacco smoke and risk of spontaneous abortion. *Epidemiology* **2006**, *17*, 500–505. [CrossRef] [PubMed]

43. Ananth, C.V.; Smulian, J.C.; Vintzileos, A.M. Incidence of placental abruption in relation to cigarette smoking and hypertensive disorders during pregnancy: A meta-analysis of observational studies. *Obstet. Gynecol.* **1999**, *93*, 622–628. [CrossRef] [PubMed]

44. Faiz, A.S.; Ananth, C.V. Etiology and risk factors for placenta previa: An overview and meta-analysis of observational studies. *J. Matern. Fetal Neonatal Med.* **2003**, *13*, 175–190. [CrossRef] [PubMed]

45. Hung, T.H.; Hsieh, C.C.; Hsu, J.J.; Chiu, T.H.; Lo, L.M.; Hsieh, T.T. Risk factors for placenta previa in an Asian population. *Int. J. Gynaecol. Obstet.* **2007**, *97*, 26–30. [CrossRef] [PubMed]

46. Kolas, T.; Nakling, J.; Salvesen, K.A. Smoking during pregnancy increases the risk of preterm births among parous women. *Acta Obstet. Gynecol. Scand.* **2000**, *79*, 644–648. [PubMed]

47. Fantuzzi, G.; Aggazzotti, G.; Righi, E.; Facchinetti, F.; Bertucci, E.; Kanitz, S.; Barbone, F.; Sansebastiano, G.; Battaglia, M.A.; Leoni, V.; et al. Preterm delivery and exposure to active and passive smoking during pregnancy: A case-control study from Italy. *Paediatr. Perinat. Epidemiol.* **2007**, *21*, 194–200. [CrossRef] [PubMed]

48. Jaddoe, V.W.; Troe, E.J.; Hofman, A.; Mackenbach, J.P.; Moll, H.A.; Steegers, E.A.; Witteman, J.C. Active and passive maternal smoking during pregnancy and the risks of low birthweight and preterm birth: The Generation R Study. *Paediatr. Perinat. Epidemiol.* **2008**, *22*, 162–171. [CrossRef] [PubMed]

49. Wisborg, K.; Kesmodel, U.; Henriksen, T.B.; Olsen, S.F.; Secher, N.J. Exposure to tobacco smoke in utero and the risk of stillbirth and death in the first year of life. *Am. J. Epidemiol.* **2001**, *154*, 322–327. [CrossRef] [PubMed]

50. Hogberg, L.; Cnattingius, S. The influence of maternal smoking habits on the risk of subsequent stillbirth: Is there a causal relation? *BJOG* **2007**, *114*, 699–704. [CrossRef] [PubMed]

51. Mitchell, E.A.; Milerad, J. Smoking and the sudden infant death syndrome. *Rev. Environ. Health* **2006**, *21*, 81–103. [CrossRef] [PubMed]

52. Sharpe, R.M.; Franks, S. Environment, lifestyle and infertility—An inter-generational issue. *Nat. Cell Biol.* **2002**, *4*, s33–s40. [CrossRef] [PubMed]

53. Zenzes, M.T.; Krishnan, S.; Krishnan, B.; Zhang, H.; Casper, R.F. Cadmium accumulation in follicular fluid of women in in vitro fertilization-embryo transfer is higher in smokers. *Fertil. Steril.* **1995**, *64*, 599–603. [CrossRef]

54. Younglai, E.V.; Foster, W.G.; Hughes, E.G.; Trim, K.; Jarrell, J.F. Levels of environmental contaminants in human follicular fluid, serum, and seminal plasma of couples undergoing in vitro fertilization. *Arch. Environ. Contam. Toxicol.* **2002**, *43*, 121–126. [CrossRef] [PubMed]

55. Neal, M.S.; Zhu, J.; Foster, W.G. Quantification of benzo[a]pyrene and other PAHs in the serum and follicular fluid of smokers versus non-smokers. *Reprod. Toxicol.* **2008**, *25*, 100–106. [CrossRef] [PubMed]

56. Van Voorhis, B.J.; Dawson, J.D.; Stovall, D.W.; Sparks, A.E.; Syrop, C.H. The effects of smoking on ovarian function and fertility during assisted reproduction cycles. *Obstet. Gynecol.* **1996**, *88*, 785–791. [CrossRef]

57. Ness, R.B.; Grisso, J.A.; Hirschinger, N.; Markovic, N.; Shaw, L.M.; Day, N.L.; Kline, J. Cocaine and tobacco use and the risk of spontaneous abortion. *N. Engl. J. Med.* **1999**, *340*, 333–339. [CrossRef] [PubMed]

58. Huang, J.; Okuka, M.; McLean, M.; Keefe, D.L.; Liu, L. Effects of cigarette smoke on fertilization and embryo development in vivo. *Fertil. Steril.* **2009**, *92*, 1456–1465. [CrossRef] [PubMed]

59. Zhao, Z.; Reece, E.A. Nicotine-induced embryonic malformations mediated by apoptosis from increasing intracellular calcium and oxidative stress. *Birth Defects Res. B Dev. Reprod. Toxicol.* **2005**, *74*, 383–391. [CrossRef] [PubMed]

60. Holloway, A.C.; Kellenberger, L.D.; Petrik, J.J. Fetal and neonatal exposure to nicotine disrupts ovarian function and fertility in adult female rats. *Endocrine* **2006**, *30*, 213–216. [CrossRef]

61. Mokhtar, N.; Rajikin, M.H.; Zakaria, Z. Role of tocotrienol-rich palm vitamin E on pregnancy and preimplantation embryos in nicotine treated rats. *Biomed. Res.* **2008**, *19*, 181–184.

62. Rajikin, M.H.; Latif, E.S.; Mar, M.R.; Mat Top, A.G.; Mokhtar, N.M. Deleterious effects of nicotine on the ultrastructure of oocytes: Role of gamma-tocotrienol. *Med. Sci. Monit.* **2009**, *15*, BR378–BR383. [PubMed]

63. Asadi, E.; Mehrdad, J.; Mohammad, J.G. Effect of vitamin E on oocytes apoptosis in nicotine-treated mice. *Iran. J. Basic Med. Sci.* **2012**, *15*, 880–884. [PubMed]

64. Kamsani, Y.S.; Rajikin, M.H.; Nor-Ashikin, M.N.K.; Nuraliza, A.S.; Chatterjee, A. Nicotine-induced cessation of embryonic development is reversed by γ-tocotrienol in mice. *Med. Sci. Monit. Basic Res.* **2013**, *19*, 87–92. [CrossRef] [PubMed]

65. Phoebe, C.J.; Julie, A.M.; Emma, L.B.; Philip, M.H.; Keith, T.J. Increased zona pellucida thickness and meiotic spindle disruption in oocytes from cigarette smoking mice. *Hum. Reprod.* **2011**, *26*, 878–884. [CrossRef]

66. Agarwal, A.; Gupta, S.; Sekhon, L.; Shah, R. Redox considerations in female reproductive function and assisted reproduction: From molecular mechanisms to health implications. *Antioxid. Redox Signal.* **2008**, *10*, 1375–1403. [CrossRef] [PubMed]

67. Ames, B.N.; Shigenaga, M.K.; Hagen, T.M. Oxidants, antioxidants, and the degenerative diseases of aging. *Proc. Natl. Acad. Sci. USA* **1993**, *90*, 7915–7922. [CrossRef] [PubMed]

68. Puglia, C.D.; Powell, S.R. Inhibition of cellular antioxidants: A possible mechanism of toxic cell injury. *Environ. Health Perspect.* **1984**, *57*, 307–311. [CrossRef] [PubMed]

69. Evans, M.D.; Dizdaroglu, M.; Cooke, M.S. Oxidative DNA damage and disease: Induction, repair and significance. *Mutat. Res.* **2004**, *567*, 1–61. [CrossRef] [PubMed]

70. Dennery, P.A. Role of redox in fetal development and neonatal diseases. *Antioxid. Redox Signal.* **2004**, *6*, 147–153. [CrossRef] [PubMed]

71. Lee, T.H.; Wu, M.Y.; Chen, M.J.; Chao, K.H.; Ho, H.N.; Yang, Y.S. Nitric oxide is associated with poor embryo quality and pregnancy outcome in in vitro fertilization cycles. *Fertil. Steril.* **2004**, *82*, 126–131. [CrossRef] [PubMed]

72. Sharma, R.K.; Agarwal, A. Role of reactive oxygen species in gynecologic diseases. *Reprod. Med. Biol.* **2004**, *3*, 177–199. [CrossRef]

73. Guerin, P.; El Mouatassim, S.; Menezo, Y. Oxidative stress and protection against reactive oxygen species in the pre-implantation embryo and its surroundings. *Hum. Reprod. Update* **2001**, *7*, 175–189. [CrossRef] [PubMed]

74. Agarwal, A.; Saleh, R.A.; Bedaiwy, M.A. Role of reactive oxygen species in the pathophysiology of human reproduction. *Fertil. Steril.* **2003**, *79*, 829–843. [CrossRef]

75. Jurisicova, A.; Varmuza, S.; Casper, R.F. Programmed cell death and human embryo fragmentation. *Mol. Hum. Reprod.* **1996**, *2*, 93–98. [CrossRef] [PubMed]

76. Walsh, S.W.; Wang, Y. Secretion of lipid peroxides by the human placenta. *Am. J. Obstet. Gynecol.* **1993**, *169*, 1462–1466. [CrossRef]

77. Myatt, L.; Cui, X. Oxidative stress in the placenta. *Histochem. Cell Biol.* **2004**, *122*, 369–382. [CrossRef] [PubMed]

78. Poston, L.; Raijmakers, M.T. Trophoblast oxidative stress, antioxidants and pregnancy outcome—A review. *Placenta* **2004**, *25* (Suppl. A), S72–S78. [CrossRef] [PubMed]

79. Wang, Y.; Walsh, S.W. Placental mitochondria as a source of oxidative stress in pre-eclampsia. *Placenta* **1998**, *19*, 581–586. [CrossRef]

80. Jauniaux, E.; Watson, A.L.; Hempstock, J.; Bao, Y.; Skepper, J.N.; Burton, G.J. Onset of maternal arterial blood flow and placental oxidative stress: A possible factor in human early pregnancy failure. *Am. J. Pathol.* **2000**, *157*, 2111–2122. [CrossRef]

81. Parman, T.; Wiley, M.J.; Wells, P.G. Free radical-mediated oxidative DNA damage in the mechanism of thalidomide teratogenicity. *Nat. Med.* **1999**, *5*, 582–585. [PubMed]

82. Burton, G.J.; Hempstock, J.; Jauniaux, E. Oxygen, early embryonic metabolism and free radical-mediated embryopathies. *Reprod. Biomed. Online* **2003**, *6*, 84–96. [CrossRef]

83. Nicol, C.J.; Zielenski, J.; Tsui, L.C.; Wells, P.G. An embryoprotective role for glucose-6-phosphate dehydrogenase in developmental oxidative stress and chemical teratogenesis. *FASEB J.* **2000**, *14*, 111–127. [CrossRef] [PubMed]

84. Murphy, A.A.; Santanam, N.; Parthasarathy, S. Endometriosis: A disease of oxidative stress? *Semin. Reprod. Endocrinol.* **1998**, *16*, 263–273. [CrossRef] [PubMed]

85. Rong, R.; Ramachandran, S.; Santanam, N.; Murphy, A.A.; Parthasarathy, S. Induction of monocyte chemotactic protein-1 in peritoneal mesothelial and endometrial cells by oxidized low-density lipoprotein and peritoneal fluid from women with endometriosis. *Fertil. Steril.* **2002**, *78*, 843–848. [CrossRef]

86. Szczepanska, M.; Kozlik, J.; Skrzypczak, J.; Mikolajczyk, M. Oxidative stress may be a piece in the endometriosis puzzle. *Fertil. Steril.* **2003**, *79*, 1288–1293. [CrossRef]

87. Bedaiwy, M.A.; Falcone, T. Peritoneal fluid environment in endometriosis. Clinicopathological implications. *Minerva Ginecol.* **2003**, *55*, 333–345. [PubMed]

88. Jackson, L.W.; Schisterman, E.F.; Dey-Rao, R.; Browne, R.; Armstrong, D. Oxidative stress and endometriosis. *Hum. Reprod.* **2005**, *20*, 2014–2020. [CrossRef] [PubMed]

89. Mier-Cabrera, M.; Jimenez-Zamudio, L.; Garcia-Latorre, E.; Cruz-Orozco, O.; Hernandez-Guerrero, C. Quantitative and qualitative peritoneal immune profiles, T-cell apoptosis and oxidative stress-associated characteristics in women with minimal and mild endometriosis. *BJOG* **2011**, *118*, 6–16. [CrossRef] [PubMed]

90. Sharma, I.; Dhaliwal, L.K.; Saha, S.C.; Sangwan, S.; Dhawan, V. Role of 8-iso-prostaglandin F 2alpha and 25-hydroxycholesterol in the pathophysiology of endometriosis. *Fertil. Steril.* **2010**, *94*, 63–70. [CrossRef] [PubMed]

91. Gupta, S.; Agarwal, A.; Banerjee, J.; Alvarez, J.G. The role of oxidative stress in spontaneous abortion and recurrent pregnancy loss: A systematic review. *Obstet. Gynecol. Surv.* **2007**, *62*, 335–347. [CrossRef] [PubMed]

92. Polak, G.; Rola, R.; Gogacz, M.; Koziol-Montewka, M.; Kotarski, J. Malonyldialdehyde and total antioxidant status in the peritoneal fluid of infertile women. *Ginekol. Pol.* **1999**, *70*, 135–140. [PubMed]

93. Polak, G.; Koziol-Montewka, M.; Tarkowski, R.; Kotarski, J. Peritoneal fluid and plasma 4-hydroxynonenal and malonyldialdehyde concentrations in infertile women. *Ginekol. Pol.* **2011**, *72*, 1316–1320.

94. Agarwal, A.; Gupta, S.; Sikka, S. The role of free radicals and antioxidants in reproduction. *Curr. Opin. Obstet. Gynecol.* **2006**, *18*, 325–332. [CrossRef] [PubMed]

95. Burton, G.J.; Jauniaux, E. Placental oxidative stress: From miscarriage to preeclampsia. *J. Soc. Gynecol. Investig.* **2004**, *11*, 342–352. [CrossRef] [PubMed]

96. Redman, C.W.; Sargent, I.L. Placental stress and pre-eclampsia: A revised view. *Placenta* **2009**, *30* (Suppl. A), S38–S42. [CrossRef] [PubMed]

97. Burton, G.J.; Yung, H.W.; Cindrova-Davies, T.; Charnock-Jones, D.S. Placental endoplasmic reticulum stress and oxidative stress in the pathophysiology of unexplained intrauterine growth restriction and early onset preeclampsia. *Placenta* **2009**, *30* (Suppl. A), S43–S48. [CrossRef] [PubMed]

98. Karowicz-Bilinska, A. Lipid peroxides concentration in women with intrauterine growth restriction. *Ginekol. Pol.* **2004**, *75*, 6–9. [PubMed]
99. Biri, A.; Bozkurt, N.; Turp, A.; Kavutcu, M.; Himmetoglu, O.; Durak, I. Role of oxidative stress in intrauterine growth restriction. *Gynecol. Obstet. Investig.* **2007**, *64*, 187–192. [CrossRef] [PubMed]
100. Hong, Y.C.; Lee, K.H.; Yi, C.H.; Ha, E.H.; Christiani, D.C. Genetic susceptibility of term pregnant women to oxidative damage. *Toxicol. Lett.* **2002**, *129*, 255–262. [CrossRef]
101. Frosali, S.; DiSimplicio, P.; Perrone, S.; DiGiuseppe, D.; Longini, M.; Tanganelli, D.; Buonocore, G. Glutathione recycling and antioxidant enzyme activities in erythrocytes of term and preterm newborns at birth. *Neonatology* **2004**, *85*, 188–194. [CrossRef] [PubMed]
102. Mustafa, M.D.; Pathak, R.; Ahmed, T.; Ahmed, R.S.; Tripathi, A.K.; Guleria, K.; Banerjee, B.D. Association of glutathione S-transferase M1 and T1 gene polymorphisms and oxidative stress markers in preterm labor. *Clin. Biochem.* **2010**, *43*, 1124–1128. [CrossRef] [PubMed]
103. Pathak, R.; Suke, S.G.; Ahmed, T.; Ahmed, R.S.; Tripathi, A.K.; Guleria, K.; Sharma, C.S.; Makhijani, S.D.; Banerjee, B.D. Organochlorine pesticide residue levels and oxidative stress in preterm delivery cases. *Hum. Exp. Toxicol.* **2010**, *29*, 351–358. [CrossRef] [PubMed]
104. Perkins, A.V. Anti-oxidants in pregnancy Endogenous anti-oxidants in pregnancy and preeclampsia. *Aust. N. Z. J. Obstet. Gynaecol.* **2006**, *46*, 77–83. [CrossRef] [PubMed]
105. Mustacich, D.; Powis, G. Thioredoxin reductase. *Biochem. J.* **2000**, *346*, 1–8. [CrossRef] [PubMed]
106. Rhee, S.G.; Chae, H.Z.; Kim, K. Peroxiredoxins: A historical overview and speculative preview of novel mechanisms and emerging concepts in cell signaling. *Free Radic. Biol. Med.* **2005**, *38*, 1543–1552. [CrossRef] [PubMed]
107. Gasdaska, J.R.; Berggren, M.; Powis, G. Cell growth stimulation by the redox protein thioredoxin occurs by a novel helper mechanism. *Cell Growth Differ.* **1995**, *6*, 1643–1650. [PubMed]
108. Saitoh, M.; Nishitoh, H.; Fujii, M.; Takeda, K.; Tobiume, K.; Sawada, Y.; Kawabata, M.; Miyazono, K.; Ichijo, H. Mammalian thioredoxin is a direct inhibitor of apoptosis signal-regulating kinase (ASK) 1. *EMBO J.* **1998**, *17*, 2596–2606. [CrossRef] [PubMed]
109. Nordberg, J.; Arner, E.S.J. Reactive oxygen species, antioxidants, and the mammalian thioredoxin system. *Free Radic. Biol. Med.* **2001**, *31*, 1287–1312. [CrossRef]
110. Aerts, L.; Van, A.F.A. Taurine and taurine-deficiency in the perinatal period. *J. Perinat. Med.* **2002**, *30*, 281–286. [CrossRef] [PubMed]
111. Uriu-Adams, J.Y.; Keen, C.L. Zinc and reproduction: Effects of zinc deficiency on prenatal and early postnatal development. *Birth Defects Res. B. Dev. Reprod. Toxicol.* **2010**, *89*, 313–325. [CrossRef] [PubMed]
112. Silvia, I.A.; Castañón, S.G.; Ruata, M.L.; Aragüés, E.F.; Terraz, P.B.; Irazabal, Y.G.; González, E.G.; Rodríguez, B.G. Updating of normal levels of copper, zinc and selenium in serum of pregnant women. *J. Trace Elem. Med. Biol.* **2007**, *21* (Suppl. 1), 49–52. [CrossRef]
113. Hess, S.Y.; King, J.C. Effects of maternal zinc supplementation on pregnancy and lactation outcomes. *Food Nutr. Bull.* **2009**, *30* (Suppl. 1), S60–S78. [CrossRef] [PubMed]
114. Levine, M.; Katz, A.; Padayatty, S.J.; Vitamin, C. *Modern Nutrition in Health and Disease*; Shils, M.E., Shike, M., Ross, A.C., Caballero, B., Cousins, R.J., Eds.; Lippincott Williams & Wilkins: Philadelphia, PA, USA, 2006; pp. 507–524.
115. Zhang, C.; Williams, M.A.; King, I.B.; Dashow, E.E.; Sorensen, T.K.; Frederick, I.O.; Thompson, M.L.; Luthy, D.A. Vitamin C and the risk of preeclampsia—Results from dietary questionnaire and plasma assay. *Epidemiology* **2002**, *13*, 409–416. [CrossRef] [PubMed]
116. Henmi, H.; Endo, T.; Kitajima, Y.; Manase, K.; Hata, H.; Kudo, R. Effects of ascorbic acid supplementation on serum progesterone levels in patients with a luteal phase defect. *Fertil. Steril.* **2003**, *80*, 459–461. [CrossRef]
117. Westphal, L.M.; Polan, M.L.; Trant, A.S.; Mooney, S.B. A nutritional supplement for improving fertility in women: A pilot study. *J. Reprod. Med.* **2004**, *49*, 289–293. [PubMed]
118. Rumiris, D.; Purwosunu, Y.; Wibowo, N.; Farina, A.; Sekizawa, A. Lower rate of preeclampsia after antioxidant supplementation in pregnant women with low antioxidant status. *Hypertens. Pregnancy* **2006**, *25*, 241–253. [CrossRef] [PubMed]
119. Traber, M.G.; Jeffrey, A. Vitamin E, antioxidant and nothing more. *Free Radic. Biol. Med.* **2007**, *43*, 4–15. [CrossRef] [PubMed]

120. Fraser, W.D.; Audibert, F.; Bujold, E.; Leduc, L.; Xu, H.; Boulvain, M.; Julien, P. The Vitamin E debate: Implications for ongoing trials of pre-eclampsia prevention. *BJOG* **2005**, *112*, 684–688. [CrossRef] [PubMed]

121. Parvin, B.; Kobra, H.; Fatemeh, A.; Nazli, N. Effects of vitamin E supplementation on some pregnancy health indices: A randomized clinical trial. *Int. J. Gen. Med.* **2011**, *4*, 461–464. [CrossRef]

122. De Vriese, S.R.; Dhont, M.; Christophe, A.B. Oxidative stability of low density lipoproteins and vitamin E levels increase in maternal blood during normal pregnancy. *Lipids* **2001**, *36*, 361–366. [CrossRef] [PubMed]

123. Chelchowska, M.; Laskowska-Klita, T.; Leibschang, J. The effect of tobacco smoking during pregnancy on concentration of vitamin E in blood of mothers and their newborns in umbilical cord blood. *Ginekol. Pol.* **2006**, *77*, 263–268. [PubMed]

124. Von Mandach, U.; Huch, R.; Huch, A. Maternal and cord serum vitamin E levels in normal and abnormal pregnancy. *Int. J. Vitam. Nutr. Res.* **1994**, *64*, 26–32. [PubMed]

125. Tamura, T.; Goldenberg, R.L.; Johnston, K.E.; Cliver, S.P.; Hoffman, H.J. Serum concentrations of zinc, folate, vitamins A and E, and proteins, and their relationships to pregnancy outcome. *Acta Obstet. Gynecol. Scand. Suppl.* **1997**, *165*, 63–70. [PubMed]

126. Cicek, N.; Eryilmaz, O.G.; Sarikaya, E.; Gulerman, C.; Genc, Y. Vitamin E effect on controlled ovarian stimulation of unexplained infertile women. *J. Assist. Reprod. Genet.* **2012**, *29*, 325–328. [CrossRef] [PubMed]

127. Traber, G.M. Vitamin E Inadequacy in Humans: Causes and Consequences. *Adv. Nutr.* **2014**, *5*, 503–514. [CrossRef] [PubMed]

128. Barrie, M.M. Vitamin E deficiency in rats: Fertility in the female. *Biochem. J.* **1938**, *32*, 2134–2137. [CrossRef] [PubMed]

129. Simsek, M.; Naziroglu, M.; Simsek, H.; Cay, M.; Aksakal, M.; Kumru, S. Blood plasma levels of lipoperoxides, glutathione peroxidase, beta carotene, vitamin A and E in women with habitual abortion. *Cell Biochem. Funct.* **1998**, *16*, 227–231. [CrossRef]

130. Ibrahim, S.A.; Abd el-Maksoud, A.; Nassar, M.F. Nutritional stunting in Egypt: Which nutrient is responsible? *East. Mediterr. Health J.* **2002**, *8*, 272–280. [PubMed]

131. Fares, S.; Sethom, M.M.; Khouaja-Mokrani, C.; Jabnoun, S.; Feki, M.; Kaabachi, N. Vitamin A, E, and D deficiencies in Tunisian very low birth weight neonates: Prevalence and risk factors. *Pediatr. Neonatol.* **2014**, *55*, 196–201. [CrossRef] [PubMed]

132. Saleh, H.; Omar, E.; Froemming, G.; Said, R. Tocotrienol rich fraction supplementation confers protection on the ovary from cyclophasphamide induced apoptosis. *Asian Pac. J. Trop. Dis.* **2014**, *4*, 234. [CrossRef]

133. Saleh, H.; Omar, E.; Froemming, G.; Said, R. Tocotrienol preserves ovarian function in cyclophosphamide therapy. *Hum. Exp. Toxicol.* **2015**, *34*, 946–952. [CrossRef] [PubMed]

134. Nasibah, A.; Rajikin, M.H.; Nor-Ashikin, M.N.K.; Nuraliza, A.S. Tocotrienol improves the quality of impaired mouse embryos induced by corticosterone. In Proceedings of the Symposium on Humanities, Science and Engineering Research (SHUSER2012), Kuala Lumpur, Malaysia, 24–27 June 2012; pp. 135–138.

135. Lee, E.; Min, S.-H.; Song, B.-S.; Yeon, J.-Y.; Kim, J.-W.; Bae, J.-H.; Park, S.-Y.; Lee, Y.-H.; Kim, S.-U.; Lee, D.-S.; et al. Exogenous γ-tocotrienol promotes preimplantation development and improves the quality of porcine embryos. *Reprod. Fertil. Dev.* **2014**. [CrossRef] [PubMed]

136. Paumgartten, F.J.R.; De-Carvalho, R.R.; Araujo, I.B.; Pinto, F.M.; Borges, O.O.; Souza, C.A.M.; Kuriyama, S.M. Evaluation of the developmental toxicity of annatto in the rat. *Food Chem. Toxicol.* **2002**, *40*, 1595–1601. [CrossRef]

137. Syairah, S.M.M.; Rajikin, M.H.; Sharaniza, A.-R.; Nor-Ashikin, M.N.K.; Anne, T.; Barrie, T. Annatto (*Bixa orellana*) derived δ-tocotrienol supplementation suppresses PIK3CA oncogene expression in 2- and 4-cell embryos of nicotine-induced mice. *Anticancer Res.* **2014**, *34*, 6064.

138. Syairah, S.M.M.; Rajikin, M.H.; Sharaniza, A.-R. Supplementation of annatto (*Bixa orellana*)-derived δ-tocotrienol produced high number of morula through increased expression of 3-phosphoinositide dependent protein kinase-1 (PDK1) in mice. *Int. J. Biol. Biomol. Agric. Food Biotechnol. Eng.* **2015**, *9*, 741–745.

139. Syairah, S.M.M.; Rajikin, M.H.; Sharaniza, A.R.; Nor-Ashikin, N.K.; Kamsani, Y.S. Chromosomal status in murine preimplantation 2-cell embryos following annatto (*Bixa orellana*)-derived pure delta-tocotrienol supplementation in normal and nicotine-treated mice. *WASJ* **2016**, *34*, 1855–1859. [CrossRef]

140. Syairah, S.M.M.; Rajikin, M.H.; Sharaniza, A.R.; Nor-Ashikin, M.N.K. Annatto (*Bixa orellana*) δ-TCT supplementation protected against embryonic DNA damages through alterations in PI3K/Akt-Cyclin D1 pathway. *Int. J. Vitam. Nutr. Res.* **2017**, accepted.

141. Sugahara, R.; Sato, A.; Uchida, A.; Shiozawa, S.; Sato, C.; Virgona, N.; Yano, T. Annatto tocotrienol induces a cytotoxic effect on human prostate cancer PC3 cells via the simultaneous inhibition of Src and Stat3. *J. Nutr. Sci. Vitaminol.* **2015**, *61*, 497–501. [CrossRef] [PubMed]

142. Olson, S.E.; Seidel, G.E., Jr. Culture of in vitro-produced bovine embryos with vitamin E improves development in vitro and after transfer to recipients. *Biol. Reprod.* **2000**, *62*, 248–252. [CrossRef] [PubMed]

143. Kitagawa, Y.; Suzuki, K.; Yoneda, A.; Watanabe, T. Effects of oxygen concentration and antioxidants on the *in vitro* developmental ability, production of reactive oxygen species (ROS), and DNA fragmentation in porcine embryos. *Theriogenology* **2004**, *62*, 1186–1197. [CrossRef] [PubMed]

144. Thiyagarajan, B.; Valivittan, K. Ameliorating effect of vitamin E on in vitro development of preimplantation buffalo embryos. *J. Assist. Reprod, Genet.* **2009**, *26*, 217–225. [CrossRef] [PubMed]

145. Natarajan, R.; Shankar, M.B.; Munuswamy, D. Effect of α-tocopherol supplementation on in vitro maturation of sheep oocytes and in vitro development of preimplantation sheep embryos to the blastocyst stage. *J. Assist. Reprod. Genet.* **2010**, *27*, 483–490. [CrossRef] [PubMed]

antioxidants

MDPI

Review

Vitamin E as a Treatment for Nonalcoholic Fatty Liver Disease: Reality or Myth?

Hamza El Hadi, Roberto Vettor and Marco Rossato *

Internal Medicine 3, Department of Medicine—DIMED, University of Padova, Via Giustiniani 2, 35128 Padova, Italy; dr.hamza.elhadi@gmail.com (H.E.H.); roberto.vettor@unipd.it (R.V.)
* Correspondence: marco.rossato@unipd.it; Tel.: +39-049-821-8747; Fax:+39-049-821-3332

Received: 1 November 2017; Accepted: 10 January 2018; Published: 16 January 2018

Abstract: Obesity is one of the major epidemics of this millennium, and its incidence is growing worldwide. Following the epidemics of obesity, nonalcoholic fatty liver disease (NAFLD) has become a disease of increasing prevalence and a leading cause of morbidity and mortality closely related to cardiovascular disease, malignancies, and cirrhosis. It is believed that oxidative stress is a main player in the development and progression of NAFLD. Currently, a pharmacological approach has become necessary in NAFLD because of a failure to modify lifestyle and dietary habits in most patients. Vitamin E is a potent antioxidant that has been shown to reduce oxidative stress in NAFLD. This review summarizes the biological activities of vitamin E, with a primary focus on its therapeutic efficacy in NAFLD.

Keywords: vitamin E; nonalcoholic fatty liver disease; obesity; oxidative stress

1. Introduction

Nonalcoholic fatty liver disease (NAFLD) is defined as the accumulation of excessive fat in the liver, as demonstrated by imaging or by histology, in the setting of no significant alcohol consumptionand the absence of any secondary cause [1].

NAFLD encompasses a broad pathological spectrum of phenotypes ranging from isolated hepatic steatosis (IHS) to nonalcoholic steatohepatitis (NASH)—the progressive form of fatty liver disease associated with inflammation and cellular injury, which can lead to NASH-related cirrhosis and hepatocellular carcinoma [2]. This pathology is now regarded as a leading cause of chronic liver diseases and liver transplantation in most countries [3]. In addition, NAFLD has been also linked to extra-hepatic morbidity, including systemic metabolic complications, chronic kidney and cardiovascular disease, and malignancies, which all contribute to a higher mortality observed in NASH patients [4].

NAFLD is strongly associated with obesity and related metabolic disorders such as insulin resistance and dyslipidemia. In the last decades, adult and childhood obesity has reached epidemic levels, and as a consequence the global prevalence of NAFLD has increased significantly. According to a recent report, prevalence estimates in the general population of Europe and the Middle East are 20–30%, with higher prevalence in Western countries' populations with obesity or diabetes (75%) and with morbid obesity (90–95%) [5].

It is acknowledged that vitamin E is the major lipid-soluble chain-breaking antioxidant found in the human body. In addition to its anti-oxidative properties, molecules of the vitamin E family exertanti-atherogenicand anti-inflammatory activities [6]. Although the pathogenesis of NAFLD and its progression to fibrosis needs to be fully clarified, it is believed that oxidative stress plays a crucial role in producing the lethal hepatocyte injury associated with NAFLD.

Therefore, by targeting oxidative stress components, vitamin E appears as a promising therapeutic approach in NASH patients.

The present review briefly discusses the biological activities of vitamin E, focusing on its potential as a treatment for NAFLD/NASH. To this aim, we also highlight the role of oxidative stress in the pathogenesis of NAFLD.

2. Vitamin E: Brief Overview of Structure, Metabolism and Function

2.1. Structural Perspectives

The discovery of vitamin E dates back to 1922 due to the observations of Herbert Evans and his associate Bishop who isolated an as-yet uncharacterized fat-soluble compound from green leafy vegetables which is required for reproduction in rats [7]. Upon isolating this dietary compound, it was termed as tocopherol (Greek: tocos—child birth; pheros—to bear; ol—alcohol). Today the term "vitamin E" encompasses a group of eight lipophilic molecules that are synthesized by plants starting from homogentisic acid. It includes four tocopherols and four tocotrienols. Tocopherols and tocotrienols are subdivided into alpha (α), beta (β), gamma (γ), and delta (δ) forms based on the methyl and hydroxyl substitution in their phenolic rings [8] (Figure 1). The tocopherols are saturated forms of vitamin E, whereas the tocotrienols are unsaturated and possess an isoprenoid side chain [8]. α-Tocopherol is considered the most abundant form in nature, and is consequently the most widely studied. Common food oils including corn, peanut, and soybean oil contain largely α-tocopherol. In contrast, tocotrienols are relatively rarer in food sources and prevail in rice bran, barley, oats, and palm oil [9].

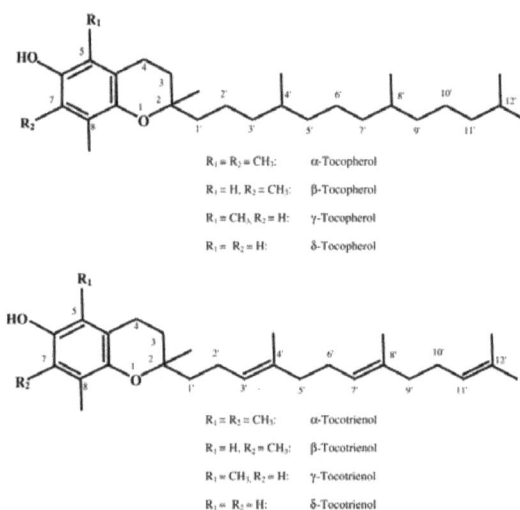

Figure 1. Structure of tocopherols and tocotrienols (with permission from reference [8]).

Currently, synthetic forms of vitamin E consist mainly of α-tocopherol, whichwas first synthesized in 1938 [10]. Unlike naturally-occurring d-RRR-α-tocopherol, synthesized α-tocopherol consists of a racemic mixture of eight stereoisomers named the all-racemic or (2RS, 4'RS, 8'RS) product [11].

2.2. Insight into Metabolism

Like other fat-soluble vitamins, the bioavailability of vitamin E depends on pancreatic function, biliary secretion, micellar formation, and penetration across intestinal membranes. After being cleaved by the enzyme esterase located in the stomach lining and partly processed enzymatically in the stomach by gastric lipase, vitamin E isoforms reach the basolateral side of enterocytes [12,13]. In the intestinal lumen, dietary tocopherols and tocotrienols appear to be similarly absorbed along with dietary fat, and are secreted in chylomicron particles together with triacylglycerol, phospholipids, and cholesterol [12,13]. The chylomicron-bound vitamin E isoforms are then transferred to peripheral tissues such as adipose tissues, bones, brain, lung, muscle, and skin via the lymphatic system. During their transport, chylomicrons are catabolized in the circulation by the endothelial-bound enzyme, which hydrolyzes triglycerides, releasing free fatty acids (FFAs) [12,13]. The resulting chylomicron remnants containing absorbed vitamin E are subsequently taken up by the liver, probably by a receptor-mediated process, and then preferentially incorporated within very low density lipoprotein (VLDL) and high-density lipoprotein (HDL) into the bloodstream [12]. After hepatic uptake, the α-tocopherol form of vitamin E is preferentially re-excreted into the circulation. α-Tocopherol transfer protein (α-TTP), a small cytoplasmic hepatic protein with differential affinity for various vitamin E forms, is responsible for the biodiscriminating process underlying the selective resecretion of α-tocopherol from the liver into plasma [14].

While α-TTP has a high affinity for α-tocopherol (100%), it has a lower affinity for other vitamin E isoforms: approximately 50% for β-tocopherol, 10–30% for γ-tocopherol, and 1% for δ-tocopherol [12]. In the liver, isoforms not bound to α-TTP will be susceptible to catabolization via cytochrome P450 (CYP4F2)-initiated ω-hydroxylation and oxidation by ω-hydroxylase, and thus vitamin E isoforms are metabolized to carboxychromanols, hydroxycarboxychromanol, and carboxyethylhydroxychroman derivatives [15]. Besides catabolism, it was estimated that metabolized vitamin E isoforms are also discarded via biliary excretion [16].

2.3. Biological Activity of Vitamin E

2.3.1. Antioxidant Activity

Oxidative stress is defined as the imbalance between the generation of reactive species and antioxidant defense, and leads to the damage of DNA and disturbances in cellular biology [17]. Vitamin E is widely accepted as one of the most potent antioxidants in nature [18]. The antioxidant property is attributed to the hydroxyl group from the aromatic ring of tocochromanols, which donates hydrogen to neutralize free radicals or reactive oxygen species (ROS). The antioxidant activity of α-, β-, γ-isoforms of tocotrienols and tocopherols is similar, but the γ-isoform showed weaker activity when tested in pyrogallolsulfonphthalein and 2,7-dichlorodihydrofluorescein diacetate methods [18,19].

In the same context, the antioxidant efficacy of vitamin E on reactive nitrogen species (RNS) has been gaining more attention recently. RNS include nitric oxide (NO), nitrogen dioxide (NO_2), and peroxynitrite ($ONOO^-$) [20–22].

On the other hand, in vitro studies have shown that vitamin E can alternatively switch to a pro-oxidant action under certain circumstances, such as a constant low-level flux of initiator free radicals and the absence of co-antioxidants as such as vitamin C [23]. In addition, current evidence from in vivo studies showed that vitamin E may produce pro-oxidant effects at high doses [24] or in cigarette smokers consuming a high polyunsaturated fat diet [25].

2.3.2. Beyond Vitamin E Antioxidant Activity

Vitamin E biological activity is not limited to antioxidant properties. In fact, vitamin E forms are involved in the regulation of inflammatory response, gene expression, membrane-bound enzymes, modulation of cellular signaling, and cell proliferation.

Over the last two decades, vitamin E has been shown to have direct and indirect effects on several enzymes involved in signal transduction, such as protein kinase C (PKC), protein phosphatase 2A (PP2A), protein tyrosine phosphatase (PTP), protein tyrosine kinase (PTK), diacylglycerol kinase (DAGK), 5-, 12- and 15-lipoxygenases (5-, 12-, and 15-LOX), phospholipase A2 (PLA2), cyclooxygenase-2 (COX-2), and the mitogen activated protein kinase (MAPK) signal transduction pathway [26,27].

The first evidence that vitamin E can modulate enzymes involved in signal transduction came from studies with PKC when α-tocopherol exerted an inhibitory effect unrelated to antioxidant action [28]. PKC inhibition by α-tocopherol is mainly correlated with the reduction of cell proliferation in many different cell types, including vascular smooth muscle cells, monocytes/macrophages, neutrophils, fibroblasts, mesangial cells, as well as various cancer cell lines [29]. In addition to interference with cell proliferation, the inhibition of PKC by α-tocopherol inhibited NADPH-oxidase assembly in monocytes, leading to lower superoxide production [30]. Additionally, PKC inhibition mediated by α-tocopherol and not by β-tocopherol suppressed endothelin secretion in endothelial cells [31].

Another evidence of a non-antioxidant function of α-tocopherol is related to its role in regulating the expression of specific genes not only coupled to oxidative stress but also involved in cholesterol homeostasis, inflammatory pathways, and cellular trafficking [32,33]. These genes include those encoding for proteins involved in inflammation and cell adhesion (such as E-selectin, intercellular adhesion molecule-1, vascular cell adhesion molecule [VCAM]-1, integrin, interleukin [IL]-1b, IL-2, IL-4, and transforming growth factor [TGF]-β), extracellular matrix formation and degradation (tropomyosin, glycoprotein IIb, collagen A1, matrix metalloproteinase [MMP]-1, MMP-19 and connective tissue growth factor), cell cycle regulation (cyclin D1, cyclin E1, and p27), transcriptional control lipoprotein (peroxisome proliferator-activated receptor [PPAR]-γ), receptors (CD36, scavenger receptor class B type 1, low density lipoprotein receptor), and metabolism (Cytochrome P450 3A4 [CYP3A4] and HMG-CoA reductase) [32,33].

3. Vitamin E and NAFLD

3.1. NAFLD/NASH Pathogenesis

NAFLD is a complex disease trait where inter-patient genetic and epigenetic variations and environmental factors are combined to define development and disease progression [34] (Figure 2).

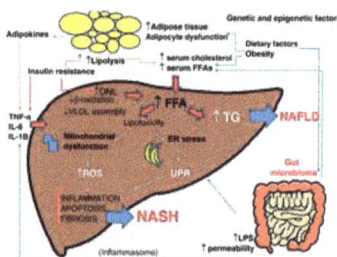

Figure 2. Multiple parallel-hit hypothesis of nonalcoholic fatty liver disease (NAFLD). Obesity together with dietary habits and environmental factors can lead to raised serum levels of free fatty acids (FFAs) and cholesterol, development of insulin resistance, adipocyte proliferation, and dysfunction in the intestinal microbiome. Insulin resistance acts on adipose tissue, worsening adipocyte dysfunction, induces hepatic de novo lipogenesis (DNL) and release of proinflammatory adipokines such as interleukin (IL)-6, IL-1β, and tumor necrosis factor (TNF)-α, which also exacerbates the insulin resistance state. The increased hepatic FFAs flux which derives from the above processes and from an altered activity of the gut microbiome leads to accumulation of triglycerides (TGs) and "toxic"levels of FFAs, free cholesterol, and other lipid metabolites which cause mitochondrial dysfunction, oxidative stress with the production of reactive oxygen species (ROS), and endoplasmic reticulum (ER) stress with the activation of the unfolded protein response (UPR), all leading to hepatic inflammation and fibrogenesis (nonalcoholic steatohepatitis, NASH). Increased intestinal permeability of gut-derived microbial products such as lipopolysaccharides (LPS) contributes to the activation of the inflammasome, ER stress, and activation of inflammatory cascades. Epigenetic factors are also involved in progression to NASH or persistence in a stable stage of disease (with permission from [34]). VLDL: very low density lipoprotein.

The generally accepted dogma in the pathogenesis of NAFLD is that insulin resistance—commonly associated with obesity—leads to the hepatic accumulation of triglycerides, a process that usually results from increased FFAs flux from adipose tissue to the liver, dietary fat via chylomicron metabolism, and increased de novo lipogenesis [35–37]. Obesity generates a state of low-grade inflammation characterized by the accumulation of immune cells in adipose tissue (particularly macrophages), and the production of proinflammatory cytokines by adipocytes contributes to the development of systemic and hepatic insulin resistance [35–37]. However, multiple pathways (multiple hits) are involved in the development of NASH and fibrosis. "Hits" that may contribute include oxidative stress, endotoxins, changes in the gut–liver axis, and mitochondrial dysfunction [34].

Oxidative stress and mitochondrial dysfunction were proposed as main triggers for the progression of steatosis to steatohepatitis [38]. FFAs catabolism in the liver takes place mainly via mitochondrial β-oxidation—a process that can lead to the generation of ROS, including superoxide, hydrogen peroxide, and hydroxyl radicals, in the case of increased FFAs delivery [38].

In the same context, impaired mitochondrial activity in NASH patients due to reduced enzymatic activities of mitochondrial electron transport chain and excessive in FA oxidation results in hepatic ATP depletion and may cause structural mitochondrial abnormalities consisting of enlargement (megamitochondria), paracrystalline inclusions, and loss of cristae [39,40].

In NAFLD, enhanced cytochrome P450 2E1 (CYP2E1) expression and activity seem to be an important source of ROS which trigger oxidative stress and perpetuate the hepatic mitochondrial dysfunction [41]. Moreover, it has been reported that upregulated microsomal CYP4A enzymes ω-hydroxylate fatty acids into dicarboxylic acids that are then preferentially oxidized by peroxisomes, thus promoting ROS production in NAFLD [42]. The ablation of a homolog of human CYP4A gene (CYP4A14) in animal models of steatohepatitis has shown to attenuate hepatic inflammation and fibrosis [43]. The abundant production of ROS induces the peroxidation of hepatic triglycerides with the release of reactive aldehydes such as 4-hydroxynonenal (4-HNE) and malondialdehyde (MDA) which can damage mitochondrial components [44].

Accumulating data have implicated the disruption of endoplasmic reticulum (ER) homeostasis (i.e., ER stress) in the development of NASH [45]. Factors that disturb ER folding capacity (e.g., excessive protein synthesis, mitochondrial dysfunction, oxidative stress) will lead to the activation of a physiologic mechanism called the "unfolded protein response" (UPR) in hepatocytes. This adaptive mechanism aims to increase the folding capacity of the ER, thus bringing the organelle and the cell into a state of equilibrium. When the activation of the UPR fails to promote cell survival, the cell is taken down the pro-apoptotic ER stress response pathway, which can ultimately lead to apoptotic cell death, inflammation, and/or fat accumulation [45].

Several studies have suggested a role of gut microbiome in NASH pathogenesis [46]. Intestinal barrier alterations cause increased intestinal permeability of bacteria, viruses, or microbial products such as lipopolysaccharide (LPS). These pathogens are recognized through specialized recognition receptors that include toll-like receptors (TLRs) and inflammasomes, inducing a signaling cascade leading to the production of inflammatory cytokines [46].

3.2. Vitamin E and NAFLD: Experimental Studies

The methionine and choline deficient (MCD) diet is a widely employed diet in NASH animal studies. The MCD diet is high in sucrose and fat (40% sucrose, 10% fat), and is deficient in methionine and choline, which are essential for hepatic beta-oxidation and the production of VLDL [47].

A trial carried out with rats fed with an MCD-diet showed that vitamin E significantly reduces the oxidative stress. In the control group, the authors observed a depletion in the liver glutathione stores and a notable increase in hepatic fibrosis, whereas vitamin E supplementation repleted hepatic glutathione, reduced steatosis, inflammation, hepatic stellate cell activation, and collagen mRNA expression, and ameliorated fibrosis [48].

In another animal NASH model, the combination of MCD diet and vitamin E significantly lowered serum transaminase levels and ameliorated hepatic steatosis and necroinflammation. These effects were associated with suppressed expression of the fibrotic genes TGF-β and MMP-2, inflammatory factor COX-2, and pro-apoptotic genes (Bax). In addition, vitamin E enhanced the activity of hepatic superoxide dismutase (SOD) and inhibited that of nuclear factor kappa B (NFkB) [49].

Similar effects were reported in arecent mouse model for NAFLD in which vitamin E therapy after partial hepatectomy significantly reduced the oxidative stress level and attenuated the progression of NAFLD [50].

In an obese (ob/ob) mouse model, α- or γ-tocopherol exerted a hepatoprotective role in alipopolysaccharide-induced NASH, as shown by suppressing hepatic malondialdehyde (MDA), tumor necrosis factor-α, and serum alanine aminotransferase levels [51].

Moreover, chickens fed ahigh-oxidant diet with the supplementation of vitamin E were able to normalize elevated hepatic transaminase levels [52]. In addition to the role of being an antioxidant, some studies have also proposed that vitamin E improves the liver integrity by down-regulating hepatic cluster of differentiation 36 protein (CD36)—a membrane transporter responsible for the uptake of fatty acids into the liver [53].

3.3. Vitamin E and NAFLD: Human Studies

Vitamin E has been used in monotherapy or with other agents in multiple clinical trials to treat NAFLD or NASH, with reported improvement in liver biochemistries and histology [54–68]. These trials varied in duration (24 weeks to >2 years) and dose (100–1200 IU/day) of vitamin E used. Long-term studies (≥2 years) are summarized in Table 1.

Treatment with vitamin E combined with vitamin C and atorvastatin was demonstrated to be effective in reducing the odds of having hepatic steatosis in individuals with computed tomography (CT)-diagnosed NAFLD after 4 years of active therapy [63].

The combination of ursodeoxycholic acid (UDCA) with vitamin E has been evaluated in a small sample size compared with UDCA alone or placebo. Improvement in steatosis and transaminases level were observed only in the group who received combination therapy with UDCA and vitamin E [64]. By evaluating the long-term (>2 years) efficacy of a similar combination (UDCA with vitamin E) in patients with NASH, Piettu et al.demonstrated an improvement in histological lesions in the majority of patients [62].

Nobili et al.studied 90 children with NAFLD who were given calorie-restricted diet and exercise. Patients were randomized to treatment with vitamin E (600 IU/day) in combination with ascorbic acid (500 mg/daily; $n = 45$) or placebo ($n = 45$) [65]. At the end of 24 months, both groups had significant improvement in steatosis, lobular inflammation, hepatocyte ballooning. However, the addition of α-tocopherol and ascorbic acid was not associated with a greater histological or biochemical improvement as compared to placebo [65,66].

Clinical trials with vitamin E supplementation in NASH patients yielded promising results. In the PIVENS (Pioglitazone, Vitamin E or Placebo for Nonalcoholic Steatohepatitis) trial, vitamin E was evaluated as a treatment for NASH. The rate of achievement of the primary outcome was higher in patients treated with high-dose vitamin E (800 IU/day) for 96 weeks compared to placebo (43% vs. 19%, $p = 0.001$), while pioglitazone did not reach statistical significance. The histological analysis showed a reduction in hepatocyte ballooning (50% vs. 29%, $p = 0.005$) and lobular inflammation (54% vs. 35%, $p = 0.02$), thus reflecting its expected effect as an antioxidant leading to a decrease of oxidative stress-mediated injury. Interestingly, it significantly reduced liver steatosis and alanine aminotransferase (ALT), but had no significant changes on fibrosis [60].

Table 1. Effects of Vitamin E administration in patients with NAFLD.

| Author | Study Design | Duration | Vitamin E Dosage | Effect of Vitamin E on NAFLD/NASH Histology and Biochemistry Compared to Placebo | | | | |
				ALT	Steatosis	Inflammation	Hepatocytes Ballooning	Fibrosis
Dufour et al., 2006 [64]	48 NASH patients randomized to vitamin E plus UDCA, UDCA versus placebo	2 years	400 IU twice daily	↓	↓	ns	NA	ns
Nobili et al., 2008 [66]	53 NAFLD patients randomized to vitamin E plus ascorbic and life style intervention versus placebo	2 years	600 IU/day	ns	ns	ns	ns	ns
Sanyal et al., 2010 [60]	247 NASH patients randomized to vitamin E, pioglitazone versus placebo	96 weeks	800 IU daily	↓		↓	↓	ns
Foster et al., 2011 [63]	1005 randomized to vitamin E combined with ascorbic acid and atorvastatin versus placebo. At baseline 80 had CT-diagnosed NAFLD	4 years	1000 IU daily	ns	↓ (Based on abdominal CT scan)	NA	NA	NA
Lavine et al., 2011 [61]	173 NAFLD patients randomized to vitamin E, metformin versus placebo	96 weeks	400 IU twice daily	ns	ns	ns	↓	ns
Pietu et al., 2012 [62]	101 patients treated with a combination of vitamin E with UDCA. 10 patients had a second liver biopsy during follow-up	4 years	500 IU daily	↓	↓ 3/10 ↑1/10 patients	↓ 3/10 ↑2/10 patients	↓ 3/10 ↑1/10patients	↓ 4/10 ↑1/10 patients

ALT: alanine aminotransferase; CT: computed tomography; NAFLD: nonalcoholic fatty liver disease; NASH: nonalcoholic steatohepatitis; ns: no significant statistical difference versus placebo; NA: not available; UDCA: ursodeoxycholic acid; ↑: worsening; ↓: improvement.

Following those studies, Lavine et al. conducted a multicenter, double-blind, double-placebo, randomized clinical trial in pediatric patients. The TONIC (Treatment of NAFLD in Children) trial involved 173 children and adolescents witha mean age of 13 years that received metformin (500 mg twice daily), vitamin E (400 IU twice daily), or placebo twice daily for 96 weeks. The primary outcome was sustained reduction in ALT level (defined as $\leq$50% baseline or $\leq$40 U/L from 48 weeks to 96 weeks of treatment). The only histologic feature of NASH that improved after treatment with both medications was the hepatocellular ballooning. Disappointingly, neither vitamin E nor metformin were superior to placebo in achieving sustained ALT reduction or in improving steatosis, lobular inflammation, or fibrosis scores [61].

In the light of these observations, today the European Association for the Study of the Liver (EASL) and the American Association for the Study of Liver Diseases (AASLD) (Alexandria, VA, USA) guidelines consider vitamin E as a potential short-term treatment for non-diabetic adults with biopsy-proven NASH [69]. Until further data supporting its effectiveness is available, vitamin E is not recommended to treat NASH in diabetic patients, NAFLD without liver biopsy, NASH cirrhosis, or cryptogenic cirrhosis [1,69].

3.4. Vitamin E: Safety Concerns

Long-term safety should be carefully discussed with NASH patients before starting therapy with vitamin E.A meta-analysis has suggested that high-dosage ($\geq$400 IU/day) vitamin E supplements may increase all-cause mortality [70]. However, this meta-analysis has been criticized because several studies with low mortality were excluded and concomitant vitamin A and the administration of other drugs as well as common factors such as smoking were not considered. Another meta-analysis where new clinical trials and updated results of mortality were included suggested that the higher mortality can be explained by a higher proportion of male patients that were included in these trials and not due to the higher dose of vitamin E supplementation [71]. On the other hand, large meta-analysis that included 57 trials showed that vitamin E supplementation appears to have no effect on all-cause mortality at doses up to 5500 IU/day [72].

Finally, a meta-analysis investigating the effect of vitamin E on the incidence of stroke reported an increase in the relative risk of hemorrhagic stroke by 22%, while the risk of ischemic stroke was reduced by 10% [73]. Another concern about vitamin E use is related to its association with a modest increase in the risk of prostate cancer [74].

Thus, patients chosen for treatment with vitamin E should be aware of these risks or be considered for alternative treatments.

4. Conclusions

Despite its high prevalence and the intensive research in the field, the treatment of NAFLD remains an unmet medical need. To date, no definite pharmacological treatment has been approved for NAFLD, and patients are often advised to engage in physical activity and lose weight, which is difficult to achieve and more difficult to maintain. However, policies to promote physical activity and the management of associated comorbidities (i.e., obesity and metabolic syndrome components) are expected to decrease both hepatic and cardiovascular-related morbidity and mortality in NASH patients.

Clinical trials for NAFLD or NASH showed a modest improvement in liver biochemistries and histology induced by vitamin E administration. However, some limitations of these trials should be noted. The duration of therapy may be not long enough to detect long-term histological change complications. Moreover, the number of randomized trials included was limited, and the number of participants in the trials was low. An additional reason for the lack of efficacy of vitamin E could be related to the use of oral preparations of varied dosage, which may not always guarantee an adequate bioavailability in patients with liver disease.

In this context, further monotherapy clinical trials and pharmacological evaluations are still needed to elucidate the underlying molecular mechanisms of prevention or therapy and possible adverse outcomes. This may also help to identify the optimal daily intake of vitamin E in both pediatric and adult patients.

In addition, more research is needed to seek novel biological activities of vitamin Emetabolites in the liver of NAFLD patients that may differ from those of parent compounds.

Finally, a better understanding of the pathophysiology of NAFLD/NASH will provide the opportunity to create trials of combination therapies that achieve high rates of therapeutic responses.

Author Contributions: Hamza El Hadi, Roberto Vettor and Marco Rossato designed the concept and wrote the manuscript.

Conflicts of Interest: The authors declare no conflict of interest.

References

1. Chalasani, N.; Younoss, Z.; Lavine, J.E.; Diehl, A.M.; Brunt, E.M.; Cusi, K.; Charlton, M.; Sanyal, A.J. The diagnosis and management of non-alcoholic fatty liver disease: Practice Guideline by the American Association for the Study of Liver Diseases, American College of Gastroenterology, and the American Gastroenterological Association. *Hepatology* **2012**, *55*, 2005–2023. [CrossRef] [PubMed]
2. Polyzos, S.A.; Kountouras, J.; Zavos, C. Nonalcoholic fatty liver disease: The pathogenetic roles of insulin resistance and adipocytokines. *Curr. Mol. Med.* **2009**, *72*, 299–314. [CrossRef]
3. Khan, R.S.; Newsome, P.N. Non-alcoholic fatty liver disease and liver transplantation. *Metabolism* **2016**, *65*, 1208–1223. [CrossRef] [PubMed]
4. Rinella, M.E. Nonalcoholic fatty liver disease: A systematic review. *JAMA* **2015**, *313*, 2263–2273. [CrossRef] [PubMed]
5. Review, T.; LaBrecque, D.R.; Abbas, Z.; Anania, F.; Ferenci, P.; Khan, A.G.; Goh, K.L.; Hamid, S.S.; Isakov, V.; Lizarzabal, M.; et al. World Gastroenterology Organisation global guidelines: Nonalcoholic fatty liver disease and nonalcoholic steatohepatitis. *J. Clin. Gastroenterol.* **2014**, *48*, 467–473.
6. El Hadi, H.; Vettor, R.; Rossato, M. Congenital Vitamin E deficiency. In *Handbook of Famine, Starvation, and Nutrient Deprivation*; Preedy, V.R., Patel, V.B., Eds.; Springer International Publishing AG: Basel, Switzerland, 2018; pp. 1–18, ISBN 978-3-319-40007-5.
7. Evans, H.M.; Bishop, K.S. On the existence of a hitherto unrecognized dietary factor essential for reproduction. *Science* **1922**, *56*, 650–651. [CrossRef] [PubMed]
8. Gee, P.T. Unleashing the untold and misunderstood observations on vitamin E. *Genes Nutr.* **2011**, *6*, 5–16. [CrossRef] [PubMed]
9. Sheppard, A.J.; Pennington, J.A.T.; Weihrauch, J.L. Analysis and distribution of vitamin E in vegetable oils and foods. In *Vitamin E in Health and Disease*; Packer, L., Fuchs, J., Eds.; Marcel Dekker Inc.: New York, NY, USA, 1993; pp. 9–31.
10. Karrer, P.; Fritzsche, H.; Ringier, B.; Salomon, H. Synthesis of alpha-tocopherol (vitamin E). *Nature* **1938**, *141*, 1057. [CrossRef]
11. Jensen, S.K.; Lauridsen, C. Alpha-tocopherol stereoisomers. *Vitam. Horm.* **2007**, *76*, 281–308. [PubMed]
12. Traber, M.G. Vitamin E regulatory mechanisms. *Annu. Rev. Nutr.* **2007**, *27*, 347–362. [CrossRef] [PubMed]
13. Kayden, H.J.; Traber, M.G. Absorption, lipoprotein transport, and regulation of plasma concentrations of vitamin E in humans. *J. Lipid Res.* **1993**, *34*, 343–358. [PubMed]
14. Kaempf-Rotzoll, D.E.; Traber, M.G.; Arai, H. Vitamin E and transfer proteins. *Curr. Opin. Lipidol.* **2003**, *14*, 249–254. [CrossRef] [PubMed]
15. Jiang, Q. Natural forms of vitamin E: Metabolism, antioxidant, and anti-inflammatory activities and their role in disease prevention and therapy. *Free Radic. Biol. Med.* **2014**, *72*, 76–90. [CrossRef] [PubMed]
16. Bardowell, S.A.; Duan, F.; Manor, D.; Swanson, J.E.; Parker, R.S. Disruption of mouse cytochrome p4504f14 (Cyp4f14 gene) causes severe perturbations in vitamin E metabolism. *J. Biol. Chem.* **2012**, *287*, 26077–26086. [CrossRef] [PubMed]

17. Birben, E.; Sahiner, U.M.; Sackesen, C.; Erzurum, S.; Kalayci, O. Oxidative stress and antioxidant defense. *World Allergy Organ J.* **2012**, *5*, 9–19. [CrossRef] [PubMed]

18. Peh, H.Y.; Tan, W.S.; Liao, W.; Wong, W.S. Vitamin E therapy beyond cancer: Tocopherol versus tocotrienol. *Pharmacol. Ther.* **2016**, *162*, 152–169. [CrossRef] [PubMed]

19. Yoshida, Y.; Saito, Y.; Jones, L.S.; Shigeri, Y. Chemical reactivities and physical effects in comparison between tocopherols and tocotrienols: Physiological significance and prospects as antioxidants. *J. Biosci. Bioeng.* **2007**, *104*, 439–445. [CrossRef] [PubMed]

20. Cooney, R.V.; Harwood, P.J.; Franke, A.A.; Narala, K.; Sundström, A.K.; Berggren, P.O.; Mordan, L.J. Products of gamma-tocopherol reaction with NO_2 and their formation in rat insulinoma (RINm5F) cells. *Free Radic. Biol. Med.* **1995**, *19*, 259–269. [CrossRef]

21. Christen, S.; Jiang, Q.; Shigenaga, M.K.; Ames, B.N. Analysis of plasma tocopherols α, γ, and 5-nitro-γ in rats with inflammation by HPLC coulometric detection. *J. Lipid Res.* **2002**, *43*, 1978–1985. [CrossRef] [PubMed]

22. Berbée, M.; Fu, Q.; Boerma, M.; Wang, J.; Kumar, K.S.; Hauer-Jensen, M. gamma-Tocotrienol ameliorates intestinal radiation injury and reduces vascular oxidative stress after total-body irradiation by an HMG-CoA reductase-dependent mechanism. *Radiat. Res.* **2009**, *171*, 596–605. [CrossRef] [PubMed]

23. Upston, J.M.; Terentis, A.C.; Stocker, R. Tocopherol-mediated peroxidation of lipoproteins: Implications for vitamin E as a potential antiatherogenic supplement. *FASEB J.* **1999**, *13*, 977–994. [PubMed]

24. Pearson, P.; Lewis, S.A.; Britton, J.; Young, I.S.; Fogarty, A. The pro-oxidant activity of high-dose vitamin E supplements in vivo. *BioDrugs* **2006**, *20*, 271–273. [CrossRef] [PubMed]

25. Weinberg, R.B.; VanderWerken, B.S.; Anderson, R.A.; Stegner, J.E.; Thomas, M.J. Pro-oxidant effect of vitamin E in cigarette smokers consuming a high polyunsaturated fat diet. *Arterioscler. Thromb. Vasc. Biol.* **2001**, *21*, 1029–1033. [CrossRef] [PubMed]

26. Zingg, J.M.; Azzi, A. Non-antioxidant activities of vitamin E. *Curr. Med. Chem.* **2004**, *11*, 1113–1133. [CrossRef] [PubMed]

27. Rimbach, G.; Minihane, A.M.; Majewicz, J.; Fischer, A.; Pallauf, J.; Virgli, F.; Weinberg, P.D. Regulation of cell signalling by vitamin E. *Proc. Nutr. Soc.* **2002**, *61*, 415–425. [CrossRef] [PubMed]

28. Boscoboinik, D.; Szewczyk, A.; Hensey, C.; Azzi, A. Inhibition of cell proliferation by alpha-tocopherol. Role of protein kinase C. *J. Biol. Chem.* **1991**, *266*, 6188–6194. [PubMed]

29. Cook-Mills, J.M. Isoforms of Vitamin E Differentially Regulate PKC α and Inflammation: A Review. *J. Clin. Cell. Immunol.* **2013**, *4*, 1000137. [CrossRef] [PubMed]

30. Venugopal, S.K.; Devaraj, S.; Yang, T.; Jialal, I. Alpha-tocopherol decreases superoxide anion release in human monocytes under hyperglycemic conditions via inhibition of protein kinase C-alpha. *Diabetes* **2002**, *51*, 3049–3054. [CrossRef] [PubMed]

31. Martin-Nizard, F.; Boullier, A.; Fruchart, J.C.; Duriez, P. Alpha-tocopherol but not beta-tocopherol inhibits thrombin-induced PKC activation and endothelin secretion in endothelial cells. *J. Cardiovasc. Risk* **1998**, *5*, 339–345. [CrossRef] [PubMed]

32. Rimbach, G.; Moehring, J.; Huebbe, P.; Lodge, J.K. Gene-regulatory activity of alpha-tocopherol. *Molecules* **2010**, *15*, 1746–1761. [CrossRef] [PubMed]

33. Ahsan, H.; Ahad, A.; Iqbal, J.; Siddiqui, W.A. Pharmacological potential of tocotrienols: A review. *Nutr. Metab.* **2014**, *11*, 52. [CrossRef] [PubMed]

34. Buzzetti, E.; Pinzani, M.; Tsochatzis, E.A. The multiple-hit pathogenesis of non-alcoholic fatty liver disease (NAFLD). *Metabolism* **2016**, *65*, 1038–1048. [CrossRef] [PubMed]

35. Hassan, K.; Bhalla, V.; El Regal, M.E.; A-Kader, H.H. Nonalcoholic fatty liver disease: A comprehensive review of a growing epidemic. *World J. Gastroenterol.* **2014**, *20*, 12082–12101. [CrossRef] [PubMed]

36. Fabbrini, E.; Mohammed, B.S.; Magkos, F.; Korenblat, K.M.; Patterson, B.W.; Klein, S. Alterations in adipose tissue and hepatic lipid kinetics in obese men and women with nonalcoholic fatty liver disease. *Gastroenterology* **2008**, *134*, 424–431. [CrossRef] [PubMed]

37. Sanyal, A.J.; Campbell-Sargent, C.; Mirshahi, F.; Rizzo, W.B.; Contos, M.J.; Sterling, R.K.; Luketic, V.A.; Shiffman, M.L.; Clore, J.N. Nonalcoholic steatohepatitis: Association of insulin resistance and mitochondrial abnormalities. *Gastroenterology* **2001**, *120*, 1183–1192. [CrossRef] [PubMed]

38. Hardwick, R.N.; Fisher, C.D.; Canet, M.J.; Lake, A.D.; Cherrington, N.J. Diversity in antioxidant response enzymes in progressive stages of human nonalcoholic fatty liver disease. *Drug Metab. Dispos.* **2010**, *38*, 2293–2301. [CrossRef] [PubMed]

39. Day, C.P. Pathogenesis of steatohepatitis. *Best Prac. Res. Clin. Gastroenterol.* **2002**, *16*, 663–678. [CrossRef]

40. Caldwell, S.H.; Chang, C.Y.; Nakamoto, R.K.; Krugner-Higby, L. Mitochondria in nonalcoholic fatty liver disease. *Clin. Liver Dis.* **2004**, *8*, 595–617. [CrossRef] [PubMed]

41. Aubert, J.; Begriche, K.; Knockaert, L.; Robin, M.A.; Fromenty, B. Increased expression of cytochrome P450 2E1 in nonalcoholic fatty liver disease: Mechanisms and pathophysiological role. *Clin. Res. Hepatol. Gastroenterol.* **2011**, *35*, 630–637. [CrossRef] [PubMed]

42. Leclercq, I.A.; Farrell, G.C.; Field, J.; Bell, D.R.; Gonzalez, F.J.; Robertson, G.R. CYP2E1 and CYP4A as microsomal catalysts of lipid peroxides in murine nonalcoholic steatohepatitis. *J. Clin. Investig.* **2000**, *105*, 1067–1075. [CrossRef] [PubMed]

43. Zhang, X.; Li, S.; Zhou, Y.; Su, W.; Ruan, X.; Wang, B.; Zheng, F.; Warner, M.; Gustafsson, J.Å.; Guan, Y. Ablation of cytochrome P450 omega-hydroxylase 4A14 gene attenuates hepatic steatosis and fibrosis. *Proc. Natl. Acad. Sci. USA* **2017**, *114*, 3181–3185. [CrossRef] [PubMed]

44. Demeilliers, C.; Maisonneuve, C.; Grodet, A.; Mansouri, A.; Nguyen, R.; Tinel, M.; Lettéron, P.; Degott, C.; Feldmann, G.; Pessayre, D.; et al. Impaired adaptive resynthesis and prolonged depletion of hepatic mitochondrial DNA after repeated alcohol binges in mice. *Gastroenterology* **2002**, *123*, 1278–1290. [CrossRef] [PubMed]

45. Dara, L.; Ji, C.; Kaplowitz, N. The contribution of endoplasmic reticulum stress to liver diseases. *Hepatology* **2011**, *53*, 1752–1763. [CrossRef] [PubMed]

46. Marra, F.; Svegliati-Baroni, G. Lipotoxicity and the gut-liver axis in NASH pathogenesis. *J. Hepatol.* **2017**. [CrossRef] [PubMed]

47. Larter, C.Z.; Yeh, M.M.; Williams, J.; Bell-Anderson, K.S.; Farrell, G.C. MCD-induced steatohepatitis is associated with hepatic adiponectin resistance and adipogenic transformation of hepatocytes. *J. Hepatol.* **2008**, *49*, 407–416. [CrossRef] [PubMed]

48. Phung, N.; Pera, N.; Farrell, G.; Leclercq, I.; Hou, J.Y.; George, J. Pro-oxidant-mediated hepatic fibrosis and effects of antioxidant intervention in murine dietary steatohepatitis. *Int. J. Mol. Med.* **2009**, *24*, 171–180. [PubMed]

49. Nan, Y.M.; Wu, W.J.; Fu, N.; Liang, B.L.; Wang, R.Q.; Li, L.X.; Zhao, S.X.; Zhao, J.M.; Yu, J. Antioxidants vitamin E and 1-aminobenzotriazole prevent experimental non-alcoholic steatohepatitis in mice. *Scand. J. Gastroenterol.* **2009**, *44*, 1121–1131. [CrossRef] [PubMed]

50. Karimian, G.; Kirschbaum, M.; Veldhuis, Z.J.; Bomfati, F.; Porte, R.J.; Lisman, T. Vitamin E Attenuates the Progression of Non-Alcoholic Fatty Liver Disease Caused by Partial Hepatectomy in Mice. *PLoS ONE* **2015**, *10*, e0143121. [CrossRef] [PubMed]

51. Chung, M.Y.; Yeung, S.F.; Park, H.J.; Volek, J.S.; Bruno, R.S. Dietary α- and γ-tocopherol supplementation attenuates lipopolysaccharide-induced oxidative stress and inflammatory-related responses in an obese mouse model of nonalcoholic steatohepatitis. *J. Nutr. Biochem.* **2010**, *21*, 1200–1206. [CrossRef] [PubMed]

52. Lu, T.; Harper, A.F.; Zhao, J.; Corl, B.A.; LeRoith, T.; Dalloul, R.A. Effects of a dietary antioxidant blend and vitamin E on fatty acid profile, liver function, and inflammatory response in broiler chickens fed a diet high in oxidants. *Poult. Sci.* **2014**, *93*, 1658–1666. [CrossRef] [PubMed]

53. Podszun, M.C.; Grebenstein, N.; Spruss, A.; Schlueter, T.; Kremoser, C.; Bergheim, I.; Frank, J. Dietary alpha-tocopherol and atorvastatin reduce high-fat-induced lipid accumulation and down-regulate CD36 protein in the liver of guinea pigs. *J. Nutr. Biochem.* **2014**, *25*, 573–579. [CrossRef] [PubMed]

54. Lavine, J.E. Vitamin E treatment of nonalcoholic steatohepatitis in children: A pilot study. *J. Pediatr.* **2000**, *136*, 734–738. [CrossRef]

55. Hasegawa, T.; Yoneda, M.; Nakamura, K.; Makino, I.; Terano, A. Plasma transforming growth factor-beta1 level and efficacy of alpha-tocopherol in patients with non-alcoholic steatohepatitis: A pilot study. *Aliment. Pharmacol. Ther.* **2001**, *15*, 1667–1672. [CrossRef] [PubMed]

56. Kugelmas, M.; Hill, D.B.; Vivian, B.; Marsano, L.; McClain, C.J. Cytokines and NASH: A pilot study of the effects of lifestyle modification and vitamin E. *Hepatology* **2003**, *38*, 413–419. [CrossRef] [PubMed]

57. Yakaryilmaz, F.; Guliter, S.; Savas, B.; Erdem, O.; Ersoy, R.; Erden, E.; Akyol, G.; Bozkaya, H.; Ozenirler, S. Effects of vitamin E treatment on peroxisome proliferator-activated receptor-alpha expression and insulin resistance in patients with non-alcoholic steatohepatitis: Results of a pilot study. *Intern. Med. J.* **2007**, *37*, 229–235. [CrossRef] [PubMed]

58. Bugianesi, E.; Gentilcore, E.; Manini, R.; Natale, S.; Vanni, E.; Villanova, N.; David, E.; Rizzetto, M.; Marchesini, G. A randomized controlled trial of metformin versus vitamin E or prescriptive diet in nonalcoholic fatty liver disease. *Am. J. Gastroenterol.* **2005**, *100*, 1082–1090. [CrossRef] [PubMed]

59. Vajro, P.; Mandato, C.; Franzese, A.; Ciccimarra, E.; Lucariello, S.; Savoia, M.; Capuano, G.; Migliaro, F. Vitamin E treatment in pediatric obesity-related liver disease: A randomized study. *J. Pediatr. Gastroenterol. Nutr.* **2004**, *38*, 48–55. [CrossRef] [PubMed]

60. Sanyal, A.J.; Chalasani, N.; Kowdley, K.V.; McCullough, A.; Diehl, A.M.; Bass, N.M.; Neuschwander-Tetri, B.A.; Lavine, J.E.; Tonascia, J.; Unalp, A.; et al. Pioglitazone, vitamin E, or placebo for nonalcoholic steatohepatitis. *N. Engl. J. Med.* **2010**, *362*, 1675–1685. [CrossRef] [PubMed]

61. Lavine, J.E.; Schwimmer, J.B.; Van Natta, M.L.; Molleston, J.P.; Murray, K.F.; Rosenthal, P.; Abrams, S.H.; Scheimann, A.O.; Sanyal, A.J.; Chalasani, N.; et al. Effect of vitamin E or metformin for treatment of nonalcoholic fatty liver disease in children and adolescents: The TONIC randomized controlled trial. *JAMA* **2011**, *305*, 1659–1668. [CrossRef] [PubMed]

62. Pietu, F.; Guillaud, O.; Walter, T.; Vallin, M.; Hervieu, V.; Scoazec, J.Y.; Dumortier, J. Ursodeoxycholic acid with vitamin E in patients with nonalcoholic steatohepatitis: Long-term results. *Clin. Res. Hepatol. Gastroenterol.* **2012**, *36*, 146–155. [CrossRef] [PubMed]

63. Foster, T.; Budoff, M.J.; Saab, S.; Ahmadi, N.; Gordon, C.; Guerci, A.D. Atorvastatin and antioxidants for the treatment of nonalcoholic fatty liver disease: The St Francis Heart Study randomized clinical trial. *Am. J. Gastroenterol.* **2011**, *106*, 71–77. [CrossRef] [PubMed]

64. Dufour, J.F.; Oneta, C.M.; Gonvers, J.J.; Bihl, F.; Cerny, A.; Cereda, J.M.; Zala, J.F.; Helbling, B.; Steuerwald, M.; Zimmermann, A. Swiss Association for the Study of the Liver. Randomized placebo-controlled trial of ursodeoxycholic acid with vitamin E in nonalcoholic steatohepatitis. *Clin. Gastroenterol. Hepatol.* **2006**, *4*, 1537–1543. [CrossRef] [PubMed]

65. Nobili, V.; Manco, M.; Devito, R.; Ciampalini, P.; Piemonte, F.; Marcellini, M. Effect of vitamin E on aminotransferase levels and insulin resistance in children with non-alcoholic fatty liver disease. *Aliment. Pharmacol. Ther.* **2006**, *24*, 1553–1561. [CrossRef] [PubMed]

66. Nobili, V.; Manco, M.; Devito, R.; Di Ciommo, V.; Comparcola, D.; Sartorelli, M.R.; Piemonte, F.; Marcellini, M.; Angulo, P. Lifestyle intervention and antioxidant therapy in children with nonalcoholic fatty liver disease: A randomized, controlled trial. *Hepatology* **2008**, *48*, 119–128. [CrossRef] [PubMed]

67. Sanyal, A.J.; Mofrad, P.S.; Contos, M.J.; Sargeant, C.; Luketic, V.A.; Sterling, R.K.; Stravitz, R.T.; Shiffman, M.L.; Clore, J.; Mills, A.S. A pilot study of vitamin E versus vitamin E and pioglitazone for the treatment of nonalcoholic steatohepatitis. *Clin. Gastroenterol. Hepatol.* **2004**, *2*, 1107–1115. [CrossRef]

68. Harrison, S.A.; Torgerson, S.; Hayashi, P.; Ward, J.; Schenker, S. Vitamin E and vitamin C treatment improves fibrosis in patients with nonalcoholic steatohepatitis. *Am. J. Gastroenterol.* **2003**, *98*, 2485–2490. [CrossRef] [PubMed]

69. European Association for the Study of the Liver (EASL); European Association for the Study of Diabetes (EASD); European Association for the Study of Obesity (EASO). EASL-EASD-EASO Clinical Practice Guidelines for the management of non-alcoholic fatty liver disease. *J. Hepatol.* **2016**, *64*, 1388–1402.

70. Miller, E.R.; Pastor-Barriuso, R.; Dalal, D.; Riemersma, R.A.; Appel, L.J.; Guallar, E. Meta-analysis: High-dosage vitamin E supplementation may increase all-cause mortality. *Ann. Intern. Med.* **2005**, *142*, 37–46. [CrossRef] [PubMed]

71. Gerss, J.; Köpcke, W. The questionable association of vitamin E supplementation and mortality–inconsistent results of different meta-analytic approaches. *Cell. Mol. Biol.* **2009**, *55* (Suppl. OL), 1111–1120.

72. Abner, E.L.; Schmitt, F.A.; Mendiondo, M.S.; Marcum, J.L.; Kryscio, R.J. Vitamin E and all-cause mortality: A meta-analysis. *Curr. Aging Sci.* **2011**, *4*, 158–170. [CrossRef] [PubMed]

73. Schurks, M.; Glynn, R.J.; Rist, P.M.; Tzourio, C.; Kurth, T. Effects of vitamin E on stroke subtypes: Meta-analysis of randomised controlled trials. *BMJ* **2010**, *341*, c5702. [CrossRef] [PubMed]

74. Klein, E.A.; Thompson, I.M.; Tangen, C.M.; Crowley, J.J.; Lucia, M.S.; Goodman, P.J.; Minasian, L.M.; Ford, L.G.; Parnes, H.L.; Gaziano, J.M.; et al. Vitamin E and the risk of prostate cancer: The Selenium and Vitamin E Cancer Prevention Trial (SELECT). *JAMA* **2011**, *306*, 1549–1556. [CrossRef] [PubMed]

antioxidants

MDPI

Review

Antioxidant Tocols as Radiation Countermeasures (Challenges to be Addressed to Use Tocols as Radiation Countermeasures in Humans)

Ujwani Nukala [1,2], Shraddha Thakkar [3], Kimberly J. Krager [1], Philip J. Breen [1,4], Cesar M. Compadre [1,4] and Nukhet Aykin-Burns [1,4,*]

[1]   Department of Pharmaceutical Sciences, College of Pharmacy, University of Arkansas for Medical Sciences, Little Rock, AR 72205, USA; uxnukala@ualr.edu (U.N.); KJKrager@uams.edu (K.J.K.); BreenPhilipJ@uams.edu (P.J.B.); CompadreCesarM@uams.edu (C.M.C.)
[2]   Joint Bioinformatics Graduate Program, University of Arkansas at Little Rock, Little Rock, AR 72204, USA
[3]   Division of Bioinformatics and Biostatistics, National Center for Toxicological Research, US Food and Drug Administration, Jefferson, AR 72079, USA; Shraddha.Thakkar@fda.hhs.gov
[4]   Tocol Pharmaceuticals, LLC, Little Rock, AR 77205, USA
*   Correspondence: NAykinBurns@uams.edu

Received: 8 January 2018; Accepted: 22 February 2018; Published: 23 February 2018

Abstract: Radiation countermeasures fall under three categories, radiation protectors, radiation mitigators, and radiation therapeutics. Radiation protectors are agents that are administered before radiation exposure to protect from radiation-induced injuries by numerous mechanisms, including scavenging free radicals that are generated by initial radiochemical events. Radiation mitigators are agents that are administered after the exposure of radiation but before the onset of symptoms by accelerating the recovery and repair from radiation-induced injuries. Whereas radiation therapeutic agents administered after the onset of symptoms act by regenerating the tissues that are injured by radiation. Vitamin E is an antioxidant that neutralizes free radicals generated by radiation exposure by donating H atoms. The vitamin E family consists of eight different vitamers, including four tocopherols and four tocotrienols. Though alpha-tocopherol was extensively studied in the past, tocotrienols have recently gained attention as radiation countermeasures. Despite several studies performed on tocotrienols, there is no clear evidence on the factors that are responsible for their superior radiation protection properties over tocopherols. Their absorption and bioavailability are also not well understood. In this review, we discuss tocopherol's and tocotrienol's efficacy as radiation countermeasures and identify the challenges to be addressed to develop them into radiation countermeasures for human use in the event of radiological emergencies.

Keywords: tocopherol; tocotrienol; tocols; radioprotectors; radiation countermeasures; radiomitigators; alpha tocopherol transfer protein

1. Introduction

The use of ionizing radiation is increasing day by day for various purposes, including its clinical uses for diagnostic purposes and cancer treatment and in non-clinical applications, such as the nuclear generated energy production, engineering, construction, and sterilization of food products [1–3]. With such widespread uses, the likelihood of an intentional or unintentional encounter with radiation is quite high. The risk of radiation exposure has been increasing mainly with increased use of ionizing radiation for nuclear power plants or nuclear weapons, both of which can result in accidental radiological emergencies. There were nearly 105 civilian and military nuclear reactor accidents between 1952 and 2015 that resulted in massive loss of human life and property [1]. Recently, on 24 August 2017 the Pennsylvania Department of Health distributed potassium iodide tablets for free to the residents who live or work within 10 miles of the Peach Bottom and Three Mile Island nuclear plants to be ready in case of an emergency. However, potassium iodide, a specific blocker of thyroid radioactive iodine uptake, only protects the thyroid gland of individuals exposed to radiation.

Exposure to ionizing radiation produces oxygen derived reactive oxygen and nitrogen species (ROS and RNS), including hydroxyl radical (OH•), superoxide ($O_2\bullet^-$), peroxynitrite ($ONOO^-$), and hydrogen peroxide (H_2O_2). Ionizing radiation-induced ROS and RNS damage DNA, proteins, and lipids as well as activate intracellular signaling pathways and stimulate cytochrome C release from mitochondria, leading to apoptosis [4–6].

The Office of Science and Technology Policy and the United States Department of Homeland Security have identified radiation countermeasure development as the highest priority for preparedness against a potential bioterrorism event [1]. As of today, amifostine is the only FDA approved drug for use in patients undergoing radiotherapy. However, because of its adverse side effects, its use is limited and the search for safe and effective radiation countermeasures continues.

Vitamin E and its derivatives have attracted the attention of researchers in recent years for their radioprotective effects, which have been heavily studied against total body irradiation [7–10], as well as partial body irradiation [11–16]. In this review, we outline the research endeavors dedicated to studying the radiation protection efficacy of vitamin E analogs. We also identify the issues that need to be addressed when using vitamin E analogues as safe and effective radiation countermeasures in humans.

2. Radiation Induced Injuries

Ionizing radiation is radiation that carries sufficient energy to liberate electrons from atoms or molecules leaving them with unpaired electrons thereby ionizing them and producing free radicals. These free radicals and ROS/RNS including, OH•, $O_2\bullet^-$, $ONOO^-$, and H_2O_2 can damage nucleic acids, proteins, and membrane lipids. Exposing an individual to ionizing radiation for a brief period can cause severe tissue injuries, referred to as Acute Radiation Syndrome (ARS). ARS can occur with doses higher than 1 Gray (Gy) that are delivered at relatively high rates. The three clinical syndromes of ARS are based on the acute whole-body dose, duration, and dose rate, including hematopoietic or bone marrow sub-syndrome, gastro-intestinal sub-syndrome, and cerebrovascular sub-syndrome. Each of these three sub-syndromes follows a 4-phase clinical pattern according to CDC Emergency Preparedness and Response, as detailed in Table 1 [17]

Table 1. Acute Radiation Syndrome.

Acute Radiation Syndrome			
	Hematopoietic Sub-Syndrome	Gastro-Intestinal Sub-Syndrome	Neuro/Cerebrovascular Sub-Syndrome
Quantity of radiation	>2–3 Gy	5–12 Gy	10–20 Gy
Prodromal stage symptoms	Anorexia, nausea and vomiting	Anorexia, severe nausea, vomiting, cramps and diarrhea	Extreme nervousness and confusion; severe nausea, vomiting, and watery diarrhea; loss of consciousness; and burning sensations of the skin.
Latent Stage symptoms	Stem cells in bone marrow are dying, although patient may appear and feel well.	Stem cells in bone marrow and cells lining GI tract are dying, although patient may appear and feel well.	Patient may return to partial functionality.
Manifest Phase/Illness Phase symptoms	Anorexia, fever, and malaise. Drop in all blood cell counts occurs for several weeks.	Malaise, anorexia, severe diarrhea, fever, dehydration, and electrolyte imbalance.	Watery diarrhea, convulsions, and coma.
Recovery or death	Bone marrow cells will begin to repopulate the marrow.	>10 Gy radiation leads to death due to gastro-intestinal syndrome	No recovery is expected.

3. Radiation Countermeasures

Effective radiation countermeasures should be safe, efficient, stable, easy to administer, and have good bioavailability. Radiation countermeasures fall under three categories, radiation protectors, radiation mitigators, and radiation therapeutics.

Radiation protectors are agents that are administered prior to radiation exposure to protect from radiation-induced injuries by many mechanisms, such as scavenging free radicals that are generated by initial radiochemical events, delaying cell cycle, promoting DNA repair, etc. Radiation protectants are given to personnel that are at risk of exposure to radiation like the military, first responders and civilians during the evacuation of disaster areas. Radiation mitigators are agents that are administered after the exposure of radiation but before the onset of symptoms. They accelerate the recovery and repair from radiation-induced injuries. Whereas radiation therapeutic agents are administered after the onset of symptoms and act by regenerating the tissues that are injured by radiation. Radiation mitigators and radiation therapeutics can be given to people who are victims of nuclear accidents or terrorist attacks or to patients undergoing radiotherapy. One among the various candidates [18–21] that are under development as radiation countermeasures are vitamin E's tocols.

Natural products with health benefits are often attractive targets for research [22]. Among the natural products, vitamins are notably considered beneficial for human health. Vitamin E is a well-known antioxidant that can scavenge free radicals produced by radiation exposure. Vitamin E is found in our diet and has an acceptable toxicity profile [23].

The vitamin E family consists of eight different naturally occurring vitamers, four saturated analogs (α, β, γ, and δ) called tocopherols, and four unsaturated analogs (α, β, γ, and δ) referred to as tocotrienols, which are collectively called tocols. Tocopherols and tocotrienols are structurally similar with the same chromanol head, except that tocotrienols have unsaturated farnesyl isoprenoid side chain at C-3', C-7', and C-11', whereas tocopherols have saturated phytyl isoprenoid side chain (Figure 1).

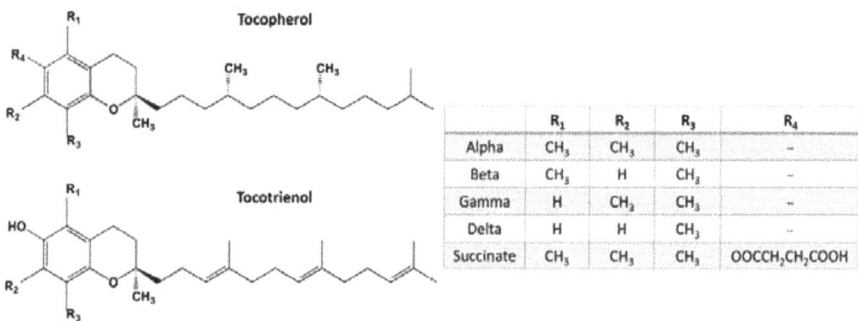

	R_1	R_2	R_3	R_4
Alpha	CH$_3$	CH$_3$	CH$_3$	--
Beta	CH$_3$	H	CH$_3$	--
Gamma	H	CH$_3$	CH$_3$	--
Delta	H	H	CH$_3$	--
Succinate	CH$_3$	CH$_3$	CH$_3$	OOCCH$_2$CH$_2$COOH

Figure 1. Chemical structures of tocopherol and tocotrienol isoforms.

3.1. Tocopherols and Tocopherol Succinate

Vitamin E components have been reported to be radioprotective in various studies [24–26]. However, there are several factors that influence the differential radioprotective efficacy of tocol analogs, including the rate of absorption after oral administration, which is found to be greater in tocotrienols than tocopherols due to higher absorption of tocotrienols by the intestinal epithelial cells [27]. Serbinova et al. [28] reported that the antioxidant potential of tocotrienols is 1600 times more than that of α-tocopherol (AT). Studies by Pearce et al. [29] and Qureshi et al. [30] suggest that tocotrienols are better radio-protectants because of their ability to inhibit HMG-CoA reductase.

AT's dose reduction factor was determined to be 1.11 when administered subcutaneously at a dose of 100 IU/kg when administered within 15 m after irradiation of 9 Gy in mice. AT significantly increased the 30-day survival of male CD2F1 mice when given one h before or within 15 m after irradiation. Combination studies of α-tocopherol with WR-3689 showed that the radio-protective efficacy of WR-3689 (150 mg/kg) significantly increased when given in combination with α-tocopherol at 100 IU/kg with a dose reduction factor of 1.49 [31]. AT also reduced the frequency of micronuclei and chromosomal aberrations in bone marrow cells when administered orally either 2 h before, immediately after, or 2 h after irradiation of 1 Gy in mice [32]. The study by Kumar et al. [33], demonstrated that AT has more radio-protective activity when administered subcutaneously than when given orally at a dose of 400 IU/kg 24 h before total body irradiation at 10.5 Gy. AT (20 IU/kg/day) in combination with pentoxifylline (100 mg/kg/day) induced a significant improvement in radiation-induced myocardial fibrosis and left ventricular diastolic dysfunction after irradiation at 9 Gy [34]. A study by Empey et al. [35] showed that AT protects gastrointestinal mucosa against radiation induced absorptive injury when administered before 10 Gy of abdominal radiation. In another study, AT protected mice from radiation injury when administered i.p. after irradiation of <10 Gy, when administered no later than 5 h after irradiation, suggesting enhancement of repair processes and antioxidant scavenging of metabolically produced radicals when produced by irradiation [36].

To study the role of hematopoietic cytokines in the radioprotective activity of tocopherol succinate and other tocols (α-tocopherol, δ-tocopherol, γ-tocopherol, γ-tocotrienol, and tocopherol acetate), Singh et al. [8] measured cytokine levels by Luminex, ELISA, and cytokine array in mice serum after administrating 400 mg/kg tocols subcutaneously 24 h before whole-body irradiation at a dose of 3 and 7 Gy. Among all the tocols studied, tocopherol succinate was most effective in stimulating granulocyte colony stimulating factor (G-CSF) and IL-6. Since G-CSF and IL-6 play an important role in hematopoietic injury, the study indicates that tocopherol succinate's radio protective activity is mediated by cytokines [8]. When tocopherol succinate injected mice were administered a neutralizing antibody to G-CSF, the protective effect of tocopherol succinate was significantly abrogated [9]. A study by Singh et al. [10] showed that tocopherol succinate protects mice against lethal doses of ionizing radiation by inhibiting radiation-induced apoptosis and DNA damage as well as by increasing cell proliferation. In a study by Singh et al. [37], the dose reduction factor of tocopherol succinate was determined to be 1.28 in mice, administered subcutaneously with 400 mg/kg of tocopherol succinate, 24 h before total body irradiation at 9, 9.5, 10, 10.5, 10.75, and 11 Gy. This study showed that tocopherol succinate stimulated high levels of G-CSF with a peak at 24 h, moderate levels of IL-6 between 24 and 48 h after treatment, and protected myeloid components from radiation injury at 3 and 7 Gy. A similar study [38] suggests that tocopherol succinate protects mice from radiation-induced gastrointestinal damage by promoting the generation of crypt cells and inhibiting apoptosis and translocation of gut bacteria to the heart, spleen, and liver in irradiated mice. Tocopherol succinate has also demonstrated radioprotection from total body irradiation by decreasing the number of CD68-positive cells, reducing DNA damage, and apoptotic cells and by increasing proliferating cells in irradiated mice [10]. Tocopherol succinate also modulates the expression of antioxidant enzymes and inhibits expression of oncogenes in irradiated mice, according to a study by Singh et al. [39].

3.2. Tocotrienols

Multiple studies have reported the antioxidant, anti-inflammatory, anticancer, hypocholesterolemic, and neuroprotective properties of tocotrienols in different cell lines, animal models, and in humans. This review will discuss their radioprotective activity studied in different animal models and in humans.

A number of studies have shown that tocotrienols are superior antioxidants compared to tocopherols [28,29,40–43]. Studies [24,44] have demonstrated that γ-tocotrienol (GT3) protects against radiation injury by increasing hematopoietic progenitors, neutrophils, platelets, white blood cells, and reticulocytes. Singh et al. [45] evaluated the protective effects of GT3 in nonhuman primates treated with 5.8, 6.5, and 7.2 Gy doses of cobalt-60 gamma radiation (0.6 Gy/min). This study reports the pharmacokinetic (PK) parameters (at 9.375, 18.75 and 37.5 mg/kg doses) and efficacy of

GT3 (37.5 mg/kg and 75 mg/kg). Their PK analysis showed increased area under the curve with increasing drug dose and half-life of GT3. Unexpectedly, t_{max} increased in a dose-dependent manner. This could be due to the slow release of GT3 from the site of injection (sub-cutaneous). The study also demonstrated that GT3's efficacy in reducing the severity of neutropenia and thrombocytopenia is dose-dependent, and 75 mg/kg treatment is more effective than 37.5 mg/kg treatment after a 5.8 Gy dose of ionizing radiation. However, there was no significant difference in animal survival at 60 days between the vehicle group and the GT3 treated groups.

The study by Li et al. [46] demonstrated that a single injection of δ-tocotrienol (DT3), given to mice 24 h before a total body irradiation, presented 100% survival, measured by 30-day post-irradiated survival, by increasing cell survival and regeneration of hematopoietic microfoci. Singh et al. [43] showed evidence that DT3 mediates its radioprotection by inducing G-CSF in irradiated mice and also showed that DT3 induces high levels of several cytokines, comparable to other tocols in mice. A study by Loose et al. [47] suggested that GT3 is more efficacious than the tocopherols in terms of radiation protection because of its greater potency to induce gene expression in human endothelial cells. This could be due to the low cellular uptake of α-tocopherol compared to GT3. When GT3 was administered s.c. at 200 mg/kg in combination with pentoxifylline, showed synergistic radioprotection activity in mice exposed to 11.5 Gy total body irradiation, and this radioprotective activity of GT3 was shown to be mediated through induction of G-CSF [48].

Studies by Satyamitra et al. [23] showed that DT3 has radiation protection and mitigation effects. DT3, when given s.c. at 150 mg/kg or 300 mg/kg 24 h before total body irradiation, showed effective radioprotection in mice. In addition, DT3 showed reduced lethality when administered at 150 mg/kg in mice 2, 6, or 12 h after irradiation. Another study by Kumar et al. [49] indicated that a GT3 and DT3 combination at 800 mg/kg given orally twice a day for six months to patients who had radiotherapy for head and neck cancer had significantly improved mouth opening and subjective symptoms due to radiation-induced fibrosis. Studies also suggested that tocols exert their biological effects not only by their antioxidant properties but also by inhibiting HMG-CoA reductase [49]. Tocotrienols accumulate in the small intestine as well as in the colon to a greater level than tocopherols, and this may aid with their ability to reduce GI injury [50]. A study by Naito et al. [51] demonstrated that GT3 concentrations in endothelial cells were 30–50 times greater than those of α-tocopherol. The effect of GT3 on tetrahydrobiopterin (BH4) bioavailability was studied by Berbee et al. [52], where GT3 counteracted the decrease in BH4. GT3 protected hematopoietic tissue by preserving the hematopoietic stem cells (HSCs) and hematopoietic progenitor cells (HPCs). The HPC numbers in GT3 treated mice recovered 90% at day seven after total body irradiation [44]. Modulation of sphingolipids leads to cellular stress and upregulation of A20, a well-established NF-kB negative regulator. It has been shown that the mechanism of GT3 radioprotection may involve the inhibition of NF-κB activation by induction of A20 [53,54]. It also has been shown that the ability of GT3 to protect against vascular injury is related to its ability to inhibit HMG-CoA reductase [55].

4. Vitamin E Tocols

Vitamin E in the diet passes through the gastrointestinal tract and gets absorbed in the small intestine. It is emulsified by bile and absorbed in the form of micelles [56]. Yap et al. [57] have reported that the dietary fat consumption plays an important role in the absorption of tocols. The administration route of vitamin E also influences its absorption [58,59].

4.1. α-Tocopherol Transfer Protein

The circulating levels of vitamin E in the body are maintained by α-tocopherol transfer protein (ATTP), which is abundantly found in the liver. This hepatic protein is a member of a lipid-binding protein family and plays a major discriminating role in the plasma and tissue retention of dietary tocols. This protein is regulated by α-tocopherol transfer gene present on chromosome 8q13 [60]. ATTP has two different conformations, closed and open [61]. The structure of ATTP has a hinge, which flips to a closed conformation after incorporating AT in the binding pocket. The fate of the tocol depends on the preferential binding of ATTP in the liver, and those that are not selected by ATTP are excreted via bile or renal excretion (Figure 2) [25,61]. ATTP has a much greater affinity AT than for all other tocols, and this would explain why the other tocols exhibit lower plasma levels over time [62–64]. Vitamin E supplements are commonly formulated with AT, which is often present as a synthetic racemic mixture. However, only the naturally occurring RRR isomer of AT has a high affinity for ATTP, causing the other isomers to be much less bioavailable [65]. Studies [25,66] have proposed that unsaturation in the side chain of tocotrienols accounts for their lower affinity for ATTP by making it impossible for the tocotrienol bend inside the ligand-binding pocket of ATTP and hindering the protein's ability to attain the closed conformation. Thus, there is a significant difference in the distribution and metabolism of tocols among various tissues [67–70].

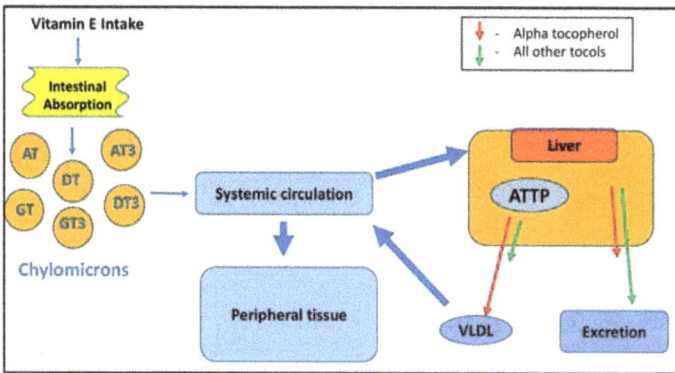

Figure 2. Major pathway for vitamin E absorption and metabolism.

There is a population, with a neurodegenerative disease known as ataxia, with vitamin E deficiency (AVED) [71]. Patients diagnosed with AVED have three frame shift mutations in the α-tocopherol transfer gene present on chromosome 8q13. AVED is an autosomal recessive defect in which the patient is not able to absorb vitamin E into the systemic circulation and becomes deficient [71]. Without a regular supply of this key antioxidant, the body becomes susceptible to damage by free radicals. The coordination of movement becomes uncontrolled and the patient experiences a loss of sensation [71]. Individuals suffering from AVED have normal intestinal absorption of vitamin E, but they are unable to maintain normal levels of vitamin E in systemic circulation. The reason for this is that AVED patients lack functional ATTP, thus the recirculation of α-tocopherol is inhibited. To maintain a normal blood concentration of vitamin E, patients suffering from AVED require very high doses of oral vitamin E throughout their lives [71]. AVED exemplifies the importance of the presence of functional ATTP. However, the fact that ATTP has a much greater affinity for α-tocopherol than for other tocols results in a much faster elimination rate and lower plasma levels over time for all the other tocopherols and for tocotrienols.

4.2. Absorption and Distribution

It is likely that the inconclusive outcomes of various vitamin E clinical trials may be due to limited understanding of the pharmacokinetics of tocols. The bioavailability of vitamin E in humans is dependent on many factors. Studies have shown that vitamin E bioavailability is greater when administered with food, particularly fat [57,58]. Reboul et al. [72] studied the role of the SR-B1 receptor in the intestinal absorption of vitamin E and showed that the expression of the SR-B1 receptor may be responsible for the inter-individual variability in vitamin E absorption and that this receptor mediates the intestinal vitamin E absorption.

Though there are several studies showing tocotrienol's radioprotective activity, their dosing, efficacy, and therapeutic concentrations still remain controversial. This is partially due to the lack of sufficient data explaining their absorption, distribution, and elimination. There is a large variability in the rate and extent of absorption of tocotrienols in different populations from healthy subjects to smokers and diseased patients. The route of administration can also play an important role in the plasma concentrations of tocotrienols [45,57,58].

The absorption of tocotrienols was negligible when administered via intraperitoneal and intramuscular route, whereas the absorption was incomplete when given orally [58]. Several studies have focused on developing a water-soluble delivery system to increase the solubility of the lipophilic tocols. The use of cyclodextrin and self-emulsifying formulations showed improved absorption and higher plasma levels in rats. The peak plasma concentration (C_{max}) and the area under the curve (AUC) of tocotrienols administered with self-emulsifying systems were increased 2–4-fold compared to non-emulsified formulations [73,74]. In a study by Yap et al. [57], comparisons were made between eight healthy male volunteers given 300 mg mixed tocotrienols (87 mg α-tocotrienol, 166 mg GT3, and 43 mg DT3) under a fasted state or after a fatty meal. The 24 h AUC of tocotrienols was increased by at least 2-fold in the fed state. The maximum plasma concentrations for α, γ, and δ tocotrienols were found to be 1.83, 2.13, and 0.34 µg/mL, respectively. While in another study, where tocotrienols were given at higher doses of 296 mg α-tocotrienol, 284 mg GT3, and 83 mg DT3, the maximum plasma concentrations were 1.55, 2.79, and 0.44 µg/mL, respectively [75]. The discrepancies in the plasma levels observed may be an indication that levels may depend on the ratios of the analogs [76]. It has been reported that tocotrienols were transported in triacylglycerol-rich fractions after administration of tocotrienol-rich fraction (TRF) at 1011 mg in healthy subjects, and tocotrienols were found in significant amounts in the plasma and lipoproteins [75]. This study [75], as well as other studies [77–79], indicate that high TRF doses from 200 to 3200 mg/d are safe for healthy human consumption. However, it was also mentioned in a review by Ju et al. [80] that the therapeutic efficacy of tocotrienols depends on dose, formulation, route of administration, and study population. Several studies [81–84] investigated the changes in lipid profile at different doses of tocotrienols.

In spite of all these studies, the dose dependent effect of TRF on the lipid profile is still inconclusive, and further studies are needed to identify the factors responsible for this disparity. For example, it was reported that reduced DNA damage was observed in people over the age of 50 years who were given tocotrienols [85]. This was supported by a later study by Chen et al. [86] who compared the absorption of tocotrienols in different age groups of subjects who were given TRF supplementations for six months. They observed that the plasma tocotrienol levels increased significantly in participants aged over 50 years but not in younger people between the ages of 35–49 years. In contrast to these studies, a study by Heng et al. [87] reported higher plasma concentrations of tocotrienols in younger people 30–34 years old, compared to people 50–54 years old after receiving TRF supplementations for six months. In study by Qureshi et al. [77], TRF doses at 125 mg/d, 250 mg/d, and 500 mg/d were given orally to healthy fed subjects ($n = 11$/dose), and dose-dependent increases in AUC and C_{max} were found. In a later study by Qureshi et al. [78], the safety and bioavailability of higher oral doses of 750 mg/d and 1000 mg/d of annatto-based tocotrienols in healthy fed subjects ($n = 3$/dose) were analyzed. This study showed a dose-dependent increase in plasma concentration (ng/mL) and plasma t_{max} of 3.33 and 4 h; elimination half-lives of 2.74 and 2.68 h for δ-tocotrienol. Similar results were reported for all other tocols, except for α-tocopherol. It was reported that a dose of δ-tocotrienol lower than 500 mg/d decreased the levels of serum total cholesterol, LDL-cholesterol, and triglycerides in a dose-dependent manner, but a higher dose of 750 mg/d increased the levels of these lipid parameters, compared to 250 mg/d [88]. Studies suggest that δ-tocotrienol has the unique dual biological property of inhibition (anti-inflammatory) and activation (pro-inflammatory), depending on its concentration [78,79,88–90]. The bioavailability of a mixture of α-tocotrienol, GT3, and DT3 was studied after administering a 300 mg capsule to fasted and fed healthy subjects ($n = 8$) and found the plasma t_{max} to be between 3 and 5 h for both fasted and fed subjects [57]. In another study, plasma t_{max}

after administering two 300 mg/capsules of a mixture of γ-tocotrienol and δ-tocotrienol was found to be 5.64 ± 1.50 h and 4.73 ± 0.90 h, respectively [91]. When a dose of 450 mg of a rich fraction of barley oil versus a rich fraction of palm oil was administered to healthy subjects (n = 7), the area under the curve (0–24 h) of total (α-, β-, γ-, δ-) tocotrienols was significantly (2.6-fold) higher in the barley oil than in the palm oil [92].

A study by Abuasal et al. [93] compared the intestinal absorption kinetics and the bioavailability of GT3 and AT administered to rats. The oral bioavailability of AT (36%) was significantly higher than GT3 (9%), and AT showed higher intestinal permeability than GT3. These results indicate that the intestinal permeability could be a contributing factor for the higher bioavailability of α-tocopherol and suggests that enhancing the permeability of γ-tocotrienol would increase its oral bioavailability. There are findings suggesting that increasing the permeability of GT3 via new approaches, like solid lipid nanoparticles, leads to an increase in its bioavailability [94]. It was reported that co-administration of tocotrienols with lipids causes a delay in the rate of gastric emptying and this leads to an increase in tocotrienol's solubility by stimulating the secretion of bile salts and phospholipids into the GI tract and therefore increasing the absorption and bioavailability of tocotrienols [95,96].

A self-emulsifying drug delivery system (SEDDS) has been used to achieve increased bioavailability of tocotrienols. GT3 dissolved in SEDDS versus commercial Tocovid was studied in vitro and in vivo in rats [97]. There was a 2-fold increase in the solubilization, higher cellular uptake in vitro, and a 2-fold increase in oral bioavailability for the SEDDS formulation [97]. Alqahtani et al. [98] also reported the bioavailability of γ and δ tocotrienols, administered to rats as SEDDS compared to commercially available UNIQUE E® tocotrienol capsules, and the results showed that SEDDS increased the bioavailability. However, bioavailability showed a progressive decrease with increased treatment dose due to nonlinear absorption kinetics [98].

Collectively these studies show that tocotrienols respond differently at different doses in different populations. The major determinant for the limited bioavailability of the tocotrienols could be their limited binding to the transporter, ATTP, which is responsible for transporting the tocols out of the liver into the systemic circulation [99]. The key role of ATTP in regulating the pharmacokinetics of vitamin E has been elegantly demonstrated in several studies [61,99–105]. Previous studies have shown that there is a good linear relationship between relative affinity of tocols to ATTP and their biological activity [106]. These studies also demonstrate the need to conduct more randomized controlled trials with large sample sizes to better understand the bioavailability mechanisms and the therapeutic window of tocotrienols. Thus, it is highly desirable to develop a new vitamin E analog with an increased half-life and improved pharmacokinetic properties. This can be achieved by increasing the affinity of the analogs to ATTP, which may be responsible for maintaining the plasma levels of these compounds and also by increasing their permeability. Multiple studies [25,63,66] have recently reported the development of such analogs, the tocoflexols (Figure 3). The tocoflexols were designed, using computer aided techniques, to behave like tocopherols in terms of their bioavailability and like tocotrienols in terms of their biological activity. To that effect, tocoflexol has been shown to have an antioxidant activity and rate of cell uptake on a par with DT3 and GT3, but it has a greater ability to bind to ATTP than the tocotrienols and thus has the potential for improved bioavailability.

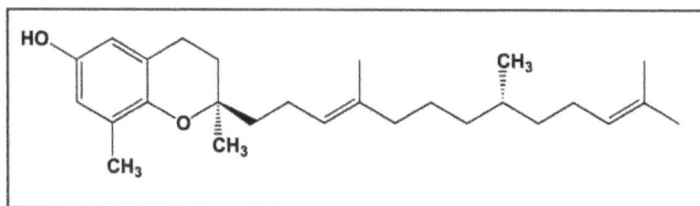

Figure 3. Chemical structure of δ-tocoflexol.

5. Conclusions and Future Directions

Given the increasing use of radioactive materials in healthcare as well as the number of nuclear reactors and the amplified nuclear terrorism risk, development of effective radioprotectors and radiomitigators that can be deployed immediately is the key to have a successful contingency strategy in case of unwanted/unexpected radiation exposures. Multiple studies have reported the mechanisms by which tocols exert their radioprotection. It has been suggested that tocols' mechanism of protection from radiation-induced hematopoietic death involves cytokines and chemokines [8]. The importance of G-CSF induction on the mechanism of radioprotection of tocotrienols was demonstrated when the protective effects of GT3 were abrogated in irradiated mice treated with G-CSF antibodies [9,40,48]. It has also been shown that tocols reduce post-irradiation GI syndrome by decreasing IL-1β and IL-6 [107]. Some studies have shown that tocotrienols have a greater radioprotective effect than tocopherols because of their ability to inhibit HMG-CoA reductase [55]. The radioprotective effects of GT3 depend not only on its antioxidant properties but also on its ability to concentrate in endothelial cells [51]. A study by Loose et al. [47] investigated the gene expression profile in human epithelial cells after treatment with GT3, γ-tocopherol (GT), and AT for 24 h. GT3 was found to be more effective in modulating changes in gene expression than GT and AT, when a genome wide analysis was performed. Several genes were affected, including those responsible for cell cycle, cell proliferation, cell death, hematopoiesis, angiogenesis, and DNA damage. The poor bioavailability of the tocotrienols is a major limiting factor for their clinical use as radioprotectants and radiomitigators [66]. In this regard, the development of the tocoflexols represents a promising approach [25].

Low doses of ionizing radiation in medical treatments and occupational exposure results in high risk for cardiovascular diseases [108]. Long distance space missions are associated with exposure to galactic cosmic rays and solar particle events that can increase the risk for cataract, cancers, and cardiovascular diseases [109–111]. Because of their effectiveness and low toxicity, the tocols may prove to be effective to protect against low-dose radiation, but further research is required to establish their potential.

Acknowledgments: This work was supported in part by the National Institutes of Health (P01AG012411-18; P20, GM109005; R15, ES022781, and P20 GM103429), the Arkansas Biosciences Institute FY13, and the Arkansas Science and Technology Authority 15-B-19.

Conflicts of Interest: The University of Arkansas has applied for patent protection on the tocoflexols and other tocotrienol containing products. A potential royalty stream to Shraddha Thakkar, Philip J. Breen, Cesar M. Compadre and Nukhet Aykin-Burns, may occur consistent with the University of Arkansas policy.

Disclaimer: The views presented in this article do not necessarily reflect current or future opinion or policy of the U.S. Food and Drug Administration. Any mention of commercial products is for clarification and not intended as an endorsement.

References

1. Singh, P.K.; Krishnan, S. Vitamin E analogs as radiation response modifiers. *Evid.-Based Complement. Altern. Med.* **2015**, *2015*, 741301. [CrossRef] [PubMed]
2. Lacroix, M.; Ouattara, B. Combined industrial processes with irradiation to assure innocuity and preservation of food products—A review. *Food Res. Int.* **2000**, *33*, 719–724. [CrossRef]
3. Pereira, E.; Antonio, A.L.; Barreira, J.C.; Barros, L.; Bento, A.; Ferreira, I.C. Gamma irradiation as a practical alternative to preserve the chemical and bioactive wholesomeness of widely used aromatic plants. *Food Res. Int.* **2015**, *67*, 338–348. [CrossRef]
4. Kobashigawa, S.; Kashino, G.; Suzuki, K.; Yamashita, S.; Mori, H. Ionizing radiation-induced cell death is partly caused by increase of mitochondrial reactive oxygen species in normal human fibroblast cells. *Radiat. Res.* **2015**, *183*, 455–464. [CrossRef] [PubMed]
5. Redza-Dutordoir, M.; Averill-Bates, D.A. Activation of apoptosis signalling pathways by reactive oxygen species. *Biochim. Biophys. Acta (BBA)-Mol. Cell Res.* **2016**, *1863*, 2977–2992. [CrossRef] [PubMed]

6. Chen, Q.; Chai, Y.C.; Mazumder, S.; Jiang, C.; Macklis, R.M.; Chisolm, G.M.; Almasan, A. The late increase in intracellular free radical oxygen species during apoptosis is associated with cytochrome c release, caspase activation, and mitochondrial dysfunction. *Cell Death Differ.* **2003**, *10*, 323–334. [CrossRef] [PubMed]

7. Singh, V.K.; Beattie, L.A.; Seed, T.M. Vitamin E: Tocopherols and tocotrienols as potential radiation countermeasures. *J. Radiat. Res.* **2013**, *54*, 973–988. [CrossRef] [PubMed]

8. Singh, V.K.; Shafran, R.L.; Jackson, W.E.; Seed, T.M.; Kumar, K.S. Induction of cytokines by radio protective tocopherol analogs. *Exp. Mol. Pathol.* **2006**, *81*, 55–61. [CrossRef] [PubMed]

9. Singh, P.K.; Wise, S.Y.; Ducey, E.J.; Brown, D.S.; Singh, V.K. Radioprotective efficacy of tocopherol succinate is mediated through granulocyte-colony stimulating factor. *Cytokine* **2011**, *56*, 411–421. [CrossRef] [PubMed]

10. Singh, V.K.; Singh, P.K.; Wise, S.Y.; Posarac, A.; Fatanmi, O.O. Radioprotective properties of tocopherol succinate against ionizing radiation in mice. *J. Radiat. Res.* **2012**, *54*, 210–220. [CrossRef] [PubMed]

11. Chitra, S.; Devi, C.S. Effect of alpha-tocopherol on pro-oxidant and antioxidant enzyme status in radiation-treated oral squamous cell carcinoma. *Indian J. Med. Sci.* **2008**, *62*, 141. [CrossRef] [PubMed]

12. Gomes, C.C.; Ramos-Perez, F.M.; Perez, D.E.; Novaes, P.D.; Boscolo, F.N.; Almeida, S.M. Radioprotective effect of vitamin E in parotid glands: A morphometric analysis in rats. *Braz. Dent. J.* **2013**, *24*, 183–187. [CrossRef] [PubMed]

13. Laurent, C.; Pouget, J.P.; Voisin, P. Modulation of DNA damage by pentoxifylline and α-tocopherol in skin fibroblasts exposed to gamma rays. *Radiat. Res.* **2005**, *164*, 63–72. [CrossRef] [PubMed]

14. Ferreira, P.R.; Fleck, J.F.; Diehl, A.; Barletta, D.; Braga-Filho, A.; Barletta, A.; Ilha, L. Protective effect of alpha-tocopherol in head and neck cancer radiation-induced mucositis: A double-blind randomized trial. *Head Neck* **2004**, *26*, 313–321. [CrossRef] [PubMed]

15. Sridharan, V.; Tripathi, P.; Sharma, S.; Corry, P.M.; Moros, E.G.; Singh, A.; Compadre, C.M.; Hauer-Jensen, M.; Boerma, M. Effects of late administration of pentoxifylline and tocotrienols in an image-guided rat model of localized heart irradiation. *PLoS ONE* **2013**, *8*, e68762. [CrossRef] [PubMed]

16. Misirlioglu, C.H.; Demirkasimoglu, T.; Kucukplakci, B.; Sanri, E.; Altundag, K. Pentoxifylline and alpha-tocopherol in prevention of radiation-induced lung toxicity in patients with lung cancer. *Med. Oncol.* **2007**, *24*, 308–311. [CrossRef] [PubMed]

17. Center for Disease Control. Acute Radiation Syndrome: A Brochure for Physicians. U.S. Department of Health and Human Services. Available online: https://emergency.cdc.gov/radiation/pdf/ars.pdf (accessed on 5 January 2018).

18. Singh, V.K.; Ducey, E.J.; Brown, D.S.; Whitnall, M.H. A review of radiation countermeasure works ongoing at the Armed Forces Radiobiology Research Institute. *Int. J. Radiat. Boil.* **2012**, *88*, 296–310. [CrossRef] [PubMed]

19. Weiss, J.F.; Landauer, M.R. History and development of radiation-protective agents. *Int. J. Radiat. Boil.* **2009**, *85*, 539–573. [CrossRef] [PubMed]

20. Dumont, F.; Roux, A.L.; Bischoff, P. Radiation countermeasure agents: An update. *Expert Opin. Ther. Pat.* **2010**, *20*, 73–101. [CrossRef] [PubMed]

21. Weiss, J.F.; Landauer, M.R. Protection against ionizing radiation by antioxidant nutrients and phytochemicals. *Toxicology* **2003**, *189*, 1–20. [CrossRef]

22. Agarwal, S.S.; Singh, V.K. Immunomodulators: A review of studies on Indian medicinal plants and synthetic peptides. Part I: Medicinal plants. *Proc. Ind. Nat. Sci. Acad. B* **1999**, *65*, 179–204.

23. Satyamitra, M.M.; Kulkarni, S.; Ghosh, S.P.; Mullaney, C.P.; Condliffe, D.; Srinivasan, V. Hematopoietic recovery and amelioration of radiation-induced lethality by the vitamin E isoform δ-tocotrienol. *Radiat. Res.* **2011**, *175*, 736–745. [CrossRef] [PubMed]

24. Ghosh, S.P.; Kulkarni, S.; Hieber, K.; Toles, R.; Romanyukha, L.; Kao, T.C.; Hauer-Jensen, M.; Kumar, K.S. Gamma-tocotrienol, a tocol antioxidant as a potent radioprotector. *Int. J. Radiat. Boil.* **2009**, *85*, 598–606. [CrossRef] [PubMed]

25. Compadre, C.M.; Singh, A.; Thakkar, S.; Zheng, G.; Breen, P.J.; Ghosh, S.; Kiaei, M.; Boerma, M.; Varughese, K.I.; Hauer-Jensen, M. Molecular dynamics guided design of tocoflexol: A new radio protectant tocotrienol with enhanced bioavailability. *Drug Dev. Res.* **2014**, *75*, 10–22. [CrossRef] [PubMed]

26. Felemovicius, I.; Bonsack, M.E.; Baptista, M.L.; Delaney, J.P. Intestinal radioprotection by vitamin E (alpha-tocopherol). *Ann. Surg.* **1995**, *222*, 504. [CrossRef] [PubMed]

27. Tsuzuki, W.; Yunoki, R.; Yoshimura, H. Intestinal epithelial cells absorb γ-tocotrienol faster than α-tocopherol. *Lipids* **2007**, *42*, 163. [CrossRef] [PubMed]

28. Serbinova, E.; Kagan, V.; Han, D.; Packer, L. Free radical recycling and intramembrane mobility in the antioxidant properties of alpha-tocopherol and alpha-tocotrienol. *Free Radic. Biol. Med.* **1991**, *10*, 263–275. [CrossRef]

29. Pearce, B.C.; Parker, R.A.; Deason, M.E.; Qureshi, A.A.; Wright, J.K. Hypocholesterolemic activity of synthetic and natural tocotrienols. *J. Med. Chem.* **1992**, *35*, 3595–3606. [CrossRef] [PubMed]

30. Qureshi, A.A.; Pearce, B.C.; Nor, R.M.; Gapor, A. Dietary alpha-tocopherol attenuates the impact of gamma-tocotrienol on hepatic 3-hydroxy-3-methylglutaryl coenzyme A reductase activity in chickens. *J. Nutr.* **1996**, *126*, 389. [CrossRef] [PubMed]

31. Srinivasan, V.; Weiss, J.F. Radioprotection by vitamin E: Injectable vitamin E administered alone or with WR-3689 enhances survival of irradiated mice. *Int. J. Radiat. Oncol. Biol. Phys.* **1992**, *23*, 841–845. [CrossRef]

32. Sarma, L.; Kesavan, P.C. Protective effects of vitamins C and E against γ-ray-induced chromosomal damage in mouse. *Int. J. Radiat. Boil.* **1993**, *63*, 759–764. [CrossRef]

33. Kumar, K.S. Nutritional approaches to radioprotection: Vitamin, E. *Mil. Med.* **2002**, *167*, 57. [PubMed]

34. Boerma, M.; Roberto, K.A.; Hauer-Jensen, M. Prevention and treatment of functional and structural radiation injury in the rat heart by pentoxifylline and alpha-tocopherol. *Int. J. Radiat. Oncol. Biol. Phys.* **2008**, *72*, 170–177. [CrossRef] [PubMed]

35. Empey, L.R.; Papp, J.D.; Jewell, L.D.; Fedorak, R.N. Mucosal protective effects of vitamin E and misoprostol during acute radiation-induced enteritis in rats. *Dig. Dis. Sci.* **1992**, *37*, 205–214. [CrossRef] [PubMed]

36. Roy, R.M.; Petrella, M.; Shateri, H. Effects of administering tocopherol after irradiation on survival and proliferation of murine lymphocytes. *Pharmacol. Ther.* **1988**, *39*, 393–395. [CrossRef]

37. Singh, V.K.; Brown, D.S.; Kao, T.C. Tocopherol succinate: A promising radiation countermeasure. *Int. Immunopharmacol.* **2009**, *9*, 1423–1430. [CrossRef] [PubMed]

38. Singh, P.K.; Wise, S.Y.; Ducey, E.J.; Fatanmi, O.O.; Elliott, T.B.; Singh, V.K. α-Tocopherol succinate protects mice against radiation-induced gastrointestinal injury. *Radiat. Res.* **2011**, *177*, 133–145. [CrossRef] [PubMed]

39. Singh, V.K.; Parekh, V.I.; Brown, D.S.; Kao, T.C.; Mog, S.R. Tocopherol succinate: Modulation of antioxidant enzymes and oncogene expression, and hematopoietic recovery. *Int. J. Radiat. Oncol. Biol. Phys.* **2011**, *79*, 571–578. [CrossRef] [PubMed]

40. Singh, V.K.; Newman, V.L.; Romaine, P.L.; Wise, S.Y.; Seed, T.M. Radiation countermeasure agents: An update (2011–2014). *Expert Opin. Ther. Pat.* **2014**, *24*, 1229–1255. [CrossRef] [PubMed]

41. Kamal-Eldin, A.; Appelqvist, L.Å. The chemistry and antioxidant properties of tocopherols and tocotrienols. *Lipids* **1996**, *31*, 671–701. [CrossRef] [PubMed]

42. Pearce, B.C.; Parker, R.A.; Deason, M.E.; Dischino, D.D.; Gillespie, E.; Qureshi, A.A.; Wright, J.K.; Volk, K. Inhibitors of cholesterol biosynthesis. 2. hypocholesterolemic and antioxidant activities of benzopyran and tetrahydronaphthalene analogs of the tocotrienols. *J. Med. Chem.* **1994**, *37*, 526–541. [CrossRef] [PubMed]

43. Singh, V.K.; Wise, S.Y.; Scott, J.R.; Romaine, P.L.; Newman, V.L.; Fatanmi, O.O. Radioprotective efficacy of delta-tocotrienol, a vitamin E isoform, is mediated through granulocyte colony-stimulating factor. *Life Sci.* **2014**, *98*, 113–122. [CrossRef] [PubMed]

44. Kulkarni, S.; Ghosh, S.P.; Satyamitra, M.; Mog, S.; Hieber, K.; Romanyukha, L.; Gambles, K.; Toles, R.; Kao, T.C.; Hauer-Jensen, M.; Kumar, K.S. Gamma-tocotrienol protects hematopoietic stem and progenitor cells in mice after total-body irradiation. *Radiat. Res.* **2010**, *173*, 738–747. [CrossRef] [PubMed]

45. Singh, V.K.; Kulkarni, S.; Fatanmi, O.O.; Wise, S.Y.; Newman, V.L.; Romaine, P.L.; Hendrickson, H.; Gulani, J.; Ghosh, S.P.; Kumar, K.S.; et al. Radio protective efficacy of gamma-tocotrienol in nonhuman primates. *Radiat. Res.* **2016**, *185*, 285–298. [CrossRef] [PubMed]

46. Li, X.H.; Fu, D.; Latif, N.H.; Mullaney, C.P.; Ney, P.H.; Mog, S.R.; Whitnall, M.H.; Srinivasan, V.; Xiao, M. δ-tocotrienol protects mouse and human hematopoietic progenitors from γ-irradiation through extracellular signal-regulated kinase/mammalian target of rapamycin signaling. *Haematologica* **2010**, *95*, 1996–2004. [CrossRef] [PubMed]

47. Loose, D.S.; Kumar, K.S.; Berbée, M.; Boerma, M.; Hauer-Jensen, M.; Fu, Q. Mechanisms underlying the radioprotective properties of γ-tocotrienol: Comparative gene expression profiling in tocol-treated endothelial cells. *Genes Nutr.* **2012**, *7*, 75.

48. Kulkarni, S.; Singh, P.K.; Ghosh, S.P.; Posarac, A.; Singh, V.K. Granulocyte colony-stimulating factor antibody abrogates radioprotective efficacy of gamma-tocotrienol, a promising radiation countermeasure. *Cytokine* **2013**, *62*, 278–285. [CrossRef] [PubMed]

49. Kumar, D.; Aggarwal, A.K.; Shukla, D.K.; Thimothy, G.; Rani, S. To study and evaluation of role of gamma and delta tocotrienol in radiation induced fibrosis. *Pharma Innov.* **2017**, *6 Pt B*, 91.

50. Yang, C.S.; Lee, M.J.; Zhao, Y.; Yang, Z. Metabolism of tocotrienols in animals and synergistic inhibitory actions of tocotrienols with atorvastatin in cancer cells. *Genes Nutr.* **2012**, *7*, 11. [CrossRef] [PubMed]

51. Naito, Y.; Shimozawa, M.; Kuroda, M.; Nakabe, N.; Manabe, H.; Katada, K.; Kokura, S.; Ichikawa, H.; Yoshida, N.; Noguchi, N.; et al. Tocotrienols reduce 25-hydroxycholesterol-induced monocyte-endothelial cell interaction by inhibiting the surface expression of adhesion molecules. *Atherosclerosis* **2005**, *180*, 19–25. [CrossRef] [PubMed]

52. Berbee, M.; Fu, Q.; Boerma, M.; Pathak, R.; Zhou, D.; Kumar, K.S.; Hauer-Jensen, M. Reduction of radiation-induced vascular nitrosative stress by the vitamin E analog γ-tocotrienol: Evidence of a role for tetrahydrobiopterin. *Int. J. Radiat. Oncol. Biol. Phys.* **2011**, *79*, 884–891. [CrossRef] [PubMed]

53. Bai, M.; Ma, X.; Li, X.; Wang, X.; Mei, Q.; Li, X.; Wu, Z.; Han, W. The accomplices of NF-κB lead to radioresistance. *Curr. Protein Pept. Sci.* **2015**, *16*, 279–294. [CrossRef] [PubMed]

54. Wang, Y.; Park, N.Y.; Jang, Y.; Ma, A.; Jiang, Q. Vitamin E γ-tocotrienol inhibits cytokine-stimulated NF-κB activation by induction of anti-inflammatory A20 via stress adaptive response due to modulation of sphingolipids. *J. Immunol.* **2015**, *195*, 126–133. [CrossRef] [PubMed]

55. Berbée, M.; Fu, Q.; Boerma, M.; Wang, J.; Kumar, K.S.; Hauer-Jensen, M. γ-Tocotrienol ameliorates intestinal radiation injury and reduces vascular oxidative stress after total-body irradiation by an HMG-CoA Reductase-dependent mechanism. *Radiat. Res.* **2009**, *171*, 596–605. [CrossRef] [PubMed]

56. Bjørneboe, A.; Bjørneboe, G.E.; Drevon, C.A. Absorption, transport and distribution of vitamin, E. *J. Nutr.* **1990**, *120*, 233–242. [CrossRef] [PubMed]

57. Yap, S.P.; Yuen, K.H.; Wong, J.W. Pharmacokinetics and bioavailability of α-, γ-and δ-tocotrienols under different food status. *J. Pharm. Pharmacol.* **2001**, *53*, 67–71. [CrossRef] [PubMed]

58. Yap, S.P.; Yuen, K.H.; Lim, A.B. Influence of route of administration on the absorption and disposition of α, γ-and δ-tocotrienols in rats. *J. Pharm. Pharmacol.* **2003**, *55*, 53–58. [CrossRef] [PubMed]

59. Abuasal, B.; Sylvester, P.W.; Kaddoumi, A. Intestinal absorption of γ-tocotrienol is mediated by Niemann-Pick C1-Like 1: In situ rat intestinal perfusion studies. *Drug Metab. Dispos.* **2010**, *38*, 939–945. [CrossRef] [PubMed]

60. Ouahchi, K.; Arita, M.; Kayden, H.; Hentati, F.; Hamida, M.B.; Sokol, R.; Arai, H.; Inoue, K.; Mandel, J.L.; Koenig, M. Ataxia with isolated vitamin E deficiency is caused by mutations in the α–tocopherol transfer protein. *Nat. Genet.* **1995**, *9*, 141–145. [CrossRef] [PubMed]

61. Meier, R.; Tomizaki, T.; Schulze-Briese, C.; Baumann, U.; Stocker, A. The molecular basis of vitamin E retention: Structure of human α-tocopherol transfer protein. *J. Mol. Biol.* **2003**, *331*, 725–734. [CrossRef]

62. Traber, M.G. Vitamin E bioavailability. In *The Encyclopedia of Vitamin E*; Preedy, V.R., Watson, R.R., Eds.; CABI: Oxon, UK, 2007; pp. 221–230.

63. Liu, X.; Gujarathi, S.; Zhang, X.; Shao, L.; Boerma, M.; Compadre, C.M.; Crooks, P.A.; Hauer-Jensen, M.; Zhou, D.; Zheng, G. Synthesis of (2R,8'S,3'E)-δ-tocodienol, a tocoflexol family member designed to have a superior pharmacokinetic profile compared to δ-tocotrienol. *Tetrahedron* **2106**, *72*, 4001–4006. [CrossRef] [PubMed]

64. Stocker, A.; Azzi, A. Tocopherol-binding proteins: Their function and physiological significance. *Antioxid. Redox Signal.* **2000**, *2*, 397–404. [CrossRef] [PubMed]

65. Traber, M.G.; Ramakrishnan, R.; Kayden, H.J. Human plasma vitamin E kinetics demonstrate rapid recycling of plasma RRR-alpha-tocopherol. *Proc. Natl. Acad. Sci. USA.* **1994**, *91*, 10005–10008. [CrossRef] [PubMed]

66. Singh, A.; Breen, P.J.; Ghosh, S.; Kumar, S.K.; Varughese, K.I.; Crooks, P.A.; Hauer-Jensen, M.; Compadre, C.M. Structural modification of tocotrienols to improve bioavailability. In *Tocotrienols: Vitamin E Beyond Tocopherols*, 2nd ed.; Tan, B., Watson, R.R., Preedy, V.R., Eds.; American Oil Chemists Society and Taylor & Francis: Urbana, IL, USA, 2012; pp. 359–370.

67. Kawakami, Y.; Tsuzuki, T.; Nakagawa, K.; Miyazawa, T. Distribution of tocotrienols in rats fed a rice bran tocotrienol concentrate. *Biosci. Biotechnol. Biochem.* **2007**, *71*, 464–471. [CrossRef] [PubMed]

68. Hayes, K.C.; Pronczuk, A.; Liang, J.S. Differences in the plasma transport and tissue concentrations of tocopherols and tocotrienols: Observations in humans and hamsters. *Proc. Soc. Exp. Biol. Med.* **1993**, *202*, 353–359. [CrossRef] [PubMed]

69. Ikeda, S.; Toyoshima, K.; Yamashita, K. Dietary sesame seeds elevate α-and γ-tocotrienol concentrations in skin and adipose tissue of rats fed the tocotrienol-rich fraction extracted from palm oil. *J. Nutr.* **2001**, *131*, 2892–2897. [CrossRef] [PubMed]

70. Okabe, Y.; Watanabe, A.; Shingu, H.; Kushibiki, S.; Hodate, K.; Ishida, M.; Ikeda, S.; Takeda, T. Effects of α-tocopherol level in raw venison on lipid oxidation and volatiles during storage. *Meat Sci.* **2002**, *62*, 457–462. [CrossRef]

71. Min, K.C.; Kovall, R.A.; Hendrickson, W.A. Crystal structure of human α-tocopherol transfer protein bound to its ligand: Implications for ataxia with vitamin E deficiency. *Proc. Natl. Acad. Sci. USA* **2003**, *100*, 14713–14718. [CrossRef] [PubMed]

72. Reboul, E.; Klein, A.; Bietrix, F.; Gleize, B.; Malezet-Desmoulins, C.; Schneider, M.; Margotat, A.; Lagrost, L.; Collet, X.; Borel, P. Scavenger receptor class B type I (SR-BI) is involved in vitamin E transport across the enterocyte. *J. Biol. Chem.* **2006**, *281*, 4739–4745. [CrossRef] [PubMed]

73. Miyoshi, N.; Wakao, Y.; Tomono, S.; Tatemichi, M.; Yano, T.; Ohshima, H. The enhancement of the oral bioavailability of γ-tocotrienol in mice by γ-cyclodextrin inclusion. *J. Nutr. Biochem.* **2011**, *22*, 1121–1126. [CrossRef] [PubMed]

74. Yap, S.P.; Yuen, K.H. Influence of lipolysis and droplet size on tocotrienol absorption from self-emulsifying formulations. *Int. J. Pharm.* **2004**, *281*, 67–78. [CrossRef] [PubMed]

75. Fairus, S.; Nor, R.M.; Cheng, H.M.; Sundram, K. Postprandial metabolic fate of tocotrienol-rich vitamin E differs significantly from that of α-tocopherol. *Am. J. Clin. Nutr.* **2006**, *84*, 835–842. [CrossRef] [PubMed]

76. Fairus, S.; Nor, R.M.; Cheng, H.M.; Sundram, K. Alpha-tocotrienol is the most abundant tocotrienol isomer circulated in plasma and lipoproteins after postprandial tocotrienol-rich vitamin E supplementation. *Nutr. J.* **2012**, *11*, 5. [CrossRef] [PubMed]

77. Qureshi, A.A.; Khan, D.A.; Saleem, S.; Silswal, N.; Trias, A.M. Pharmacokinetics and bioavailability of annatto δ-tocotrienol in healthy fed subjects. *J. Clin. Exp. Cardiol.* **2015**, *6*. [CrossRef]

78. Qureshi, A.A.; Khan, D.A.; Silswal, N.; Saleem, S.; Qureshi, N. Evaluation of Pharmacokinetics, and Bioavailability of Higher Doses of Tocotrienols in Healthy Fed Humans. *J. Clin. Exp. Cardiol.* **2016**, *7*, 434. [CrossRef] [PubMed]

79. Springett, G.M.; Husain, K.; Neuger, A.; Centeno, B.; Chen, D.T.; Hutchinson, T.Z.; Lush, R.M.; Sebti, S.; Malafa, M.P. A Phase I Safety, Pharmacokinetic, and Pharmacodynamic Presurgical Trial of Vitamin E δ-tocotrienol in Patients with Pancreatic Ductal Neoplasia. *eBioMedicine* **2015**, *2*, 1987–1995. [CrossRef] [PubMed]

80. Fu, J.Y.; Che, H.L.; Tan, D.M.; Teng, K.T. Bioavailability of tocotrienols: Evidence in human studies. *Nutr. Metab.* **2014**, *11*, 5. [CrossRef] [PubMed]

81. Wahlqvist, M.L.; Krivokuca-Bogetic, Z.; Lo, C.S.; Hage, B.; Smith, R.; Lukito, W. Differential serum responses of tocopherols and tocotrienols during vitamin supplementation in hypercholesterolaemic individuals without change in coronary risk factors. *Nutr. Res.* **1992**, *12*, S181–S201. [CrossRef]

82. Rasool, A.H.; Yuen, K.H.; Yusoff, K.; Wong, A.R.; Rahman, A.R. Dose dependent elevation of plasma tocotrienol levels and its effect on arterial compliance, plasma total antioxidant status, and lipid profile in healthy humans supplemented with tocotrienol rich vitamin, E. *J. Nutr. Sci. Vitaminol.* **2006**, *52*, 473–478. [CrossRef] [PubMed]

83. Rasool, A.H.; Rahman, A.R.; Yuen, K.H.; Wong, A.R. Arterial compliance and vitamin E blood levels with a self-emulsifying preparation of tocotrienol rich vitamin, E. *Arch. Pharm. Res.* **2008**, *31*, 1212–1217. [CrossRef] [PubMed]

84. Qureshi, A.A.; Sami, S.A.; Salser, W.A.; Khan, F.A. Dose-dependent suppression of serum cholesterol by tocotrienol-rich fraction (TRF 25) of rice bran in hypercholesterolemic humans. *Atherosclerosis* **2002**, *161*, 199–207. [CrossRef]

85. Chin, S.F.; Hamid, N.A.; Latiff, A.A.; Zakaria, Z.; Mazlan, M.; Yusof, Y.A.; Karim, A.A.; Ibahim, J.; Hamid, Z.; Ngah, W.Z. Reduction of DNA damage in older healthy adults by Tri E® Tocotrienol supplementation. *Nutrition* **2008**, *24*, 1–10. [CrossRef] [PubMed]

86. Chin, S.F.; Ibahim, J.; Makpol, S.; Hamid, N.A.; Latiff, A.A.; Zakaria, Z.; Mazlan, M.; Yusof, Y.A.; Karim, A.A.; Ngah, W.Z. Tocotrienol rich fraction supplementation improved lipid profile and oxidative status in healthy older adults: A randomized controlled study. *Nutr. Metab.* **2011**, *8*, 42. [CrossRef] [PubMed]

87. Heng, E.C.; Karsani, S.A.; Rahman, M.A.; Hamid, N.A.; Hamid, Z.; Ngah, W.Z. Supplementation with tocotrienol-rich fraction alters the plasma levels of Apolipoprotein AI precursor, Apolipoprotein E precursor, and C-reactive protein precursor from young and old individuals. *Eur. J. Nutr.* **2013**, *52*, 1811–1820. [CrossRef] [PubMed]
88. Qureshi, A.A.; Khan, D.A.; Mahjabeen, W.; Qureshi, N. Dose-dependent Modulation of Lipid Parameters, Cytokines and RNA by [delta]-tocotrienol in Hypercholesterolemic Subjects Restricted to AHA Step-1 Diet. *Br. J. Med. Med. Res.* **2015**, *6*, 351. [CrossRef]
89. Khor, H.T.; Chieng, D.Y.; Ong, K.K. Tocotrienols inhibit liver HMG CoA reductase activity in the guinea pig. *Nutr. Res.* **1995**, *15*, 537–544. [CrossRef]
90. Husain, K.; Francois, R.A.; Yamauchi, T.; Perez, M.; Sebti, S.M.; Malafa, M.P. Vitamin E δ-Tocotrienol Augments the Anti-tumor Activity of Gemcitabine and Suppresses Constitutive NF-κB Activation in Pancreatic Cancer. *Mol. Cancer Ther.* **2011**. [CrossRef] [PubMed]
91. Meganathan, P.; Jabir, R.S.; Fuang, H.G.; Bhoo-Pathy, N.; Choudhury, R.B.; Taib, N.A.; Nesaretnam, K.; Chik, Z. A new formulation of Gamma Delta Tocotrienol has superior bioavailability compared to existing Tocotrienol-Rich Fraction in healthy human subjects. *Sci. Rep.* **2015**, *5*. [CrossRef] [PubMed]
92. Drotleff, A.M.; Bohnsack, C.; Schneider, I.; Hahn, A.; Ternes, W. Human oral bioavailability and pharmacokinetics of tocotrienols from tocotrienol-rich (tocopherol-low) barley oil and palm oil formulations. *J. Funct. Foods* **2014**, *7*, 150–160. [CrossRef]
93. Abuasal, B.S.; Qosa, H.; Sylvester, P.W.; Kaddoumi, A. Comparison of the intestinal absorption and bioavailability of γ-tocotrienol and α-tocopherol: In vitro, in situ and in vivo studies. *Biopharm. Drug Dispos.* **2012**, *33*, 246–256. [CrossRef] [PubMed]
94. Abuasal, B.S.; Lucas, C.; Peyton, B.; Alayoubi, A.; Nazzal, S.; Sylvester, P.W.; Kaddoumi, A. Enhancement of intestinal permeability utilizing solid lipid nanoparticles increases γ-tocotrienol oral bioavailability. *Lipids* **2012**, *47*, 461–469. [CrossRef] [PubMed]
95. Pouton, C.W. Formulation of poorly water-soluble drugs for oral administration: Physicochemical and physiological issues and the lipid formulation classification system. *Eur. J. Pharm. Sci.* **2006**, *29*, 278–287. [CrossRef] [PubMed]
96. Charman, S.A.; Charman, W.N.; Rogge, M.C.; Wilson, T.D.; Dutko, F.J.; Pouton, C.W. Self-emulsifying drug delivery systems: Formulation and biopharmaceutic evaluation of an investigational lipophilic compound. *Pharm. Res.* **1992**, *9*, 87–93. [CrossRef] [PubMed]
97. Alqahtani, S.; Alayoubi, A.; Nazzal, S.; Sylvester, P.W.; Kaddoumi, A. Enhanced solubility and oral bioavailability of γ-Tocotrienol using a self-emulsifying drug delivery system (SEDDS). *Lipids* **2014**, *49*, 819–829. [CrossRef] [PubMed]
98. Alqahtani, S.; Alayoubi, A.; Nazzal, S.; Sylvester, P.W.; Kaddoumi, A. Nonlinear absorption kinetics of self-emulsifying drug delivery systems (SEDDS) containing tocotrienols as lipophilic molecules: In vivo and in vitro studies. *AAPS J.* **2013**, *15*, 684–695. [CrossRef] [PubMed]
99. Lim, Y.; Traber, M.G. Alpha-tocopherol transfer protein (α-TTP): Insights from alpha-tocopherol transfer protein knockout mice. *Nutr. Res. Pract.* **2007**, *1*, 247–253. [CrossRef] [PubMed]
100. Gohil, K.; Schock, B.C.; Chakraborty, A.A.; Terasawa, Y.; Raber, J.; Farese, R.V.; Packer, L.; Cross, C.E.; Traber, M.G. Gene expression profile of oxidant stress and neurodegeneration in transgenic mice deficient in α-tocopherol transfer protein. *Free Radic. Biol. Med.* **2003**, *35*, 1343–1354. [CrossRef]
101. Jishage, K.I.; Arita, M.; Igarashi, K.; Iwata, T.; Watanabe, M.; Ogawa, M.; Ueda, O.; Kamada, N.; Inoue, K.; Arai, H.; Suzuki, H. α-Tocopherol transfer protein is important for the normal development of placental labyrinthine trophoblasts in mice. *J. Biol. Chem.* **2001**, *276*, 1669–1672. [CrossRef] [PubMed]
102. Leonard, S.W.; Terasawa, Y.; Farese, R.V.; Traber, M.G. Incorporation of deuterated RRR-or all-rac-α-tocopherol in plasma and tissues of α-tocopherol transfer protein–null mice. *Am. J. Clin. Nutr.* **2002**, *75*, 555–560. [CrossRef] [PubMed]
103. Schock, B.C.; Van der Vliet, A.; Corbacho, A.M.; Leonard, S.W.; Finkelstein, E.; Valacchi, G.; Obermueller-Jevic, U.; Cross, C.E.; Traber, M.G. Enhanced inflammatory responses in α-tocopherol transfer protein null mice. *Arch. Biochem. Biophys.* **2004**, *423*, 162–169. [CrossRef] [PubMed]
104. Terasawa, Y.; Ladha, Z.; Leonard, S.W.; Morrow, J.D.; Newland, D.; Sanan, D.; Packer, L.; Traber, M.G.; Farese, R.V. Increased atherosclerosis in hyperlipidemic mice deficient in α-tocopherol transfer protein and vitamin, E. *Proc. Natl. Acad. Sci. USA* **2000**, *97*, 13830–13834. [CrossRef] [PubMed]

105. Yokota, T.; Igarashi, K.; Uchihara, T.; Jishage, K.I.; Tomita, H.; Inaba, A.; Li, Y.; Arita, M.; Suzuki, H.; Mizusawa, H.; et al. Delayed-onset ataxia in mice lacking α-tocopherol transfer protein: Model for neuronal degeneration caused by chronic oxidative stress. *Proc. Natl. Acad. Sci. USA* **2001**, *98*, 15185–15190. [CrossRef] [PubMed]

106. Hosomi, A.; Arita, M.; Sato, Y.; Kiyose, C.; Ueda, T.; Igarashi, O.; Arai, H.; Inoue, K. Affinity for α-tocopherol transfer protein as a determinant of the biological activities of vitamin E analogs. *FEBS Lett.* **1997**, *409*, 105–108. [CrossRef]

107. Li, X.H.; Ghosh, S.P.; Ha, C.T.; Fu, D.; Elliott, T.B.; Bolduc, D.L.; Villa, V.; Whitnall, M.H.; Landauer, M.R.; Xiao, M. Delta-tocotrienol protects mice from radiation-induced gastrointestinal injury. *Radiat. Res.* **2013**, *180*, 649–657. [CrossRef] [PubMed]

108. Little, M.P.; Tawn, E.J.; Tzoulaki, I.; Wakeford, R.; Hildebrandt, G.; Paris, F.; Tapio, S.; Elliott, P. A systematic review of epidemiological associations between low and moderate doses of ionizing radiation and late cardiovascular effects, and their possible mechanisms. *Radiat. Res.* **2008**, *169*, 99–109. [CrossRef] [PubMed]

109. Cucinotta, F.A.; Manuel, F.K.; Jones, J.; Iszard, G.; Murrey, J.; Djojonegro, B.; Wear, M. Space radiation and cataracts in astronauts. *Radiat. Res.* **2001**, *156*, 460–466. [CrossRef]

110. Cucinotta, F.A.; Schimmerling, W.; Wilson, J.W.; Peterson, L.E.; Badhwar, G.D.; Saganti, P.B.; Dicello, J.F. Space radiation cancer risks and uncertainties for Mars missions. *Radiat. Res.* **2001**, *156*, 682–688. [CrossRef]

111. Boerma, M.; Nelson, G.A.; Sridharan, V.; Mao, X.W.; Koturbash, I.; Hauer-Jensen, M. Space radiation and cardiovascular disease risk. *World J. Cardiol.* **2015**, *7*, 882. [CrossRef] [PubMed]

antioxidants

MDPI

Article

The Subcellular Distribution of Alpha-Tocopherol in the Adult Primate Brain and Its Relationship with Membrane Arachidonic Acid and Its Oxidation Products

Emily S. Mohn [1,*], Matthew J. Kuchan [2], John W. Erdman Jr. [3], Martha Neuringer [4], Nirupa R. Matthan [1], Chung-Yen Oliver Chen [1] and Elizabeth J. Johnson [1]

[1]   Jean Mayer US Department of Agriculture Human Nutrition Research Center on Aging, Tufts University, Boston, MA 02111, USA; Nirupa.Matthan@tufts.edu (N.R.M.); Oliver.Chen@tufts.edu (C.-Y.O.C.); Elizabeth.Johnson@tufts.edu (E.J.J.)
[2]   Discovery Research, Abbott Nutrition, Columbus, OH 43229, USA; Matthew.Kuchan@abbott.com
[3]   Department of Food Science and Human Nutrition, University of Illinois at Urbana-Champaign, Urbana, IL 61802, USA; jwerdman@illinois.edu
[4]   Oregon National Primate Research Center, Oregon Health and Science University, Beaverton, OR 97006, USA; neuringe@ohsu.edu
*   Correspondence: Emily.Mohn@tufts.edu; Tel.: +1-617-556-3204

Received: 3 November 2017; Accepted: 23 November 2017; Published: 26 November 2017

Abstract: The relationship between α-tocopherol, a known antioxidant, and polyunsaturated fatty acid (PUFA) oxidation, has not been directly investigated in the primate brain. This study characterized the membrane distribution of α-tocopherol in brain regions and investigated the association between membrane α-tocopherol and PUFA content, as well as brain PUFA oxidation products. Nuclear, myelin, mitochondrial, and neuronal membranes were isolated using a density gradient from the prefrontal cortex (PFC), cerebellum (CER), striatum (ST), and hippocampus (HC) of adult rhesus monkeys ($n = 9$), fed a stock diet containing vitamin E (α-, γ-tocopherol intake: ~0.7 µmol/kg body weight/day, ~5 µmol/kg body weight/day, respectively). α-tocopherol, PUFAs, and PUFA oxidation products were measured using high performance liquid chromatography (HPLC), gas chromatography (GC) and liquid chromatography-gas chromatography/mass spectrometry (LC-GC/MS) respectively. α-Tocopherol (ng/mg protein) was highest in nuclear membranes ($p < 0.05$) for all regions except HC. In PFC and ST, arachidonic acid (AA, µg/mg protein) had a similar membrane distribution to α-tocopherol. Total α-tocopherol concentrations were inversely associated with AA oxidation products (isoprostanes) ($p < 0.05$), but not docosahexaenoic acid oxidation products (neuroprostanes). This study reports novel data on α-tocopherol accumulation in primate brain regions and membranes and provides evidence that α-tocopherol and AA are similarly distributed in PFC and ST membranes, which may reflect a protective effect of α-tocopherol against AA oxidation.

Keywords: α-tocopherol; brain; membranes; arachidonic acid; isoprostanes; rhesus monkey

---

1. Introduction

Vitamin E is an essential, fat-soluble nutrient, obtained from nuts, oils, and green leafy vegetables [1]. Vitamin E has eight structural isomers: 4 (α-, β-, γ- and δ-)tocopherols and 4 (α-, β-, γ- and δ-)tocotrienols; however, mammals preferentially take up and use α-tocopherol [2,3]. α-Tocopherol is crucial for proper brain function, and deficiency can result in a number of neurologic symptoms, such as ataxia, peripheral neuropathy, myopathy, and retinopathy, which can be reversed with supplementation [1]. Although deficiency is rare, there is considerable evidence linking lower α-tocopherol intake levels with cognitive decline and neurodegenerative disease. Epidemiological studies have reported that lower tocopherol intake is associated with poorer cognitive function, compared to individuals with higher intakes [4], and reduced levels of α-tocopherol in plasma [5,6] and cerebrospinal fluid [7,8] have been reported in patients with mild cognitive impairment and Alzheimer's Disease (AD), the most common form of dementia. Post-mortem brain concentrations of α-tocopherol have also been shown to be positively related to cognitive test scores in a centenarian population [6].

The brain is one of the most metabolically active organs in the body, which causes the production of reactive oxygen species (ROS). As a result, it is highly susceptible to ROS attacks on lipid, protein, and DNA [9]. This susceptibility is accentuated by the high concentrations of polyunsaturated fatty acids (PUFA) found in brain cell membranes. Enhanced oxidative stress is thought to contribute to the pathogenesis of cognitive impairment and dementia [10]. α-Tocopherol's function as a lipid antioxidant may underlie its association with cognition. Therefore, it is possible that α-tocopherol maintains the integrity of cell membranes through inhibiting oxidation of PUFAs [11], particularly arachidonic acid (AA) and docosahexaenoic acid (DHA), the two major PUFAs found in the brain [12], and thereby combats the cell damage and neurodegeneration underlying cognitive impairment. In addition, emerging evidence indicates this vitamin may also possess non-antioxidant functions, including the regulation of gene expression [13,14]. In vitro and ex vivo studies have reported that treatment with α-tocopherol can down-regulate the expression of pro-inflammatory genes [15,16] and modulate the expression of several cell signaling and cell cycle genes [15–18]. Thus, the mechanism of action underlying the inverse association between α-tocopherol and cognitive impairment/dementia in humans is of considerable interest.

It is well established that α-tocopherol accumulates within membranes [19], and its subcellular distribution has been investigated in rat brains [20,21]. However, its distribution among different types of membranes and its direct association with membrane PUFA content and brain PUFA oxidation products have not been determined in primates. Determining the subcellular localization of α-tocopherol in brain tissue may provide insight into understanding its potential function(s). The objective of the present study was to determine the distribution of α-tocopherol in membranes from different brain regions and to characterize the relationship between membrane-specific α-tocopherol, membrane PUFA content, and brain PUFA oxidation products (isoprostanes [IsoP] and neuroprostanes [NP]) in primates. This study was performed in rhesus monkeys because they are a well-accepted model for human brain physiology [22].

2. Materials and Methods

2.1. Animals and Diet

Rhesus monkey (*Macaca mulatta*) brains, analyzed in the present study, were from a subgroup of a larger parent study sample, which investigated the effect of lutein supplementation on lutein accumulation in brain regions and membranes of primates [23]. Our study samples included serum and brain tissue from 6 female and 3 male adult monkeys (mean age: 11.7 ± 3.3 years; mean body weight: 7.87 ± 2.24 kg) that were obtained through the Oregon National Primate Research Center (ONPRC) (Beaverton, Oregon) Tissue Distribution Program; animals were not euthanized specifically for this study, but were euthanized for other projects or for veterinary reasons. All animals had consumed a standard stock diet (Monkey Diet Jumbo 5037, Lab Diet, St. Louis, MO, USA) at least twice a day, along with daily supplements of a variety of fruits and vegetables. The stock diet was measured for vitamin E (α- and γ-tocopherol), using previously reported methods [24]. The n-6 PUFA content of the diet was predominantly linoleic acid (1.7% of ration), with less than 0.01% of each ration being arachidonic acid (AA). The n-3 PUFA content was 0.13% of ration, mainly contributed by linolenic acid (0.10% of ration), but also containing DHA (~0.01% of ration). Our previous study [23] also included tissue from 4 monkeys receiving lutein supplementation, but these animals were not included in the current study, due to potential confounding effects on measures of oxidation.

The study was in compliance with all institutional and federal regulations on the use of laboratory animals as well as the Guide for the Care and Use of Laboratory Animals [25]. Procedures were approved by the Institutional Animal Care and Use Committee (IACUC) of Oregon Health and Science University (Protocol IS00003766). Professional care was provided by the ONPRC Division of Comparative Medicine and all animals were observed at least twice a day by trained veterinary technicians. In addition to rotating dietary supplements of fruits and vegetables, animals were provided with environmental enrichments, including a changing variety of toys. Euthanasia was conducted by a veterinary pathologist; animals were sedated with ketamine, and then deeply anesthetized with sodium pentobarbital, according to the Guidelines of the American Veterinary Medical Association. IACUC approval from Tufts University was also obtained for biological sample receipt, storage, and analysis.

2.2. Serum and Brain Collection

Fasting blood was drawn from the saphenous vein at the time of euthanasia. Because the monkeys were not euthanized exclusively for this study, blood samples were not collected in every case but were available for 7 out of 9 animals. Blood was processed for serum ($1000 \times g$, 10 min, 4 °C) and then stored at -80 °C prior to analysis.

Brains were removed immediately after euthanasia and the prefrontal cortex (PFC), cerebellum (CER), striatum (ST), and hippocampus (HC) were dissected from the right and left hemispheres. The regions consisted of both gray and white matter but excluded major white matter tracts. The dissected regions were immediately placed on dry ice and then stored at -80 °C. The right and left hemispheres of each region were pooled, pulverized in liquid nitrogen, aliquoted, and stored at -80 °C.

2.3. Preparation of Brain Membranes

Nuclear, myelin, mitochondrial, and neuronal plasma membranes were isolated from the PFC, CER, ST, and HC, using differential centrifugation with a Ficoll density gradient [26,27]. Pulverized brain samples were homogenized in aqueous buffer (10 mM hydroxyethyl piperazineethanesulfonic acid (HEPES), 0.25 mM ethylenediaminetetraacetic acid EDTA, 0.32 M sucrose, pH 7.2), containing protease inhibitors (cOmpleteTM protease inhibitor cocktail, Roche, Basel, Switzerland). The homogenates were subsequently subjected to low-speed centrifugation ($1000 \times g$, 4 °C) to isolate the crude nuclear membrane pellet. The resulting supernatant was removed to a new tube and the protocol was repeated with the pellet. Supernatant from the second low-speed centrifugation was combined with the first. The combined supernatants were then centrifuged ($17,000 \times g$, 4 °C) to obtain the "crude membrane pellet", containing myelin, mitochondrial, and neuronal plasma membranes. The crude membrane pellet was re-homogenized in buffer containing protease inhibitors but no sucrose (10 mM HEPES, 0.25 mM EDTA, pH 7.2) and was applied to a Ficoll density gradient (consisting of 14% and 7% Ficoll solutions). The homogenate was then subjected to high-speed centrifugation ($87,000 \times g$, 4 °C), to separate myelin, mitochondrial, and neuronal plasma membranes. All membranes, including the crude nuclear membrane, were purified via centrifugation at $17,000 \times g$, 4 °C. Pure membranes were aliquoted and stored at -80 °C for tocopherol and fatty acid analyses. Membrane recovery was 76 ± 1%, as determined by measuring the sum of α-tocopherol levels in all membranes and supernatants and comparing this to total α-tocopherol in each brain region.

2.4. α-Tocopherol Extraction from Brain Regions, Membranes, and Serum

The process for the extraction of α- and γ-tocopherol from brain regions and membranes was adapted from Park et al. [28] and has been previously described in detail [29]. Briefly, after the additions of an internal standard (echinenone) and antioxidant (sodium ascorbate), samples were saponified with 5% sodium hydroxide and α-tocopherol was extracted using hexane. Extracts were then injected into a reverse-phase HPLC system to separate and quantify α- and γ-tocopherol.

Both α- and γ-tocopherol were extracted and analyzed from serum, using a previously published Folch method [30]. The lower limit of detection was 2.7 pmol, and the interassay coefficient of variation (CV) was 4%. Concentrations are expressed as ng/mg protein.

2.5. Fatty Acid Extraction and Protein Determination in Brain Regions and Membranes

Lipids were extracted overnight from homogenates of brain regions and membranes using a modified Folch method [31]. Fatty acids were analyzed using an established gas chromatography method [32]. Peaks of interest were identified by comparison with authentic fatty acid standards (Nu-Chek Prep, Inc., Elysian, MN, USA) and expressed as μg/mg protein and mole percent (%). The interassay CV ranged from 0.5 to 4.3% for fatty acids present at levels >5% of total fatty acids (TFAs), 1.8–10.1% for fatty acids present at levels between 1–5% TFAs, and 9.8–25.1% for fatty acids present at levels <1% TFAs. For the present analysis, we focused on the content of total PUFA, total n-6, total n-3, AA, and DHA.

Delipidated brain tissue and membranes from the overnight lipid extraction were digested in 1N sodium hydroxide for the determination of protein, using the bicinchoninic acid (BCA) assay, as per the manufacturer's instructions (Pierce Inc., Rockford, IL, USA). Brain regions and membranes were digested for 8 and 5 days, respectively.

2.6. PUFA Oxidation Determination in Brain Tissue

Total NP and IsoP, formed from the oxidation of DHA and AA, respectively, were extracted from the PFC, CER and ST, and quantified using published methods [33,34] with modifications. Due to the small size of HC, the amount of tissue from this region was insufficient for analysis. Briefly, lipids were extracted from homogenized brain samples using the Folch method. The lipid extract was then saponified to release esterified NP and IsoP. Neutral lipids were removed from the resulting mixture, using hexane, and samples were acidified to pH 3 to protonate NP and IsoP carboxylic acid groups. An internal standard, $[^2H_4]$ 15-F_{2t}-IsoP (Cayman Chemicals, Ann Arbor, MI, USA), was added prior to extraction with ethyl acetate. NP and IsoP were then derivatized to form pentafluorobenzyl (PFB) esters and subjected to HPLC (Agilent 1050, Santa Clara, CA, USA), using the method described by Walter et al. [33]. NP and IsoP fractions were collected, converted to trimethylsilyl ether derivatives, and quantified using GC/MS [33]. Selective ion monitoring was used for analysis at m/z 593 for NP, m/z 569 for IsoP and m/z 573 for the internal standard, $[^2H_4]$ 15-F_{2t}-IsoP. The inter-assay CV was 10%.

2.7. Statistical Analysis

α- and γ-tocopherol and fatty acid data are expressed as mean ± standard deviation. A one-way analysis of variance (ANOVA) with Tukey's honest significant difference (HSD) was performed to determine differences in tocopherol concentrations across brain regions. Due to significant differences in brain region tocopherol concentrations ($p < 0.05$), membrane data from each region was analyzed separately. A one-way ANOVA with Tukey's HSD was performed for α- and γ-tocopherol as well as PUFAs (total, n-6, n-3, AA, and DHA), to determine their membrane distributions within each region. Pearson correlations were performed to determine whether region and membrane-specific α-tocopherol, membrane PUFAs, and brain region PUFA oxidation products were related, adjusting for age. We considered $r = 0.30$–0.49 to be a weak correlation, $r = 0.50$–0.69, a moderate correlation and $r = 0.70$–0.99 a strong correlation [35]. All analyses were performed using Statistical Analysis Software, SAS 9.4, with significance set at the 0.05 level.

3. Results

3.1. Tocopherol Concentrations in Rhesus Monkey Stock Diet and Serum

Tocopherol concentrations in the monkey diet and serum are presented in Table 1. In the stock diet, γ-tocopherol concentrations were ~7.5 times greater than the α-tocopherol content. The average amount of α-tocopherol consumed from the stock diet was ~2 mg/day or 0.70 μmol/kg body weight, while average γ-tocopherol intakes were ~16 mg/day or 5.3 μmol/kg body weight. Conversely, concentrations of α-tocopherol in serum were ~8 times greater than γ-tocopherol concentrations.

Table 1. Tocopherol Concentrations (mean $\pm$ SD) in Monkey Stock Diet * and Serum ($n = 7$).

	α-Tocopherol	γ-Tocopherol
Stock Diet (μmol/kg)	23 ± 1.2	171 ± 34
Serum (μmol/L)	20.8 ± 8.3	2.5 ± 0.6

* Sample run in triplicate.

3.2. Distribution of α-Tocopherol in Brain Regions and Membranes of Adult Rhesus Monkeys

The α-tocopherol contents in the PFC, CER, ST, and HC of rhesus monkeys are presented in Figure 1. Concentrations of α-tocopherol were significantly lower in the PFC, CER and ST than in HC ($p = 0.02$), and also significantly lower in the CER than in the ST ($p < 0.05$). α-Tocopherol concentrations did not significantly differ between the ST and PFC. Concentrations of γ-tocopherol in the HC were significantly greater than in all other regions ($p < 0.05$, Figure S1). As in serum, α-tocopherol concentrations were 7–10 times greater than γ-tocopherol concentrations in the brain regions (Figure S2).

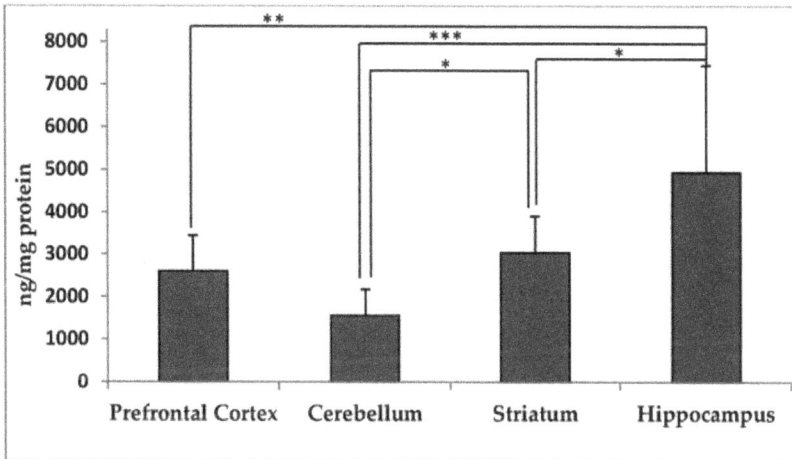

Figure 1. α-Tocopherol concentrations (ng/mg protein, mean $\pm$ SD) in different regions of the brain from adult rhesus macaques ($n = 9$). Asterisks indicate significant differences between brain regions according to one-way analysis of variance (ANOVA) followed by Tukey's honest significance difference (HSD) test; * $p < 0.05$, ** $p < 0.01$, *** $p < 0.001$.

Membrane concentrations of α-tocopherol in the PFC, CER, ST, and HC of rhesus monkeys are presented in Figure 2A–D. The distribution of α-tocopherol among membrane types differed in a region-specific manner. In the PFC, α-tocopherol concentrations were significantly greater in nuclear membranes than in mitochondrial membranes ($p < 0.05$). In the CER, α-tocopherol concentrations were significantly greater in nuclear membranes than in all other membranes, and significantly greater in myelin than in mitochondrial membranes ($p < 0.05$). In the ST, α-tocopherol was again significantly greater in nuclear membranes than in all other membrane types ($p < 0.05$). In the HC, however, there were no significant differences in α-tocopherol concentrations across membrane types. Membrane concentrations of γ-tocopherol in brain regions were considerably lower than α-tocopherol (Figure S3A–D), and there were no significant differences among membrane types in the PFC and HC. However, in the CER and ST, γ-tocopherol concentrations were significantly greater in nuclear membranes than in all other membrane types ($p \leq 0.02$).

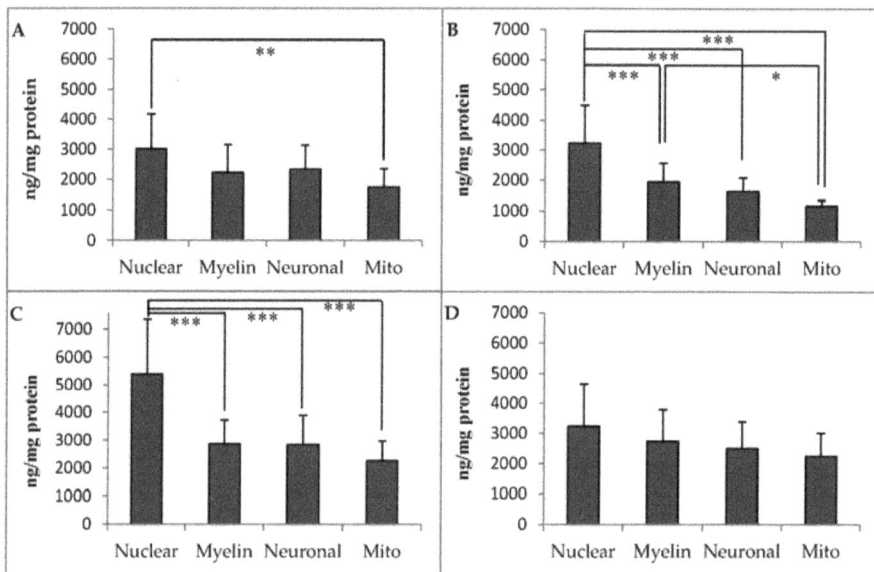

Figure 2. Membrane α-tocopherol concentrations (ng/mg protein, mean $\pm$ SD) in the (**A**) prefrontal cortex; (**B**) cerebellum; (**C**) striatum; and (**D**) hippocampus of adult rhesus macaques ($n = 9$). Asterisks indicate significant differences between brain regions according to one-way analysis of variance (ANOVA) followed by Tukey's honest significant difference (HSD) test; * $p < 0.05$, ** $p < 0.01$, *** $p < 0.001$. Nuclear: nuclear membrane; myelin: myelin membranes; Neuronal: neuronal plasma membrane; Mito: mitochondrial membranes.

3.3. Distribution of Membrane PUFAs in Different Brain Regions of Adult Rhesus Macaques

Membrane concentrations of total PUFA, n-6, n-3, AA, and DHA in the PFC, CER, ST, and HC of rhesus monkeys are presented in Table 2. In the PFC and ST, AA and total n-6 PUFA concentrations were highest in nuclear membranes, intermediate in myelin and neuronal membranes, and lowest in mitochondrial membranes ($p < 0.05$). However, in the CER, AA was similar among all membranes, except mitochondrial membranes, where concentrations were lower than in all other membrane fractions. In the HC, mean values were highest in myelin and nuclear membranes, followed by neuronal and mitochondrial membranes ($p < 0.05$).

Table 2. Membrane polyunsaturated fatty acid concentration (µg/mg protein, mean ± SD) in the prefrontal cortex, cerebellum, striatum, and hippocampus of adult rhesus macaques ($n = 9$).

Region		Nuclear	Myelin	Neuronal	Mitochondrial
Prefrontal Cortex	Arachidonic Acid	123.5 ± 30.0 [a]	99.3 ± 30.5 [b]	96.44 ± 20.0 [b]	28.0 ± 16.3 [c]
	PUFA n-6	291.7 ± 77.5 [a]	204.9 ± 48.9 [b]	201.6 ± 63.5 [b]	49.2 ± 27.6 [c]
	Docosahexaenoic acid	151.9 ± 37.2 [a]	180.6 ± 48.9 [a]	175.8 ± 54.8 [a]	35.3 ± 19.2 [b]
	PUFA n-3	164.2 ± 44.6 [a]	187.6 ± 51.7 [a]	182.9 ± 56.8 [a]	36.2 ± 19.8 [b]
	Total PUFA	455.9 ± 120.6 [a]	392.4 ± 99.9 [a]	384.5 ± 120.1 [a]	85.4 ± 47.1 [b]
Cerebellum	Arachidonic Acid	91.4 ± 19.1 [a]	84.4 ± 32.8 [a]	95.9 ± 20.4 [a]	15.1 ± 5.8 [b]
	PUFAn-6	213.0 ± 40.2 [a]	166.1 ± 59.3 [b]	187.6 ± 38.0 [a,b]	30.3 ± 11.4 [c]
	Docosahexaenoic acid	136.0 ± 26.5 [a]	178.2 ± 59.9 [b]	199.8 ± 39.7 [b]	27.8 ± 10.2 [c]
	PUFA n-3	148.8 ± 30.0 [a]	185.9 ± 63.3 [b]	207.3 ± 40.9 [b]	28.7 ± 10.5 [c]
	Total PUFA	361.8 ± 69.4 [a]	352.0 ± 121.4 [a]	394.9 ± 76.7 [a]	59.0 ± 21.9 [b]
Striatum	Arachidonic Acid	221.5 ± 65.4 [a]	156.2 ± 20.5 [b]	143.9 ± 26.5 [b]	30.5 ± 13.3 [c]
	PUFA n-6	525.2 ± 177.7 [a]	315.5 ± 46.4 [b]	289.0 ± 47.6 [b]	52.9 ± 22.1 [c]
	Docosahexaenoic acid	196.9 ± 36.1 [a]	257.7 ± 35.0 [b]	201.0 ± 37.2 [a]	37.3 ± 17.4 [c]
	PUFA n-3	224.0 ± 43.9 [a]	268.7 ± 36.7 [b]	211.7 ± 38.5 [a]	38.4 ± 17.8 [c]
	Total PUFA	749.2 ± 217.5 [a]	584.2 ± 81.6 [b]	500.7 ± 85.1 [c]	91.4 ± 39.6 [d]
Hippocampus	Arachidonic Acid	208.0 ± 54.6 [a,b]	247.1 ± 119.7 [a]	170.6 ± 50.4 [b]	21.6 ± 6.6 [c]
	PUFA n-6	452.0 ± 126.0 [a]	519.7 ± 226.7 [a]	340.0 ± 104.9 [b]	37.0 ± 10.9 [c]
	Docosahexaenoic acid	144.2 ± 40.9 [a]	325.2 ± 118.7 [b]	201.5 ± 63.4 [c]	16.7 ± 5.5 [d]
	PUFA n-3	191.6 ± 54.2 [a]	347.7 ± 127.5 [b]	219.4 ± 69.7 [a]	18.2 ± 5.5 [c]
	Total PUFA	643.6 ± 177.5 [a,b]	867.5 ± 351.2 [a]	559.5 ± 173.8 [b]	55.2 ± 16.1 [c]

Means with different superscripts (a, b, c, d) in each row are significantly different, according to one-way analysis of variance (ANOVA) followed by Tukey's honest significance difference (HSD) test ($p < 0.05$). PUFA: polyunsaturated fatty acids.

In the PFC, DHA, n-3 PUFAs, and the total PUFA concentration were lowest in mitochondrial membranes ($p < 0.05$), but did not differ in the other membranes. In the CER, DHA and n-3 PUFAs were highest in neuronal membranes and myelin, followed by nuclear membranes, and lowest in mitochondrial membranes ($p < 0.05$). In this region, the total PUFA content was lowest in mitochondrial membranes compared to the other membranes ($p < 0.05$). In both the ST and HC, PUFA n-3 and DHA were highest in myelin membranes, intermediate in neuronal and nuclear membranes, and lowest in mitochondrial membranes ($p < 0.05$). In the ST, total PUFA concentrations differed across all membranes (nuclear > myelin > neuronal > mitochondrial, $p < 0.05$). In the HC, total PUFAs were highest in myelin, intermediate in neuronal and nuclear membranes, and lowest in mitochondrial membranes ($p < 0.05$).

While the absolute concentration of fatty acids provides a measurement of the amount of each fatty acid, independent of other fatty acids, mole % data measures the relative importance of a fatty acid against the total fatty acid concentration. Thus, the membrane distribution of fatty acids differs, depending on whether fatty acids are expressed as mole % or µg/mg protein. Membrane fatty acid profiles among brain regions, expressed as mole %, are reported in the supplementary material (Tables S1–S4). Briefly, in the PFC, CER and ST, the mole % AA was highest in mitochondrial membranes, intermediate in neuronal and myelin membranes, and lowest in nuclear membranes (Tables S1–S3, $p < 0.05$). In the PFC, DHA was distributed similarly to AA among membranes (Table S1, $p < 0.05$). In the CER, the mole % DHA was highest in mitochondrial membranes, followed by neuronal membranes, followed by myelin membranes, and lowest in nuclear membranes (Table S2, $p < 0.05$). In the ST, the mole % DHA was highest in mitochondrial and myelin membranes, intermediate in neuronal membranes, and lowest in nuclear membranes (Table S3, $p < 0.05$). In the HC, the mole % AA was highest in mitochondrial membranes, but did not differ among the other membranes (Table S4, $p < 0.05$). The mole % DHA in this region was lowest in nuclear membranes but did not differ among other membrane types ($p < 0.05$).

3.4. Relationship between Membrane α-Tocopherol and PUFA Concentrations in Brain Regions

The association between the membrane-specific α-tocopherol and PUFA concentrations in the PFC, CER, ST, and HC is presented in Table 3. In all regions, except the HC, α-tocopherol concentrations were positively associated with all PUFA concentrations—including AA, DHA, n-6 and n-3 PUFA and total PUFA—in nuclear membranes ($p < 0.05$), with the single exception of DHA in the CER. No other significant associations were observed in the CER. In the PFC, α-tocopherol concentrations were also associated with all PUFAs in neuronal plasma membranes ($p < 0.05$). In the ST, α-tocopherol was significantly associated with all PUFAs in all membrane types ($p < 0.05$). In the HC, α-tocopherol was associated with all PUFAs in myelin, but only DHA and total PUFA n–3 in mitochondrial membranes ($p < 0.05$). No significant associations were observed in nuclear and neuronal plasma membranes from this region.

Table 3. Partial correlations between membrane α-tocopherol (ng/mg protein) and PUFA concentrations (μg/mg protein) from different brain regions of rhesus macaques ($n = 9$).

Region		Nuclear	Myelin	Neuronal	Mitochondrial
Prefrontal Cortex	Arachidonic acid	0.68 **	0.69 **	0.78 **	–
	PUFA n-6	0.70 **	0.61	0.82 **	–
	Docosahexaenoic acid	0.70 **	0.55	0.94 ***	–
	PUFA n-3	0.71 **	0.54	0.94 ***	–
	Total PUFA	0.71 **	0.58	0.91 ***	–
Cerebellum	Arachidonic acid	0.78 **	0.43	0.52	0.53
	PUFA n-6	0.85 ***	0.47	0.58	0.57
	Docosahexaenoic acid	0.61	0.40	0.63	0.56
	PUFA n-3	0.70 **	0.41	0.62	0.56
	Total PUFA	0.76 **	0.44	0.60	0.57
Striatum	Arachidonic acid	0.89 ***	0.81 **	0.86 ***	0.80 **
	PUFA n-6	0.81 **	0.76 **	0.84 ***	0.75 **
	Docosahexaenoic acid	0.89 ***	0.81 **	0.80 **	0.80 **
	PUFA n-3	0.91 ***	0.80 **	0.79 **	0.79 **
	Total PUFA	0.85 ***	0.79 **	0.83 **	0.77 **
Hippocampus	Arachidonic acid	0.60	0.71 **	0.48	0.61
	PUFA n-6	0.64	0.79 **	0.50	0.58
	Docosahexaenoic acid	0.61	0.70 **	0.57	0.79 **
	PUFA n-3	0.67	0.70 **	0.56	0.77 **
	Total PUFA	0.66	0.76 **	0.53	0.66

Values are partial correlation coefficients (r values) adjusted for age; ** $p < 0.05$, *** $p < 0.01$. –: r values below 0.40 are not shown.

3.5. Relationship between Membrane α-Tocopherol and PUFA Oxidation Products in Brain Regions

Concentrations of DHA oxidation products (NP) and AA oxidation products (IsoP) among the PFC, CER, and ST are presented in Figure 3. NP and IsoP were significantly lower in the CER compared to both the PFC and ST ($p < 0.01$). The ratio of IsoP/NP was also significantly lower in the CER, compared to the other two regions ($p < 0.05$). NP, IsoP, and the ratio of IsoP/NP did not differ between the PFC and ST.

The cross-sectional relationship between region and membrane-specific α-tocopherol concentrations and brain PUFA oxidation products in the PFC, CER, and ST is presented in Table 4. NP concentrations (pg/μg DHA), measured in whole tissue, were not significantly associated with α-tocopherol concentrations (ng/μg DHA) in any membrane type or brain region. Similarly, IsoP concentrations (pg/μg AA) were not associated with α-tocopherol (ng/μg AA) in any individual membrane type. However, AA oxidation products were inversely associated with concentrations of α-tocopherol in whole tissue, in both the PFC and ST ($p < 0.05$). PUFA oxidation products were not associated with γ-tocopherol concentrations in any membrane or region (data not shown).

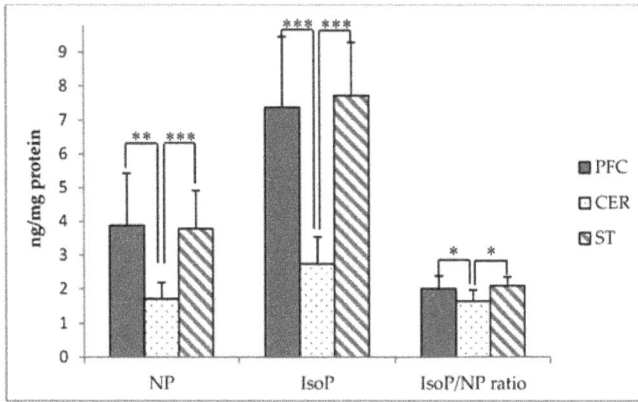

Figure 3. Neuroprostane (NP) and isoprostane (IsoP) concentrations (ng/mg protein) in the prefrontal cortex (PFC), cerebellum (CER), and striatum (ST) of adult rhesus macaques ($n = 9$). Asterisks indicate significant differences between brain regions according to one-way ANOVA followed by Tukey's HSD test; * $p < 0.05$, ** $p < 0.01$, *** $p < 0.001$.

Table 4. Partial correlations between membrane α-tocopherol and neuroprostanes and isoprostanes from different brain regions of rhesus macaques ($n = 9$).

	Total	Nuclear	Myelin	Neuronal	Mitochondrial
	Neuroprostanes				
Prefrontal Cortex	−0.66 *	–	–	–	–
Cerebellum	–	–	–	–	–
Striatum	−0.43	−0.62 *	–	–	–
	Isoprostanes				
Prefrontal Cortex	−0.74 **	−0.47	–	–	–
Cerebellum	–	−0.66 *	–	–	−0.61 *
Striatum	−0.73 **	–	−0.64 *	-0.50	–

Values are partial correlation coefficients (*r* values) adjusted for age; ** $p < 0.05$, * $p < 0.1$. –: *r* values below 0.40 are not shown.

4. Discussion

This study is the first to report the distribution of α-tocopherol and γ-tocopherol in multiple subcellular membrane types (nuclear, myelin, neuronal, and mitochondrial) for multiple brain regions in the primate brain. We also directly investigated the relationship between membrane α-tocopherol concentration and both membrane PUFA content and brain PUFA oxidation. We report that α-tocopherol was differentially distributed among the membranes, with the highest concentrations found in nuclear membranes for all regions tested, except the HC. In the PFC and ST, AA and total n-6 PUFA distribution among membrane types were similar to that of α-tocopherol, and total α-tocopherol concentrations in these regions were inversely related to AA oxidation products.

4.1. α-Tocopherol Distribution in Brain Regions of Adult Rhesus Monkeys

Although no other studies have reported on the concentration of α-tocopherol in monkey brain regions, our results are similar to α-tocopherol concentrations reported in different regions of the adult human brain [6,36], but slightly greater than those reported for human infants [37] and rats [20]. In the current study, the CER had the lowest α-tocopherol concentrations among the brain regions analyzed. This is consistent with previous findings in brain tissue from human adults and centenarians [6,36] as well as rats [20]. In the rat brain, researchers have also observed that the uptake of α-tocopherol is greatest in the CER, despite this region having the lowest steady-state amounts of α-tocopherol, suggesting a more rapid turnover of α-tocopherol in the CER, compared to other regions [20]. Alternatively, this observation in primates may be due to differences in α-tocopherol transfer protein (α-TTP) and/or tocopherol-associated protein (TAP) expression among brain regions [38,39]. In human infants, there were no differences in α-tocopherol concentrations across brain regions tested [37], suggesting that α-tocopherol distribution in the brain might differ across the lifespan. However, that study did not analyze the CER. Future studies investigating the mechanisms underlying the differential distribution of α-tocopherol across the brain tissue of adult primates are warranted.

Our detection of γ-tocopherol at significantly lower concentrations than α-tocopherol is also consistent with findings from studies in human brain tissue [36,37], but differs from rodent studies, in which only α-tocopherol has been detected [20,40]. The distribution pattern of γ-tocopherol among brain regions and membrane fractions was similar to that of α-tocopherol, indicating that tocopherol delivery to brain regions may depend on both serum levels and brain region-specific factors.

4.2. Accumulation of α-Tocopherol in Brain Membranes of Adult Rhesus Monkeys

The distribution of α-tocopherol among membrane types differed in a region-specific manner. However, α-tocopherol concentrations were generally greater in nuclear membranes, compared to other membranes, in each region, except for the HC. The enrichment of α-tocopherol in nuclear membranes suggests that α-tocopherol may play a role in nuclear-associated functions in the brain. This role may be particularly important in the ST and CER, where preferential accumulation of α-tocopherol in the nuclear membrane was most apparent. Accumulating evidence from a number of in vitro and ex vivo studies indicates that α-tocopherol can modulate the expression of cell signaling, cell cycle regulation, and pro-inflammatory genes [15–18,41]. In mice and rats, dietary and supplemental α-tocopherol or the combination of α- and γ-tocopherol has been shown to modulate the expression of genes involved in apoptosis, lipid biosynthesis, adenosine triphosphate (ATP) biosynthesis, and immune/inflammatory responses in the aging brain [42,43]. Additionally, the expression of genes encoding synaptic proteins, protein kinase C family members, and myelin proteins were decreased in transgenic mice deficient in α-TTP [44]. These results may have important implications for the potential role of α-tocopherol in protecting against cognitive impairment and dementia, given that microarray studies have demonstrated that dysregulation of genes involved in inflammation, synaptic signaling, and neuronal apoptosis alter brain cell function and contribute to the pathogenesis of these conditions [45–47]. Future studies investigating the effect of α-tocopherol enrichment in nuclear membranes and changes in gene expression in the primate brain are needed to gain a better understanding of why α-tocopherol may accumulate in this membrane.

Membrane α-tocopherol and PUFA concentrations were found to be positively associated with one another, particularly in the PFC and ST, and to a lesser extent in the CER and HC. This finding is consistent with previous studies, which demonstrated that α-tocopherol accumulates in PUFA-rich membrane domains [48,49]. No significant relationship was observed between α-tocopherol and brain PUFA oxidation products in individual membrane fractions. However, concentrations of both DHA and AA oxidation products (as measured by NP and IsoP, respectively), as well as the relative amount of AA oxidation products to DHA oxidation products (IsoP/NP ratio) were lower in the CER compared to other brain regions, similar to α-tocopherol. Additionally, α-tocopherol concentrations in whole tissue were inversely associated with AA oxidation products (IsoP) in both the PFC and ST. These are the only two brain regions where concentrations of both α-tocopherol and AA were highest in nuclear membranes, intermediate in myelin and neuronal membranes, and lowest in mitochondrial membranes. In contrast, α-tocopherol concentrations were not correlated with DHA oxidation products, as measured by NP levels. These results suggest that (1) α-tocopherol may accumulate in regions with higher PUFA oxidation, particularly AA oxidation relative to DHA oxidation; and (2) α-tocopherol concentrations within these regions of relative higher accumulation may be associated with lower AA oxidation. This is consistent with previous rodent studies, which demonstrated that supplementation with α-tocopherol is more effective at decreasing AA oxidation than DHA oxidation [50,51]. Taken together, our findings indicate that, although α-tocopherol accumulates in membranes, rich in both n-6 and n-3 PUFAs, it may be associated with preferentially protecting AA from oxidative damage. Our observation that only concentrations of α-tocopherol in whole tissue were associated with AA oxidation suggests that the free-radical scavenging function of this nutrient may not be specific to a particular membrane type. Therefore, the contribution of total α-tocopherol concentrations within regions may have the strongest relationship to total AA oxidation in brain regions compared to individual membranes. However, only whole tissue concentrations of AA and DHA oxidation products were determined in these studies as it is currently not feasible to measure membrane-specific concentrations, due to limitations in methodology. Therefore, the relationship between α-tocopherol and PUFA oxidation products within each membrane type remains unknown. Our α-tocopherol result differs from our previous findings that membrane-specific concentrations of the antioxidant, lutein, are associated with total DHA oxidation products, but not total AA oxidation products in rhesus monkey brains [23]. One limitation of both studies is that neither accounts for the potential synergistic effects of other antioxidants present in the brain and their influence on DHA and AA oxidation. Therefore, future studies measuring a more comprehensive profile of antioxidants in the brain and their associations with PUFA oxidation, are needed. Another limitation of this study is the small sample size. Our results need to be replicated in a larger sample population. However, our study provides an important first step in characterizing previously unknown relationships between membrane-associated antioxidants and PUFA oxidation in the primate brain and can guide future investigations into the antioxidant functions of α- and γ-tocopherol in the brain.

Our findings support an antioxidant-associated function of α-tocopherol towards AA, but we cannot rule out the possibility that non-antioxidant functions underlie the association between these two nutrients. Previous studies have demonstrated that α-tocopherol can significantly inhibit the activity of phospholipase A2 [41,52], which is primarily responsible for cleavage and release of AA from membrane phospholipids. Therefore, it is possible that the similarity in membrane distribution between α-tocopherol and AA may reflect not only a role of α-tocopherol in inhibiting AA oxidation, but also a role in the modulation of AA cleavage and release from membranes. Future studies, investigating the relationship between membrane α-tocopherol levels and phospholipase A2 activity in regions of the primate brain, are needed, to better understand the contribution of this potential mechanism to the overall function of α-tocopherol in the primate brain.

5. Conclusions

In conclusion, we found that dietary α-tocopherol was higher in the HC than in the PFC, CER or ST in the non-human primate brain. Within membrane types, α-tocopherol showed preferential accumulation in nuclear membranes compared to other membrane types, except in the HC. We speculate that this observation may be indicative of a role of α-tocopherol in nuclear functions in brain cells. Additionally, we observed that α-tocopherol is positively related to PUFA concentrations in membranes, but only concentrations of AA were distributed similarly to α-tocopherol among membrane types. Finally, whole tissue, but not membrane, α-tocopherol concentrations were significantly associated with AA oxidation products, in both the PFC and ST. Thus, correlations between α-tocopherol and PUFA concentrations may be associated with a protective role of α-tocopherol against AA oxidation, but may also reflect a role of α-tocopherol in inhibiting AA release from membrane phospholipids. Collectively, our study provides insight into the accumulation of dietary α-tocopherol in the primate brain, which may have important implications regarding its functions in this tissue.

Supplementary Materials: The following are available online at www.mdpi.com/2076-3921/6/4/97/s1, Figure S1. Mean γ-tocopherol concentrations (ng/mg protein, ±SD) in different regions of the brain from adult rhesus macaques (*n* = 9), Figure S2. Mean (±SD) ratio of α-tocopherol/γ-tocopherol in different regions of the brain from adult rhesus macaques (*n* = 9). Figure S3. Membrane γ-tocopherol concentrations (ng/mg protein, mean ± SD) in (A) prefrontal cortex (B) cerebellum (C) striatum (D) hippocampus of adult rhesus macaques (*n* = 9), Table S1. Mean (±SD) mole percent fatty acids in prefrontal cortex membranes of adult rhesus macaques (*n* = 9), Table S2. Mean (±SD) mole percent fatty acids in cerebellar membranes of adult rhesus macaques (*n* = 9), Table S3. Mean (±SD) mole percent fatty acids in striatal membranes of adult rhesus macaques (*n* = 9), Table S4. Mean (±SD) mole percent fatty acids in hippocampal membranes of adult rhesus macaques (*n* = 9).

Acknowledgments: Emily E. Johnson and the ONPRC Veterinary Pathology Service assisted with brain tissue collection. Jean Gallucio, Audrey Goldbaum, and Kathryn Baldyga at Tufts University performed brain region and membrane fatty acid and protein analysis. This work was supported by a grant from Abbott Nutrition through the Center for Nutrition, Learning, and Memory (CNLM) at the University of Illinois at Urbana-Champaign, DSM Nutritional Products, and USDA under grant 8050-51000-095-01S.

Author Contributions: John W. Erdman Jr, Elizabeth J. Johnson, Matthew J. Kuchan, and Martha Neuringer conceived and designed the experiments; Emily S. Mohn, Nirupa R. Matthan, and Martha Neuringer performed the experiments; Emily S. Mohn analyzed the data; Elizabeth J. Johnson, Nirupa R. Matthan, Martha Neuringer, Chun-Yen. Oliver Chen contributed reagents/materials/analysis tools; Emily S. Mohn wrote the paper; all authors edited the paper.

Conflicts of Interest: This work was partially funded by Abbott Nutrition. Matthew J. Kuchan is employed by Abbott Nutrition. All other authors declare no conflict of interest.

References

1. Institute of Medicine (US) Panel on Dietary Antioxidants and Related Compounds. *Dietary Reference Intakes for Vitamin C, Vitamin E, Selenium, and Carotenoids*; National Academies Press (US): Washington, DC, USA, 2000; Available online: http://www.ncbi.nlm.nih.gov/books/NBK225483/ (accessed on 1 February 2017).

2. Traber, M.G.; Arai, H. Molecular mechanisms of vitamin E transport. *Annu. Rev. Nutr.* **1999**, *19*, 343–355. [CrossRef] [PubMed]

3. Rigotti, A. Absorption, transport, and tissue delivery of vitamin E. *Mol. Asp. Med.* **2007**, *28*, 423–436. [CrossRef] [PubMed]

4. Grodstein, F.; Chen, J.; Willett, W.C. High-dose antioxidant supplements and cognitive function in community-dwelling elderly women. *Am. J. Clin. Nutr.* **2003**, *77*, 975–984. [PubMed]

5. Mangialasche, F.; Xu, W.; Kivipelto, M.; Costanzi, E.; Ercolani, S.; Pigliautile, M.; Cecchetti, R.; Baglioni, M.; Simmons, A.; Soininen, H.; et al. Tocopherols and tocotrienols plasma levels are associated with cognitive impairment. *Neurobiol. Aging* **2012**, *33*, 2282–2290. [CrossRef] [PubMed]

6. Johnson, E.J.; Vishwanathan, R.; Johnson, M.A.; Hausman, D.B.; Davey, A.; Scott, T.M.; Green, R.C.; Miller, L.S.; Gearing, M.; Woodard, J.; et al. Relationship between Serum and Brain Carotenoids, α-Tocopherol, and Retinol Concentrations and Cognitive Performance in the Oldest Old from the Georgia Centenarian Study. *J. Aging Res.* **2013**, *2013*, 951786. [CrossRef] [PubMed]

7. Hensley, K.; Barnes, L.L.; Christov, A.; Tangney, C.; Honer, W.G.; Schneider, J.A.; Bennett, D.A.; Morris, M.C. Analysis of Postmortem Ventricular Cerebrospinal Fluid from Patients with and without Dementia Indicates Association of Vitamin E with Neuritic Plaques and Specific Measures of Cognitive Performance. *J. Alzheimers Dis.* **2011**, *24*, 767–774. [CrossRef] [PubMed]

8. Jiménez-Jiménez, F.J.; de Bustos, F.; Molina, J.A.; Benito-León, J.; Tallón-Barranco, A.; Gasalla, T.; Ortí-Pareja, M.; Guillamón, F.; Rubio, J.C.; Arenas, J.; et al. Cerebrospinal fluid levels of alpha-tocopherol (vitamin E) in Alzheimer's disease. *J. Neural Transm.* **1997**, *104*, 703–710. [CrossRef] [PubMed]

9. Uttara, B.; Singh, A.V.; Zamboni, P.; Mahajan, R. Oxidative Stress and Neurodegenerative Diseases: A Review of Upstream and Downstream Antioxidant Therapeutic Options. *Curr. Neuropharmacol.* **2009**, *7*, 65–74. [CrossRef] [PubMed]

10. Kim, G.H.; Kim, J.E.; Rhie, S.J.; Yoon, S. The Role of Oxidative Stress in Neurodegenerative Diseases. *Exp. Neurobiol.* **2015**, *24*, 325–340. [CrossRef] [PubMed]

11. Traber, M.G.; Atkinson, J. Vitamin E, antioxidant and nothing more. *Free Radic. Biol. Med.* **2007**, *43*, 4–15. [CrossRef] [PubMed]

12. Qi, K.; Hall, M.; Deckelbaum, R.J. Long-chain polyunsaturated fatty acid accretion in brain. *Curr. Opin. Clin. Nutr. Metab. Care* **2002**, *5*, 133–138. [CrossRef] [PubMed]

13. Engin, K.N. Alpha-tocopherol: Looking beyond an antioxidant. *Mol. Vis.* **2009**, *15*, 855–860. [PubMed]
14. Quinn, P.J. Is the distribution of alpha-tocopherol in membranes consistent with its putative functions? *Biochemistry* **2004**, *69*, 58–66. [PubMed]
15. Azzi, A.; Gysin, R.; Kempná, P.; Munteanu, A.; Negis, Y.; Villacorta, L.; Visarius, T.; Zingg, J.M. Vitamin E Mediates Cell Signaling and Regulation of Gene Expression. *Ann. N. Y. Acad. Sci.* **2004**, *1031*, 86–95. [CrossRef] [PubMed]
16. Han, S.N.; Pang, E.; Zingg, J.-M.; Meydani, S.N.; Meydani, M.; Azzi, A. Differential effects of natural and synthetic vitamin E on gene transcription in murine T lymphocytes. *Arch. Biochem. Biophys.* **2010**, *495*, 49–55. [CrossRef] [PubMed]
17. Han, S.N.; Adolfsson, O.; Lee, C.-K.; Prolla, T.A.; Ordovas, J.; Meydani, S.N. Age and vitamin E-induced changes in gene expression profiles of T cells. *J. Immunol.* **2006**, *177*, 6052–6061. [CrossRef] [PubMed]
18. Han, S.N.; Adolfsson, O.; Lee, C.-K.; Prolla, T.A.; Ordovas, J.; Meydani, S.N. Vitamin E and gene expression in immune cells. *Ann. N. Y. Acad. Sci.* **2004**, *1031*, 96–101. [CrossRef] [PubMed]
19. Wang, X.; Quinn, P.J. Vitamin E and its function in membranes. *Prog. Lipid Res.* **1999**, *38*, 309–336. [CrossRef]
20. Vatassery, G.T.; Angerhofer, C.K.; Knox, C.A.; Deshmukh, D.S. Concentrations of vitamin E in various neuroanatomical regions and subcellular fractions, and the uptake of vitamin E by specific areas, of rat brain. *Biochim. Biophys. Acta* **1984**, *792*, 118–122. [CrossRef]
21. Vatassery, G.T.; Smith, W.E.; Quach, H.T. α-Tocopherol in Rat Brain Subcellular Fractions Is Oxidized Rapidly during Incubations with Low Concentrations of Peroxynitrite. *J. Nutr.* **1998**, *128*, 152–157. [PubMed]
22. Perretta, G. Non-Human Primate Models in Neuroscience Research. *Scand. J. Lab. Anim. Sci.* **2009**, *36*, 77–85.
23. Mohn, E.S.; Erdman, J.W.; Kuchan, M.J.; Neuringer, M.; Johnson, E.J. Lutein accumulates in subcellular membranes of brain regions in adult rhesus macaques: Relationship to DHA oxidation products. *PLoS ONE* **2017**, *12*, e0186767. [CrossRef] [PubMed]
24. Muzhingi, T.; Yeum, K.-J.; Russell, R.M.; Johnson, E.J.; Qin, J.; Tang, G. Determination of carotenoids in yellow maize, the effects of saponification and food preparations. *Int. J. Vitam. Nutr. Res.* **2008**, *78*, 112–120. [CrossRef] [PubMed]
25. National Research Council (US) Committee for the Update of the Guide for the Care and Use of Laboratory Animals. *Guide for the Care and Use of Laboratory Animal*, 8th ed.; National Academies Press (US): Washington, DC, USA, 2011; Available online: http://www.ncbi.nlm.nih.gov/books/NBK54050/ (accessed on 15 February 2017).
26. Sun, G.Y.; Sun, Y. Phospholipids and acyl groups of synaptosomal and myelin membranes isolated from the cerebral cortex of squirrel monkey (*Saimiri sciureus*). *Biochim. Biophys. Acta* **1972**, *280*, 306–315. [CrossRef]
27. Sun, G.Y. Phospholipids and acyl groups in subcellular fractions from human cerebral cortex. *J. Lipid Res.* **1973**, *14*, 656–663. [PubMed]
28. Park, J.; Hwang, H.; Kim, M.; Lee-Kim, Y. Effects of dietary fatty acids and vitamin E supplementation on antioxidant vitamin status of the second generation rat brain sections. *Korean J. Nutr.* **2001**, *34*, 754–761.
29. Vishwanathan, R.; Neuringer, M.; Snodderly, D.M.; Schalch, W.; Johnson, E.J. Macular lutein and zeaxanthin are related to brain lutein and zeaxanthin in primates. *Nutr. Neurosci.* **2013**, *16*, 21–29. [CrossRef] [PubMed]
30. Johnson, E.J.; Neuringer, M.; Russell, R.M.; Schalch, W.; Snodderly, D.M. Nutritional manipulation of primate retinas, III: Effects of lutein or zeaxanthin supplementation on adipose tissue and retina of xanthophyll-free monkeys. *Investig. Ophthalmol. Vis. Sci.* **2005**, *46*, 692–702. [CrossRef] [PubMed]
31. Folch, J.; Lees, M.; Sloane Stanley, G.H. A simple method for the isolation and purification of total lipides from animal tissues. *J. Biol. Chem.* **1957**, *226*, 497–509. [PubMed]
32. Lichtenstein, A.H.; Matthan, N.R.; Jalbert, S.M.; Resteghini, N.A.; Schaefer, E.J.; Ausman, L.M. Novel soybean oils with different fatty acid profiles alter cardiovascular disease risk factors in moderately hyperlipidemic subjects. *Am. J. Clin. Nutr.* **2006**, *84*, 497–504. [PubMed]
33. Walter, M.F.; Blumberg, J.B.; Dolnikowski, G.G.; Handelman, G.J. Streamlined F2-isoprostane analysis in plasma and urine with high-performance liquid chromatography and gas chromatography/mass spectroscopy. *Anal. Biochem.* **2000**, *280*, 73–79. [CrossRef] [PubMed]
34. Arneson, K.O.; Roberts, L.J., 2nd. Measurement of products of docosahexaenoic acid peroxidation, neuroprostanes, and neurofurans. *Methods Enzymol.* **2007**, *433*, 127–143. [CrossRef] [PubMed]
35. Mukaka, M. A guide to appropriate use of Correlation coefficient in medical research. *Malawi Med. J.* **2012**, *24*, 69–71. [PubMed]

36. Craft, N.E.; Haitema, T.B.; Garnett, K.M.; Fitch, K.A.; Dorey, C.K. Carotenoid, tocopherol, and retinol concentrations in elderly human brain. *J. Nutr. Health Aging* **2004**, *8*, 156–162. [PubMed]
37. Kuchan, M.J.; Jensen, S.K.; Johnson, E.J.; Lieblein-Boff, J.C. The naturally occurring α-tocopherol stereoisomer RRR-α-tocopherol is predominant in the human infant brain. *Br. J. Nutr.* **2016**, *116*, 126–131. [CrossRef] [PubMed]
38. Copp, R.P.; Wisniewski, T.; Hentati, F.; Larnaout, A.; Ben Hamida, M.; Kayden, H.J. Localization of alpha-tocopherol transfer protein in the brains of patients with ataxia with vitamin E deficiency and other oxidative stress related neurodegenerative disorders. *Brain Res.* **1999**, *822*, 80–87. [CrossRef]
39. Zimmer, S.; Stocker, A.; Sarbolouki, M.N.; Spycher, S.E.; Sassoon, J.; Azzi, A. A novel human tocopherol-associated protein: cloning, in vitro expression, and characterization. *J. Biol. Chem.* **2000**, *275*, 25672–25680. [CrossRef] [PubMed]
40. Podda, M.; Weber, C.; Traber, M.G.; Packer, L. Simultaneous determination of tissue tocopherols, tocotrienols, ubiquinols, and ubiquinones. *J. Lipid Res.* **1996**, *37*, 893–901. [PubMed]
41. Zingg, J.-M.; Azzi, A. Non-antioxidant activities of vitamin E. *Curr. Med. Chem.* **2004**, *11*, 1113–1133. [CrossRef] [PubMed]
42. Rota, C.; Rimbach, G.; Minihane, A.M.; Stoecklin, E.; Barella, L. Dietary vitamin E modulates differential gene expression in the rat hippocampus: potential implications for its neuroprotective properties. *Nutr. Neurosci.* **2005**, *8*, 21–29. [CrossRef] [PubMed]
43. Park, S.-K.; Page, G.P.; Kim, K. Differential effect of α- and γ-tocopherol supplementation in age-related transcriptional alterations in heart and brain of B6/C3H F1 mice. *J. Nutr.* **2008**, *138*, 1010–1018. [PubMed]
44. Gohil, K.; Schock, B.C.; Chakraborty, A.A.; Terasawa, Y.; Raber, J.; Farese, R.V., Jr.; Packer, L.; Cross, C.E.; Traber, M.G. Gene expression profile of oxidant stress and neurodegeneration in transgenic mice deficient in alpha-tocopherol transfer protein. *Free Radic. Biol. Med.* **2003**, *35*, 1343–1354. [CrossRef]
45. Berchtold, N.C.; Coleman, P.D.; Cribbs, D.H.; Rogers, J.; Gillen, D.L.; Cotman, C.W. Synaptic genes are extensively downregulated across multiple brain regions in normal human aging and Alzheimer's disease. *Neurobiol. Aging* **2013**, *34*, 1653–1661. [CrossRef] [PubMed]
46. Berchtold, N.C.; Sabbagh, M.N.; Beach, T.G.; Kim, R.C.; Cribbs, D.H.; Cotman, C.W. Brain gene expression patterns differentiate Mild Cognitive Impairment from normal Aged and Alzheimer Disease. *Neurobiol. Aging* **2014**, *35*, 1961–1972. [CrossRef] [PubMed]
47. Simen, A.A.; Bordner, K.A.; Martin, M.P.; Moy, L.A.; Barry, L.C. Cognitive Dysfunction with Aging and the Role of Inflammation. *Ther. Adv. Chronic Dis.* **2011**, *2*, 175–195. [CrossRef] [PubMed]
48. Atkinson, J.; Harroun, T.; Wassall, S.R.; Stillwell, W.; Katsaras, J. The location and behavior of alpha-tocopherol in membranes. *Mol. Nutr. Food Res.* **2010**, *54*, 641–651. [CrossRef] [PubMed]
49. Raederstorff, D.; Wyss, A.; Calder, P.C.; Weber, P.; Eggersdorfer, M. Vitamin E function and requirements in relation to PUFA. *Br. J. Nutr.* **2015**, *114*, 1113–1122. [CrossRef] [PubMed]
50. Montine, T.J.; Montine, K.S.; Reich, E.E.; Terry, E.S.; Porter, N.A.; Morrow, J.D. Antioxidants significantly affect the formation of different classes of isoprostanes and neuroprostanes in rat cerebral synaptosomes. *Biochem. Pharmacol.* **2003**, *65*, 611–617. [CrossRef]
51. Reich, E.E.; Montine, K.S.; Gross, M.D.; Roberts, L.J., 2nd; Swift, L.L.; Morrow, J.D.; Montine, T.J. Interactions between Apolipoprotein E Gene and Dietary α-Tocopherol Influence Cerebral Oxidative Damage in Aged Mice. *J. Neurosci.* **2001**, *21*, 5993–5999. [PubMed]
52. Lebold, K.M.; Traber, M.G. Interactions between alpha-tocopherol, polyunsaturated fatty acids, and lipoxygenases during embryogenesis. *Free Radic. Biol. Med.* **2014**, *66*, 13–19. [CrossRef] [PubMed]

MDPI

St. Alban-Anlage 66

4052 Basel

Switzerland

Tel. +41 61 683 77 34

Fax +41 61 302 89 18

www.mdpi.com

Antioxidants Editorial Office

E-mail: antioxidants@mdpi.com

www.mdpi.com/journal/antioxidants

www.ingramcontent.com/pod-product-compliance
Lightning Source LLC
Chambersburg PA
CBHW041218220326
41597CB00033BA/6033